T0343422

THE CLAUSTRUM

Structural, Functional, and Clinical Neuroscience

Edited by

JOHN R. SMYTHIES

LAWRENCE R. EDELSTEIN

VILAYANUR S. RAMACHANDRAN

AMSTERDAM • BOSTON • HEIDELBERG • LONDON
NEW YORK • OXFORD • PARIS • SAN DIEGO
SAN FRANCISCO • SINGAPORE • SYDNEY • TOKYO

Academic Press is an imprint of Elsevier

Academic Press is an imprint of Elsevier
525 B Street, Suite 1800, San Diego, CA 92101-4495, USA
32 Jamestown Road, London NW1 7BY, UK
225 Wyman Street, Waltham, MA 02451, USA

Notice
No responsibility is assumed by the publisher for any injury and/or damage to persons,
or property as a matter of products liability, negligence or otherwise, or from any use or
operation of any methods, products, instructions or ideas contained in the material herein.
Because of rapid advances in the medical sciences, in particular, independent verification of
diagnoses and drug dosages should be made.

British Library Cataloguing-in-Publication Data
A catalogue record for this book is available from the British Library

Library of Congress Cataloging-in-Publication Data
A catalog record for this book is available from the Library of Congress

ISBN: 978-0-12-404566-8

For information on all Academic Press publications
visit our website at elsevierdirect.com

Printed and bound by CPI Group (UK) Ltd, Croydon, CR0 4YY

Contents

13. Hypotheses Relating to the Function of the Claustrum   299

JOHN R. SMYTHIES, LAWRENCE R. EDELSTEIN, AND
VILAYANUR S. RAMACHANDRAN

14. What is it to be Conscious?   353

DENIS NOBLE, RAYMOND NOBLE, AND
JAMES SCHWABER

15. Selected Key Areas for Future Research on the Claustrum   365

HIROYUKI OKUNO, JOHN R. SMYTHIES, AND LAWRENCE R. EDELSTEIN

Index   377

List of Contributors

Joan Susan Baizer Department of Physiology and Biophysics, University of Buffalo, NY, USA

W. Ted Brown Department of Human Genetics, NYS Institute for Basic Research in Developmental Disabilities, Staten Island, NY, USA

Nicola G. Cascella Neuropsychiatry Program, Sheppard Pratt Hospital, Baltimore, MD, USA

Michael E. Corcoran Department of Anatomy, and Cell Biology, University of Saskatchewan, Saskatoon, Canada

Rastislav Druga Institute of Anatomy, 2nd Medical Faculty, Charles University, Prague, Czech Republic

Lawrence R. Edelstein Medimark Corporation, Del Mar, CA, USA

Brian A. Fenske Department of Radiology, Division of Human Anatomy, Michigan State University, East Lansing, MI, USA

Juan C. Fernandez-Miranda Department of Neurological Surgery, University of Pittsburgh Medical Center, Pittsburgh, PA, USA

Michael Flory Department of Infant Development, NYS Institute for Basic Research in Developmental Disabilities, Staten Island, NY, USA

Alex O. Holcombe School of Psychology, University of Sydney, NSW, Australia

Humi Imaki Department of Developmental Neurobiology, NYS Institute for Basic Research in Developmental Disabilities, Staten Island, NY, USA

John Irwin Johnson Department of Radiology, Division of Human Anatomy, Michigan State University, East Lansing, MI, USA

Michail E. Kalaitzakis Neuropathology Unit, Division of Brain Sciences, Department of Medicine, Imperial College London, UK

Przemysław Kowiański Department of Anatomy and Neurobiology, Medical University of Gdansk, Poland

Izabela Kuchna Department of Developmental Neurobiology, NYS Institute for Basic Research in Developmental Disabilities, Staten Island, NY, USA

Shuang Yong Ma Department of Developmental Neurobiology, NYS Institute for Basic Research in Developmental Disabilities, Staten Island, NY, USA

Janusz Moryś Department of Anatomy and Neurobiology, Medical University of Gdansk, Poland

Denis Noble Department of Physiology, Anatomy and Genetics, University of Oxford, Oxford, UK

Raymond Noble Institute for Women's Health, University College London, London, UK

Krzysztof Nowicki Department of Developmental Neurobiology, NYS Institute for Basic Research in Developmental Disabilities, Staten Island, NY, USA

Hiroyuki Okuno Medical Innovation Center, Kyoto University Graduate School of Medicine, Tokyo, Japan

Sudhir Pathak Department of Psychology, Learning Research and Development Center, University of Pittsburgh, PA, USA

Luis Puelles Department of Human Anatomy and Psychobiology, School of Medicine, University of Murcia, Spain

Vilayanur S. Ramachandran Center for Brain and Cognition, Department of Psychology, UC San Diego, La Jolla, CA, USA

Akira Sawa Department of Psychiatry and Behavioral Sciences, Johns Hopkins University School of Medicine, Baltimore, MD, USA

James Schwaber Department of Pathology, Anatomy and Cell Biology, Thomas Jefferson University, Philadelphia, PA, USA

Michael Shanks Academic Unit of Psychiatry, University of Sheffield, UK

Helen Sherk Department of Biological Structure, University of Washington, Seattle, WA, USA

John R. Smythies Center for Brain and Cognition, Department of Psychology, UC San Diego, La Jolla, CA, USA

Klaus M. Stiefel University of Western Sydney, MARCS Institute, Bioelectronics and Neuroscience, University of Western Sydney, NSW, Australia

Annalena Venneri Department of Neuroscience, University of Sheffield, UK

Jarek Wegiel Department of Developmental Neurobiology, NYS Institute for Basic Research in Developmental Disabilities, Staten Island, NY, USA

Jerzy Wegiel Department of Developmental Neurobiology, NYS Institute for Basic Research in Developmental Disabilities, Staten Island, NY, USA

Thomas Wisniewski Department of Psychiatry, New York University School of Medicine, New York, NY, USA

Acknowledgments

As with most any book, especially one of a technical nature, there exist motivational and experiential forces beyond the ken of the author or editor; this book is no exception. In this regard, we wish to express our sincere thanks to several individuals who truly helped us to make it so.

Were it not for the steadfast focus and requisite prodding of Elsevier's able-bodied acquisition editors, Mica Haley and April Graham, our journey from concept to publication would have undoubtedly taken a longer and less predictable route. We are most grateful to Professor Doctor Stan Gielen, Dean of the Faculty of Science at Radboud University in Nijmegen, for his cogent insights with respect to the intricacies of spike timing and oscillations. Walter Freeman, Professor Emeritus at UC Berkeley, was most kind to have reviewed an early draft of our Hypotheses chapter, thus affording us a modicum of comfort in knowing that we were heading in the right direction. Last and foremost, we are beholden to Francis Crick and Christof Koch for illuminating the path while throwing down the mother of all gauntlets (wink and all).

Introduction

The editors of this volume, who also wrote the chapter on hypotheses relating to the function of the claustrum, came from different backgrounds in neuroscience. This facilitated the multidisciplinary approach so necessary for such an enterprise. In their own words, they tell the story of their collaboration that developed.

Lawrence ('Larry') Edelstein

Spring semester 1974, State University of New York at Stony Brook, "Introduction to Physiological Psychology," Professor John Stamm (mentor and frontal lobe physiologist *par excellence*). "Essentials of Physiological Psychology," by Sebastian P. Grossman. As a significant percentage of our final grade, we were tasked with submitting a term paper on the brain structure of our choosing, to be selected from those mentioned in our textbook. Owing to an already burdensome semester with numerous finals in the air, I immediately went to work looking for that part of the brain to which was paid the least attention, in essence, the Rodney Dangerfield of the CNS. Hippocampus? Fuhgeddaboudit. Amygdala? Fight or flight; I chose the latter. Claustrum – function unknown; barely made the index. Clearly, it got no respect. The seed was planted, eventually sprouting into a doctoral thesis: "The anatomy of the claustrum: A light and electron-microscopic analysis in rat and monkey incorporating the technique of HRP cytochemistry."

I moved to San Diego in late 2001, the claustrum something I once dabbled with in the distant past. Cut to September 27, 2004 and the live-streamed public memorial held at The Salk Institute for Francis Crick. Although I was long into a new career and deskbound at the time, I felt the need to somehow be a part of this event, if only as a virtual observer. Thick with notables, Nobelists (and those to be) and molecular biology, I watched and listened with rapt attention as V.S. Ramachandran took the podium and proceeded to eloquently honor a close friend and esteemed colleague. Somewhere in the middle of his tribute I learned of Francis' interest in the claustrum, at which point my jaw dropped rather precipitously, and a few choice unmentionables were uttered upon return to its normal position. A comprehensive review paper followed shortly thereafter, co-authored with my long-time friend and fellow claustrophile, Professor Frank Denaro. For some time prior to his passing, Francis Crick, along with his brilliant Caltech colleague-in-arms, Christof Koch, had honed in on the very same slab of grey matter as I did thirty years prior, and found it to be as salient a stimulus, with mysteries yet to be revealed.

Nearly forty years on, I'm still working on that term paper. However, thanks to Francis, Christof, my fortuitously catching that live-stream, and subsequently befriending my esteemed UCSD colleagues John Smythies and Rama, I can safely say that I chose wisely. And the title of what was fated to be Francis' final publication? "What is the function of the claustrum?"

Vilayanur ('Rama') Ramachandran

It is my belief that the same strategy used to crack the genetic code might prove successful in cracking the "neural code" of consciousness and self. It's a long shot, but worth considering. The basis of this strategy is the ability to grasp analogies, seeing the difference between deep and superficial ones. Crick and Watson's own solution to the riddle of heredity ranks with natural selection as biology's most elegant discovery. Will a solution of similar elegance emerge for the problem of consciousness?

In biology, knowledge of structure often *leads* to knowledge of function – one need look no further than the whole of medical history. Inspired by Griffith and Avery, Crick and Watson realized that the answer to the problem of heredity lay in the structure of DNA. Localization was critical, as indeed it may prove to be for brain function. I believe there are similar correlations between brain structure and mind function, between neurons and consciousness.

After his triumph with heredity, Crick turned to what he called the "second great riddle" in biology – consciousness. Crick did not, in my opinion, succeed in solving consciousness (whatever that might mean). Nonetheless, I believe he was headed in the right direction. He had been richly rewarded earlier in his career for grasping the analogy between biological complementarities, the notion that the structural logic of the molecule *dictates* the functional logic of heredity. Given his phenomenal success using the strategy of structure–function analogy, it is hardly surprising that he imported the same style of thinking to study consciousness. He and his colleague Christof Koch did so by focusing on a relatively obscure structure called the claustrum.

The claustrum is a thin sheet of cells underlying the insular cortex of the brain, one on each hemisphere. It is histologically more homogeneous than most brain structures, and intriguingly, unlike most brain structures (which send and receive signals to and from a small subset of other structures), the claustrum is reciprocally connected with almost every cortical region. The structural and functional streamlining might ensure that, when waves of information come through the claustrum, its neurons will be exquisitely sensitive to the timing of the inputs. What does this have to do with consciousness? Instead of focusing on pedantic philosophical issues, Crick and Koch began with their naive intuitions. "Consciousness" has many attributes – continuity in time, a sense of agency or free will, recursiveness or "self-awareness," etc. But one attribute that stands out is subjective unity: you experience all your diverse sense impressions, thoughts, willed actions and memories as being a unity – not jittery or fragmented. This attribute of consciousness, with the accompanying sense of the immediate "present" or "here and now," is so obvious that we don't usually think about it; we regard it as axiomatic.

So a central feature of consciousness is its unity – and here is a brain structure that sends and receives signals to and from practically all other brain structures, including the right parietal (involved in polysensory convergence and embodiment) and anterior cingulate (involved in the experience of "free will"). Thus the claustrum seems to unify everything anatomically, and consciousness does so mentally. Crick and Koch recognized that this may not be a coincidence: the claustrum may be central to consciousness. It is this kind of childlike reasoning that often leads to great discoveries. Obviously, such analogies don't replace rigorous science, but they're a good place to start. I think Crick and Koch may be right or wrong, but their idea is elegant. If they're right, they've paved the way to solving one of the great mysteries of biology. Even if they are wrong, students entering the field would do well to emulate their style, as we have attempted to do ourselves. Crick has been right too often to ignore.

I visited Francis at his home in La Jolla in July of 2004. He saw me to the door as I was leaving and as we parted, gave me a sly, conspiratorial wink: "I think it's the claustrum, Rama; it's where the secret is." A week later he passed away.

John Smythies

My background is in neuroanatomy, neuropsychiatry and neurophilosophy. Latterly I had been working in Rama's lab on problems related to de- and re-afferentation plasticity, when one afternoon in late August 2011, I received an email from Rama, just then departing on a visit to India. It said something like this "John – you should really have a look at the claustrum – you know Francis's idea that it plays a key role in consciousness. The fact that it is connected with every other part of the brain must mean something. Get together with Larry on this. See you when I get back" – Larry being Lawrence Edelstein, a leading authority on the claustrum, who was working down the road from the UCSD campus. His daughter also happened to be a graduate student in Rama's lab.

At first I was puzzled. At that time the claustrum was of not much interest to anyone, except a few devoted claustrophiles, of which Larry was one. In 2004 Francis Crick and Christof Koch had taken up its cause and claimed that its interesting connections made it an excellent candidate to be the "conductor" of the consciousness "orchestra." The general response from the neuroscience community had been somewhat lacking in enthusiasm. Most neuroscientists tend to shy like startled horses at the very mention of the word "consciousness." This did not deter Francis one iota and he continued to work on the claustrum until the day he died.

So I fished the paper by Crick and Koch out of the archives and re-read it. I soon became intrigued with its contents. I had read it when it came out, but had put it on one side like almost every one else had done. So I read it this time more attentively. I was soon struck by the cogency of their arguments. When Rama got back from India we got together with Larry and decided to work together to try and take up the problem of what the claustrum is doing from where Crick and Koch had left it six years before. The general results of this enterprise led to this volume: the specific result was the hypothesis presented in our chapter.

Cover Image

High-definition fiber tractography (HDFT) reconstruction of the right claustrocortical projection fibers in a 35-year-old healthy male volunteer. Such reconstructions have shown accurate fiber tracking in crossing regions and robust end-point connectivity between cortical and subcortical regions (Fernandez-Miranda et al. (2012) *Neurosurgery, 71*, 430–453). Freesurfer-based segmented cortical regions of interest were used to subsegment claustrum connections and DSI-Studio was used to calculate the volume of fibers terminating at each cortical region: superior frontal gyrus (*cyan*) 21.3 ml; rostral middle frontal gyrus (*purple*) 4.1 ml; precentral gyrus (*red*) 5.6 ml; postcentral gyrus (*yellow*) 7.3 ml; superior parietal lobule (*blue*) 5.9 ml. The cover image was provided courtesy of Juan Fernandez-Miranda, M.D., Department of Neurological Surgery, University of Pittsburgh Medical Center.

History of the Study and Nomenclature of the Claustrum

John Irwin Johnson and Brian A. Fenske

Department of Radiology, Division of Human Anatomy,
Michigan State University, East Lansing, MI, USA

THE FIRST ERA: 1780–1820

Something New in the Brain: A Product of Revolutions in Science and Society

The history of the study and nomenclature of the claustrum of the mammalian brain begins with the remarkable and revolutionary career of the Frenchman Félix Vicq d'Azyr (see Parent, 2007, for a fascinating account of this spectacular, tragically brief career, and its lasting imprint on biological science and medical education).

Americans will note with interest that this entry into the scientific literature was published in 1786, three years following the recognition of the United States as an independent nation, and three years before the adoption of the equally revolutionary Constitution of the United States of America, distinctively French in its design and principles.

By 1789 the fame of Vicq d'Azyr, through his work in the foundation of comparative biology as a field of study and his applications of this approach to solve important matters of public health, had led to his appointment as physician to Queen Marie Antoinette. Three years later, both he and the queen were dead, but his influence was profound and lives on in many ways, including in our continued attention to the structure we now know as the claustrum.

This young and superbly talented medical student was one of the early enthusiasts who pursued the anatomical study of animals as well as humans, attending lectures on the anatomy of both. He won fame by a series of lectures, on the same topics, which he initiated when he was

J. Smythies, L. Edelstein and V. Ramachandran (Eds):
The Claustrum.

1

DOI: http://dx.doi.org/10.1016/B978-0-12-404566-8.00001-5

only 25 years old. Based on these he was awarded his medical degree, and in recognition of their merit was elected to the Academy of Sciences.

He discovered that, by hardening internal tissues in a special fixative, he was able to take thin sections without disrupting the organization of the cells. From these superb sections he proceeded to make exquisite pictures of the internal organs of animals and humans with hitherto unattainably detailed and accurate macroscopic and microscopic views. His depiction of the claustrum showing extended points of it protruding outward into the white matter of the overlying insular gyri can still be viewed today, online, courtesy of the Bibliothèque Interuniversitaire de Santé (see below).

This is prime evidence of the care and accuracy in observation that was employed in the creation of these illustrations. This feature was rarely, if ever, seen in illustrations of claustrum in the subsequent 230 years, but is present in the brains of humans, dolphins, and other large-brained mammals. It has figured in recent speculations about how claustrum achieves its distinctive morphology in different mammalian species (Buchanan and Johnson, 2011).

Even in photographs (Figure 1.1) of sections through llama, dolphin and human brains, it takes careful observation of several adjacent sections to determine the extent and configuration of these gyral core invasions. Their appearance in the illustration by Vicq d'Azyr is much clearer, paradoxically, than in photographs of stained sections. This is a reminder that his illustrations are in fact paintings, although exquisitely accurate paintings. Generations of contemporary medical students know well the advantage of learning from paintings of tissue rather than from photographs or the viewing of actual specimens. The painting isolates and clarifies the spatial relationships through the intrusion of the mind of the painter; however, rather than distorting reality it actually gives a better depiction than the more direct illustrations. Even photographic evidence has lost its advantages now that photo editing can effectively change the depiction of the reality to conform to the abstractions, or many other purposes, of transmitting such views.

At the age of 38, Vicq d'Azyr gathered his superb pictures into a grand compendium, the *Traité d'anatomie et de physiologie avec des planches coloriées représentant au naturel les divers organes de l'homme et des animaux* ["Treatise on anatomy and physiology with plates in color representing, in their natural condition, the diverse organs of men and animals"], (Vicq d'Azyr, 1786).

This was the first major reference book in the field that was thereby brought into published existence: comparative biology. The key to the lasting importance of this work is indicated by the term "au naturel" in the title. Great care was taken in the figures to reproduce what was actually seen in the sections, rather than what might be believed should be in the sections. Such concern was paramount in Vicq d'Azyr's recommendations, 5 years later, for reforming French medical education so as

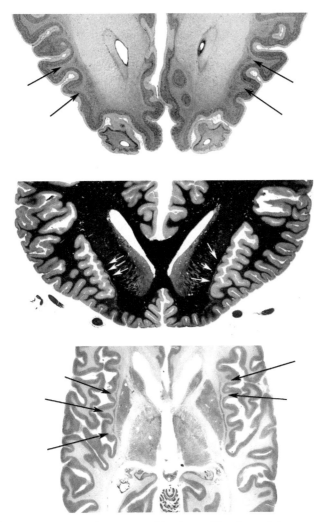

FIGURE 1.1 Invasion of white matter of insular gyri by claustral tissue, as seen in sections through brains of: llama, *Lama glama* (top); bottlenose dolphin, *Tursiops truncatus* (center); and human, *Homo sapiens* (bottom). This feature was depicted in a section of human brain by Vicq d'Azyr (1786) in the first publication that mentioned the existence of the claustrum. The llama section was stained using thionine in the Nissl method to show cell bodies; the dolphin section was stained using iron hematoxylin in the Loyez version of the Weigert method to show myelinated fibers; the human section was stained using cresyl violet in the Nissl method. *Llama section is from the Welker Wisconsin Collection; dolphin and human sections are from the Yakovlev-Haleem Collection, all now at the National Library of Health and Medicine, Silver Spring, Md.*

to base it on data rather than on theory. Although his recommendations were dismissed by the embattled authorities at that turbulent time, they remained available in the form of his published report, and eventually did shape the reform of French medical education into a system that was a world leader for a generation.

Thus it was with the *Treatise on anatomy and physiology* – its reliance on actual pictures of structures has proved of lasting value. Virtually all contributors to the rest of the history of research on the claustrum, and of many other topics as well, make reference to these illustrations as the first authoritative accounts.

This work contains five remarkable pictures that clearly depict the claustrum and constitute its earliest appearance in the scientific literature. Thanks to the generosity of the Bibliothèque numérique Medic of the Bibliothèque Interuniversitaire de Santé in Paris, readers can view these pictures from one of the few surviving copies. Accessing Volume 2 of the *Traité* via http://bit.ly/1ONIEI8 brings up Plate IX. The illustration has numerical labels; there is no reference to the name claustrum nor to any other name. The structure we now know as the claustrum is labeled '28' on this and subsequent plates.

There are explanatory texts in Volume 1 of the *Traité*, and the text for Plate IX can be accessed at http://bit.ly/ZqQ7Mz. Paging down from there leads to an explanation for the structure labeled '28', which is identified as follows :

> "28, 28, 28, ces chiffres indiquent de chaque coté une trace légère de substance corticale placée longitudinalement entre la portion supérieure et externe des stries 26, 26, 27, et le bord interne 6,6,5, des circonvolutions cérébrales qui composent la division postérieure de la scissure de Sylvius." (on page 27)
>
> [28, 28, 28, these figures indicate on each side a light trace of cortical substance placed longitudinally between the superior external portion of the striatum 26, 26, 27, and the internal edge 6, 6, 5, of the cerebral convolutions that make up the posterior division of the fissure of Sylvius.]

Number '28' was as far as Vicq d'Azyr appears to have gone in pursuing the naming of this structure that was new both to science and to the understanding of brain architecture.

With the stabilization of society, after the years of turmoil and disruption stemming from the Revolution, many additional papers by Vicq d'Azyr were collected from their scattered refuges, and several were published at later dates.

Moreau (1805) published a compendium of the writings of Vicq d'Azyr, at least part of which is now generally available as an online publication. In a 2012 reprinting, there is only a brief text followed by 500 pages of explanations of the plates and the incorporated 'Figures'. No graphic pictures are included. In the explanations, those of Plates

IX, X, and XI from the 1786 volumes are included, but are represented as explanations of Plates VII, VIII, and IX. The texts of these explanations are identical to those published in 1786, including the description of the structure labeled '28', which corresponds to the structure now known as the claustrum. There is mention of neither a *nucleus taeniaeformis* nor any other name, just the description of a trace of cortical substance lying between the striatum and the insular cortical convolutions.

Hippolyte Cloquet organized a series of works, published as jointly authored by Vicq d'Azyr and Cloquet, or vice versa, and finally by Cloquet alone. Four of these have been reprinted in whole or in part, all with the same title, but each with different content, by Nabu press. The third volume in a series entitled *Système Anatomique et de Physiologie* was apparently started by Vicq d'Azyr but completed by Cloquet (Vicq d'Azyr, 1813a). In addition, A book with the title *Traité de l'anatomie du cerveau*, and claiming the long-deceased Vicq d'Azyr as sole author, was also published in 1813 (Vicq d'Azyr, 1813b) and was available in an online sale in 2013. The circumstances of its content and publication are obscure. One of these two books may be the source of the date 1813 in many historical timelines for the discovery of the claustrum by Vicq d'Azyr.

Sorting out and analyzing the contents of all of these posthumous publications will entail a major, and worthwhile, effort, but are beyond the scope of this review.

THE SECOND ERA: 1820–1870

Naming the Claustrum: And Initiating it into the Canon of the Components of Cerebral Neuroanatomy

The German physiologist Karl Burdach published another grand compendium, this one restricted to the structure and function of the brain. The book appeared in two volumes, published 3 years apart (Burdach, 1819/2012, 1822/2012). This work included copious acknowledgments to the work of Vicq d'Azyr, as well as to that of Reil, Gall, and others. There was a separate section on the claustrum in volume II, which is the earliest text devoted specifically to this structure that the authors have been able to find in current generally available resources, and this text is presented below.

Vormauern

An der äussern Fläche der äussern Capsel, und an Der inner Fläche der Belegungs masse, welche als die Seitenfläche des Stammlappens oder als Insel erscheint, liegt die Vormauer (claustrum)* als eine Shicht grauer Substanz, welche dem Linsenkerne parallel sich erstreckt. Sie ist nämlich 1 Zoll 6 Linien lang; ihre Lage entspricht der Länge nach den Linsenkerne, und sie ist ihm entsprechend in die Länge etwas gekrümt, nach aussen gewölbt, nach innen ausgehölt; so ist sie auch ziemlich von gleicher Höhe mit demsselben. Oben schärft sie sich, wie derselbe, zu; unten breitet sie sich in eine a bis 3 Linien breite Basis auf; so dass sie auf dem senkrechten Querdurchshnitte wie ein aufrecht stehendes Dreyeck erscheint. Hier beugt sie sich aber auch an einer Stelle nach innen um, oder ihre Grundfläche verlängert sich in einen inner Arm, welcher unter dem Linsenkerne, an dem wagerechten Markblatte der äussern Capsul nach innen sich erstreckt. Sie scheint eine zur äussern Capsel gehörige Ganglienmasse zu seyn. *(Burdach, 1822/2012, §. 181)*

[On the outer face of the external capsule, and on the inner side of the occupancy mass which appears as the lateral face of the stem or insula, lies the front wall (*claustrum*)* as a layer of gray substance that stretches along the length of the lentiform nucleus. It is 1 inch and 6 lines [about one-and-a-half-inches] long; its extent equals that of the lentiform nucleus and in that extent it is correspondingly somewhat curved, convex on the outside, concave on the inside; so they are about the same height. At the top it is sharpened; below it widens into a base 3 lines [about a quarter of an inch] wide, such that in coronal section it appears as an upright triangle. But here at one point it is bent inwardly, or its base is extended in an inner arm, which under the lentiform nucleus is extended inward in the horizontal margins of the external capsule. It appears to be a ganglion mass belonging to the external capsule].

The asterisk in Burdach's text refers to a footnote listing the pictures in Vicq d'Azyr's, 1786 *Traité* wherein lie each occurrence of the structure numbered 28, indicating that this was indeed the structure he was naming *Vormauern* and "claustrum." So although, as can be seen, Burdach referred to the structure as *Vormauern*, but included the word claustrum in parentheses and italics, although only on this one occasion. It is possible that this is the "introduction" of the term claustrum reported by later historians. The Dejerines (Dejerine and Dejerine-Klumpke, 1895) were of the opinion that in this work Burdach was introducing *Vormauern* as a synonym for "claustrum." It seems strange that such an introduction or coinage of two completely new terms would include no mention of the reasons for use of such new terms, nor from whence they came.

The Dejerines also said that Reil (1809) described the gray matter of the forebrain separated by the internal, external, and extreme capsules as *caudatus, nucleus lentiformis,* and *nucleus taeniaeformis.* Our reading of this article by Reil found the description of the three capsules but no mention of the word "taeniaeformis."

Meanwhile the term *Vormauern* continued to be in use into the 20th century. *Avant-mur* is the equivalent term used in French literature, including by the Dejerines in their 1895 great French compendium of neuroanatomy, and its usage has continued in medical school courses in France into the 21st century. The Italian counterpart, *antimuro,* appears in the Italian literature at least through 1922.

According to previous histories (Edelstein and Denaro, 2004; Meyer, 1971; Olry and Haines, 2001; Rae, 1954), Vicq d'Azyr himself named the structure "nucleus taeniaeformis." This may have been done in one or another of his posthumous publications, but we did not find the source containing that information. It was not in the self-published 1786 *Traité*.

In 1820 Cloquet published a compendium of the works by himself and with Vicq d'Azyr, constituting a major anatomical resource for its age. This attracted admiration and respect such that it was soon translated into English and published, by the redoubtable Robert Knox, in several editions. The edition available to us (Knox, 1831) contains a description of the external and internal anatomy of the cerebrum, with several references to the nomenclature of Vicq d'Azyr on pp. 412 to 426, but it makes no mention of a nucleus taeniaeformis, nor of the claustrum under any other name.

In 1838, one of the many volumes on anatomy published during the 1830s by Friedrich Arnold, the *Tabula Anatomica*, in Latin, served later authors as a source of official terms. The names are accompanied by anatomical illustrations of the referents for the names. Here we find the designation in the list and in the illustration, *N. taeniaeformis* s. *claustrum*. We can see this today thanks to the library of the University of Heidelberg, who, like their counterparts in Paris, have placed online the entire book. These offerings show the wonderful illustrations along with the texts and can be viewed at http://bit.ly/13jCKOT.

The earliest mention of the claustrum that we could find, in a published scientific investigation rather than an anatomical text, was in a review of cerebral morphology by Turner (1866), which was presented at their invitation to the Royal Medical Society in London. His reference was incidental to his main subject, the cerebral convolutions, and he called our subject *Arnold's N. taeniaeformis*, indicating that that name was, to some degree, in general use at that time.

The full range of neuroanatomical designations, including *N. taeniaeformis, Vormauern,* and *claustrum* is listed in German textbooks through at least 1904 (e.g. Broesike, 1904; Wundt, 1904). The years 1889 through 1895 saw recommendations from the following societies that *"claustrum"* be the preferred term over all of the synonyms:

- 1889 – Report of the Committee of the Association of American Anatomists, adopted unanimously at Philadelphia, December 28.
- 1895 – Report of the Committee of the Anatomische Gesellschaft, adopted at Basle.
- 1896 – American Neurological Association at Philadelphia, June 5.

(From: Scientific Notes and News, Science, N. S. Vol. IV, No. 81, July 17, 1896)

Meanwhile the next era of research on the claustrum was underway.

THE THIRD ERA: 1870–1950

The Chromatic Epiphany: New Stains Show Distinctions Among Major Regions of Brain Tissues, a Major Step Towards Productive Investigations

Our third era commences with the development, in rapid succession, of four new staining methods, around 1873 and 1885. Three of these rapidly became the standard techniques, still in use today, and they bear the names of their inventors. The metallic deposition method of Camillo Golgi was able to show an entire neuronal cell with all of its processes (Golgi, 1873). Thus the morphology of the entire cell could be seen, including the size and shape of the dendrites, and of course the branching pattern of the axon.

Next, Franz Nissl devised a chemical reaction that colored nucleoli and perikarya through reactions with, it was learned much later, RNA molecules clustered in ribosomes. In this way total numbers and densities of cell bodies could be viewed. While the Nissl method was discovered in 1884 and widely used, his first publication of the method did not appear until a decade later (Nissl, 1894).

Carl Weigert, who was working his method during the same time period and at the same location as Nissl, the Städtische Irrenanstalt in Frankfurt-am-Main, published his method of attaching chelated iron to myelin sheaths throughout their cellular extent, allowing views of the extent, length, and grouping of myelinated fibers (Weigert, 1885).

In combination these methods revealed most of what we now know about the fine structure and connectivity of the brain and spinal cord.

Separately each of them reveals a quite different set of features than do the other two, as can be seen in the famous and oft-repeated depiction of layers of cerebral cortex, which can be identified in very different images. Examples can be seen on many websites, including:

- www.benbest.com/science/anatmind/anatmd5.html
- en.wikibooks.org/wiki/Consciousness_Studies/Neuroscience_1
- www.humanneurophysiology.com/cerebralcortex.htm
- www.cixip.com/index.php/page/content/id/1180 (all accessible in June 2013).

Golgi Stain

Golgi stains allow a view of only occasional cells, and the selection process is still unknown. But the cells that are selected are shown in their entirety, including all of the dendrites and often the axon, as well as the cell body. Taking full advantage of these properties, Ramón y Cajal produced an informative, classical series of studies of the architectural arrangement, with beautiful pictures of the shapes of neurons, and much of their connectivity, in most parts of the brain.

One of these stained regions was the claustrum (Ramón y Cajal, 1900, p. 178–183). Meynert (1885, p. 71–75) had asserted that claustrum was the deepest layer of cerebral cortex in the region of the insula, although he recognized that it was not coterminous with the insula, areas of claustrum extending into opercular cortices and even on to the external surface of the hemisphere. These features can be seen in the section of the dolphin brain in our Figure 1.1.

Ramón y Cajal concluded that the claustrum was a not a detached portion of cortex, nor of the corpus striatum, since the size, shape and arrangement of nearly all the neurons in the claustrum were quite distinct from the corresponding features in the neighboring structures: corpus striatum internally and insular cortex externally (Figure 1.2).

Striatum and claustrum have, as their majority constituents, sizeable neurons with radiating dendrites thickly covered with dendritic spines, and axons that project out of the nucleus; but the striatum has, in addition, numerous small cells, some with either bushy dendrites or short axons that do not leave the nucleus (or with both). These cells are spaced fairly evenly with no apparent pattern to their arrangement. In cerebral cortex, in sharp distinction, the majority of cells are pyramidal with a single apical dendrite stretching some distance radially from internal to more external levels in the cortex, along with bushes of basal dendrites very close to the cell bodies, all reaching in tangential directions at right angles to the path of the apical dendrites. Furthermore these fields of apical dendrites and cell bodies are arranged in consistent and distinctive laminas, such that the outstanding feature of cortical cytoarchitecture is its pronounced layering.

Nissl Stain

Nissl stains have the advantage that they can be magnified many times to reveal details of cellular organization in gray matter, since all cell bodies are stained. In particular the regular layers of cortex are revealed when considering large populations of cell bodies. For over a decade, Brodmann extensively and carefully studied Nissl stained sections from an amazing variety of mammalian brains, from the collections at the Vogt Institute for Brain Research. From these he produced his system of classification of cortical regions, based on the variations in sizes, shapes, and arrangements of layers of cell bodies in the different regions of cortex (Brodmann, 1909/2006).

In agreement with Meynert's (1885) original opinion, Brodmann concluded, and stated consistently and frequently, that the claustrum was a subdivision of the sixth layer, the deepest layer of cerebral cortex, that often became separated from the rest of that layer by the passage through

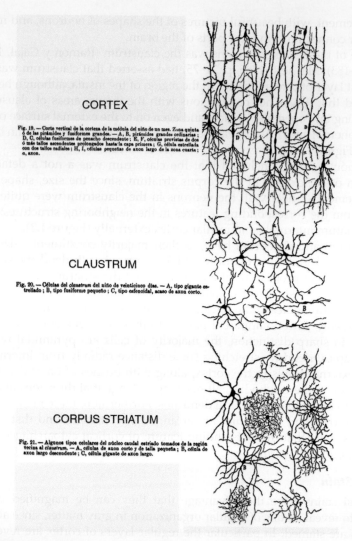

CORTEX

Fig. 19. — Corte vertical de la corteza de la médula del niño de un mes. Zona quinta ó de las pirámides y fusiformes grandes. — A, B, pirámides grandes ordinarias ; D, C, células fusiformes de penacho descendente ; E, F, células provistas de dos ó más tallos ascendentes prolongados hasta la capa primera ; G, célula estrellada con dos tallos radiales ; H, I, células pequeñas de axon largo de la zona cuarta ; a, axon.

CLAUSTRUM

Fig. 20. — Células del *claustrum* del niño de veinticinco días. — A, tipo gigante estrellado ; B, tipo fusiforme pequeño ; C, tipo esfenoidal, acaso de axon corto.

CORPUS STRIATUM

Fig. 21. — Algunos tipos celulares del núcleo caudal estriado tomados de la región vecina al *claustrum*. — A, células de axon corto y de talla pequeña ; B, célula de axon largo descendente ; C, célula gigante de axon largo.

FIGURE 1.2   Depictions of the cell types seen in human cerebral cortex, claustrum, and corpus striatum, revealed by the Golgi stain (in Spanish). *Source: Ramón y Cajal (1900). Made available in the public domain in the USA by the Hathi Trust Digital Library.*

the layer of fibers of the extreme capsule. He classified layer 6 in this region as having three sublayers: 6a, 6b, and 6c. Layer 6c was the claustrum, and this additional sublayer, in his view, was the defining feature of insular cortex. Thus in the Brodmann system, the presence of a claustrum meant that the overlying cortex was classified as insular cortex, and this was regardless of whether or not this region of cortex was covered by folded over operculae of neighboring lobes. This scheme led Brockhaus

(1940) to propose renaming the functional region known vaguely as insular cortex as instead, claustrocortex, with precise boundaries.

Thus Brodmann and Ramón y Cajal each undertook a detailed study of claustrum, with dramatically different conclusions as to its origins, structure and function. The difference is due entirely to the different staining method used in the preparation of their material.

Although far from being the only one to take up one or another sides of these conflicting arguments, Landau (1919) rendered a powerful array of data in support of Cajal's conclusion. The argument has resulted in a persistent flow of data and writings, going on in one form or another up to our own time. Many instances are included in this book, and a particularly pertinent recent study, providing a possible consilience of the two opinions, is the work by Mathur et al. (2009) that we discuss at length in our presentation of the fifth era of this history.

Weigert Stain

Weigert stains color only myelin sheathing, and yield little information about cell size, distribution, and number. But the high degree of contrast of the very black myelinated axons against the pale coloring of other tissue, of the Weigert stain, including its many later variations, gives the clearest definition of what we know as white matter. This can be seen in Figure 1.3, which shows in contrast the results of Nissl and Weigert staining, compared with unstained tissue.

With the greater contrast of the Weigert stains, they have been the usual choice in illustrating low magnification levels in much more economical monochrome publications, especially in older works (e.g. Dejerine and Dejerine-Klumpke, 1895; Ariëns Kappers, 1920).

Marchi Stain

The fourth stain, again from those remarkable years 1884 and 1885, was that invented by Vittorio Marchi (1886). In this method, osmium tetroxide is attached preferentially to myelin sheaths in the process of Wallerian degeneration. This was the earliest means of tracking the course of axons to their destination and was widely used for 70 years. Bianchi (1922) reported results of experiments done in 1897, a very early study of experimental tract-tracing. His main finding was evidence for connections between claustrum and distant regions of cerebral cortex. While he regarded this as a very tentative conclusion, it has been amply verified by subsequent studies up until our own time.

Combined Techniques

Nissl stains could also reveal information about projections from degeneration of cell bodies following the separation or destruction of their synapses in target regions. Both Marchi and Nissl methods were

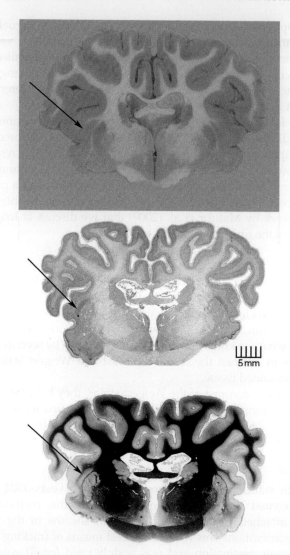

FIGURE 1.3 Comparison of claustrum in sheep (*Ovis aries*) brain before and after staining. Top: unstained fixed brain – this is a photograph of the block face as sections were being cut from the brain before staining. This is the type of material that would have been available to Vicq d'Azyr, Burdach, Arnold, etc. Center: a section stained using the Nissl method using thionine, which colors ribosomes and renders cell bodies and their distribution visible. Bottom: a section stained by the Sanides-Woelcke-Heidenhain version of the Weigert iron hematoxylin stain (Axer et al., 2003), which colors myelin black, rendering such structures as the internal and external capsules in vivid contrast. *Arrows* point to the claustrum in each section. *Images are from specimens at Neuroscience Associates, Knoxville, TN and from the Welker Wisconsin collection at the National Museum of Health and Medicine in Silver Spring MD.*

used in a long series of studies, all showing apparent connections of cells in the claustrum with cerebral cortex (e.g. Berke, 1960; Bianchi, 1922; Mettler, 1935; Narkiewicz, 1964). None, however, owing to the limitations of these techniques, were able to show definitively which cells of the claustrum projected to which cells of the cortex, and vice versa, nor what effects were produced by the connections between these regions.

Proposed Subdivisions of Claustrum and their Nomenclature

Also in this third era of studying just the anatomy of claustrum, sub-divisions of claustrum into dorsal and ventral portions were introduced.

The dorsal portion lies superior to a *rhinal sulcus* (termed the *collateral sulcus* in humans), which, in general, separates six-layered *neocortex* from the alternative laminar patterns of *allocortex* (including those regions known as *paleocortex* and *archicortex*). Intermediate laminar patterns have been called *mesocortex*, and these are in the close vicinity of the rhinal sulcus.

The cortex of the insula includes all three variants, allo-, meso-, and neo-cortex (Mesulam and Mufson, 1982). The dorsal claustrum underlies the cortex of the insular region. The ventral portion continues below the rhinal sulcus and underlies the olfactory region of allocortex (paleocortex) receiving olfactory axons from the *olfactory bulb* via the *lateral olfactory tract*. The qualification here is that, in well-differentiated brains, insular cortex itself is, in part, olfactory paleocortex marked by three layers in its anteroventral corner and receiving terminals from the lateral olfactory tract; and going progressively more posterosuperiorly, more layers are added making the central insula a form of mesocortex, and at the posterolateral end it is indistinguishable from standard six-layered neocortex. Thus ventral claustrum underlies three-layered paleocortex (allocortex), but so does the ventral end of dorsal claustrum. Dorsal claustrum proceeds to underlie all possible types of cortex classified by number of layers.

Loo in 1931 introduced the name *endopiriform nucleus* as an alternative to "ventral claustrum." The name recognizes its position under, and connections with, the *piriform* or olfactory region of paleocortex (allocortex). Dorsal claustrum has no immediately obvious connections with its overlying insular cortex, although these are often assumed to exist. This difference, and the addition of many more differences between the structure and connectivity of the two regions of claustrum, has gradually led to the abandonment of the term "ventral claustrum" in favor of recognizing the *endopiriform* region as altogether distinct from claustrum in structure, connections, and function. But the two regions always have a continuity with one another at a certain point along the fundus of the rhinal sulcus, where there is a bridge of cells in which it is not possible to identify a distinct boundary between endopiriform and claustral cell groups.

These relationships were well expressed by Narkiewicz (1964, p. 336) using the terms *claustrum insulare* for the dorsal claustrum, and *claustrum prepyriforme* for the ventral claustrum:

> Even at a casual glance one can distinguish to a dorsal part, which is composed of comparatively large cells underlying the insular granular (Ig) and agranular (Ia) cortex, and a ventral part containing smaller cells which are intercalated between prepyriform cortex on one side and putamen and amygdaloid complex on the other. These two parts will be referred to as claustrum insulare (Ci) and claustrum prepyriforme (Cp), respectively. Transition between these two sectors occurs usually at the bottom of the rhinal sulcus. However, the claustrum prepyriforme is continuous not only with the claustrum insulare but also with the deepest layers (V and VI) of the neocortex. Therefore, it must remain an open question as to what extent, if at all, claustrum insulare and claustrum prepyriforme do, in fact, form one morphological and functional unit.

This and similar questions, raised especially by the admirably thorough connectional studies of Behan and Haberly (1999) have led most recent accounts and atlases (e.g. Druga, see Chapter 2; Paxinos and Watson, 1998) to use the term *endopiriform* nucleus or region rather than ventral claustrum for the cell group lying inferior to the bottom of the rhinal sulcus. We will follow this consensus and use *endopiriform nucleus* and *claustrum* in place of ventral and dorsal claustrum, respectively, as illustrated in Figure 1.4.

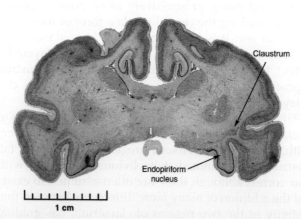

FIGURE 1.4   Nissl stained section, using thionine, through the brain of a Western gray kangaroo *Macropus fuliginosus* showing the continuity between the claustrum and the endopiriform nucleus around the fundus of the rhinal sulcus. The sulcus separates the distinctive allocortex of the piriform olfactory region from the typical 6-layered neocortex that covers most of the hemisphere. *The section is from the Welker Wisconsin Brain Collection at the National Museum of Health and Medicine in Silver Spring, MD.*

THE FOURTH ERA: 1950–2000

A Grand Flowering of Diverse Technology: Revealing Interconnections in Detail and Providing Some Intimations of Function

This next era, like the previous one, was ushered in by an even larger spate of technological advances that markedly changed the nature and the productivity of research into the claustrum. These included the introduction of:

1. **Silver degeneration stains** showing individual neurons of an investigator's choice, the route of their axons, and their synapses. (e.g. Fink and Heimer, 1967; Nauta and Gygax, 1954).
2. **Coloring substances** naturally taken up and distributed to all parts of a neuron, or inserted via micropipettes. These included radioactive solutions that could be identified later with autoradiographic exposure of radiation sensitive film (e.g. Schwab et al., 1978); enzymes to which chromophores could be attached, including horseradish peroxidase HRP (LaVail and LaVail, 1972) and *Phaseolus vulgaris* leucoagglutinin PhaL (Wouterlood et al., 1987); and a number of substances fluorescent under appropriate illumination (e.g. Kuypers et al., 1977).
3. **Microelectrodes**, enabling recording of the activity of individual neurons in the brain, as well as injecting into the interior of nerve cells various substances to mark individual cells, even the ability to record the activity of the cell, then to mark it for eventual localization (e.g. Brock et al., 1952; Graybiel and Devor, 1974; Hubel, 1957).
4. **Electron microscopy**, allowing the visualization of individual neurons, their processes, and synapses (Gray, 1959). In combination with injections into cells, this enabled such feats as identification of individual processes and synapses involved in previously recorded activity.

The power of coordinated use of these varied new techniques was demonstrated in a classical series of highly informative studies (LeVay, 1986; LeVay and Sherk 1981a, 1981b; Sherk and LeVay, 1981a, 1981b, 1983) into the cat claustrum. They provide an excellent model of the type of research needed to provide significant information about the existence and function of the claustrum.

LeVay and Sherk found that at the posterior end of what we here call the "superior pyramidoid puddle" (see Figure 1.5), extending out from the claustral lamina, is a region specialized for processing information coming from, and going back to, visual regions of the cerebral cortex. Using the large three-dimensional volume of the puddle, they were able

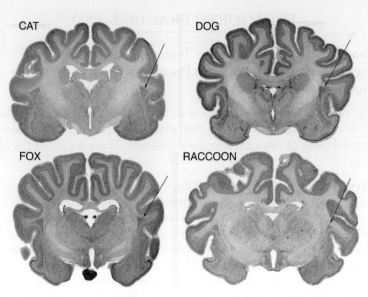

FIGURE 1.5   Superior pyramidoid claustral puddles (*arrows*) spilling upward from the canonical narrow layer of claustrum in sections from brains of four carnivores, stained with the Nissl method using thionine. The *green dots* in the left side of the claustrum of the domestic cat, *Felis catus*, show the approximate location in the posterior portion of the puddle where visual, somatotropic, and auditory specialized regions were found. The dog specimen is from a Basenji hound (*Canis familiaris*), fox from a red fox (*Vulpes vulpes*), and raccoon from a North American raccoon (*Procyon lotor*). *All specimens are from the Welker Wisconsin Collection at the National Museum of Health and Medicine in Silver Spring, MD.*

to provide a comprehensive picture of the cell types present in the claustrum, as well as the connectivity and physiological responsiveness of individual cells in relation to that of neighboring groups of cells. These in turn were related to similar analyses of cerebral cortical cells receiving projections from and sending projections to claustrum.

To achieve this impressively complete account of claustral cells and their role in brain function, it was necessary to coordinate the employment of a variety of different techniques, old and new, including:

- *Golgi staining*, to show the morphology and types of neurons and their frequency in the claustrum;
- *single unit electrophysiological recording*, to show responsiveness of individual cells to stimulus properties;
- *anterograde and retrograde tracers*, to identify not only targets of claustral efferent projections but which cells, where, were sending projections to claustrum;
- *electron microscopy*, to show the location on which cell bodies or dendrites these projections terminated.

They were thereby able to show the following:

1. The great majority of claustral cells have large spiny dendrites.
2. The dendritic spines are specialized to receive synaptic contact almost entirely from cells in the visual cortex.
3. The axons of the large spiny cells go directly and exclusively to cells in the visual cortex.
4. There are neither axons nor axon collaterals terminating in other regions of the claustrum.
5. The size of the puddle permitted mapping of individual retinal field projections (via cortex): Both the claustrum visual regions, and those in cerebral cortex with which they are connected, are retinotopically organized. This allowed placement of tracer injections at specific locations in the visual receptive field in claustrum as well as in cortex.
6. Claustro-cortical axons terminate in layer IV of the cortex; cortico-claustral axons originate in layer VI (as is the case with thalamocortical circuits).

These findings regarding the claustro-visual system in cats were replicated in part by Olson and Graybiel (1980), who were carrying out their study at the same time as the early ones of LeVay and Sherk, but these experiments were far less comprehensive. They did, however, have one additional finding: just anterior to the visual region of the cat claustrum, and also in the superior pyramidoid puddle, was a corresponding region of cortico-claustral reciprocal connections between somatic sensory cortical regions and claustrum. During the same year, claustral reciprocal connections with auditory association areas were located (Irvine and Brugge, 1980). Interestingly, Olson and Graybiel (1980) reported one auditory response in the same small region. This auditory claustral area appears to be inferior to the somatosensory claustral region, in the pyramidoid puddle, but like the somatosensory region is anterior to the visual region. In this respect the relative locations of somatosensory, auditory, and visual regions mirror the spatial relations of the corresponding areas in the cerebral cortex. It would be useful if further experiments could be undertaken to determine these relative locations, in the same way that Olson and Graybiel found that the somatosensory field was definitely anterior to the visual field.

In previous work Graybiel (1978) and Sherk (1979) had studied the parabigeminal nucleus in the midbrain, which has all of its inputs and outputs coming from and going to the superior colliculi, similar to the claustro-cortical relationship. This led to an intriguing hypothesis, phrased well by Olson and Graybiel: the claustrum is a satellite of cerebral cortex and performs some special task in the communication between one cortical region and another within a single sensory processing system.

There were a great many other studies involving the claustrum in some way during this period, most of which are covered in Chapter 2. The most coherent story to emerge from this era is in the findings of LeVay and Sherk and the two supporting studies, which suggest that those experimenting in this area should:

1. maximize the number and variety of techniques used in attacking a particular problem, and
2. maximize the chances of significant findings by working in a region with a good three-dimensional expanse.

THE FIFTH ERA: 2000–INTO THE FUTURE

The Power of Proteomics, the Promising Puddles, the Beckoning Unexplored World of Astrocytes, and More

With the coming of the new millennium, research on the claustrum welcomes new methods, notably the advent of advanced proteomics and new and revolutionary concepts.

An early beginning of the proteomic approach was the most useful finding of Druga et al. (1993) that calbindin and parvalbumin could provide fairly reliable evidence of the presence of, and difference between, the now securely named claustrum on the one hand, and endopiriform nuclei on the other.

Even more useful, and titled as such, was the work of Mathur et al. (2009) that claustral cells could be labeled in distinction to cerebral cortical cells. In rats, the conflict of opinion as to whether claustral cells were displaced cortical cells received a new twist: they found that both types of cells coexist in the deepest layer of rat insular cortex. Here, a fairly solid mass of claustral cells was peppered with a number of cortical cells, and with a thin layer of cortical cells on the internal surface abutting the external capsule overlying the corpus striatum. The claustrum is sandwiched into layer 6 and not separated at all from the overlying insular cortex, validating to some degree the opening argument of Meynert and Brodmann.

However, the lack of separation by an extreme capsule calls into question whether rats are indeed proper models for the general conclusions about mammalian claustrums. This may not be a function of body size and development of an extreme capsule – manatees also present the appearance of a claustrum-like layer of darkly staining cells on the internal surface of the cerebral cortex, not separated at all from overlying cortex; theirs is a quite large brain, but with no cortical convolutions and no identifiable insula (Figure 1.6).

Rats (and mice for that matter) may possess at best a vestigial claustrum that never developed into its full potential as a brain part, or one

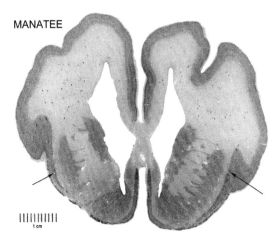

FIGURE 1.6   The possible location of a claustrum (*arrows*) in a coronal section from the brain of a Florida manatee, *Trichechus manatus latirostris*. This large brain shows that a distinct extreme capsule is not a universal property of very large mammalian brains. *This section, Nissl stained with thionine is from the Welker Wisconsin Brain Collection, and the specimen was obtained with the assistance of Dr. Roger L. Reep of the University of Florida.*

that has been lost through specialization in their complex evolutionary history – in generational terms the longest and most eventful such history among mammals, which has left them at a pinnacle of the mammalian evolutionary tree. The rate of evolutionary innovation and adaptation is a function of the number of generations, each one offering an opportunity for novel combinations of genes and the occurrence of mutations. With their rapid reproduction and maturation rates, the small rodents have become the most numerous of the mammals, accounting for about half the number of mammals and species in existence. Thus they can be considered the zenith of evolutionary specialization of mammals. However, this specialization can consist in jettisoning brain parts that are not particularly useful in the adaptational and survival wars that species undergo (e.g. Catania et al., 1999).

This reminds us of the relative decline of olfactory central neural systems in primates, and their complete eradication in some dolphin species, in the course of their evolutionary development. Either way, despite their convenience as laboratory animals (owing to their rapid reproduction rates and the consequent ease with which they can be genetically manipulated), it is being increasingly recognized that rats and mice have severe limitations as model neural systems. On the other hand, the rudimentary nature of the murine rodent claustrum probably facilitated the identification of the claustrum as being enclosed in cerebral cortex. Once the proteomic markers had been developed, the presence of cortical cells internal to the claustrum was quite evident, and might never have been noticed if studies were restricted to animals with more segregated claustrums, because there the internal cortical cells, even when marked, are so sparse that they would have escaped detection without specially looking for them. Taking the necessary precaution of testing their concept in a species with an undoubted claustrum, separated from insular cortex by a well-defined extreme capsule, Mathur et al. (2009) found that in green vervet monkeys, *Chlorocebus sabaeus*, there was a thin scattering of cortical cells on the inner surface of the claustrum, up against the external capsule. So it may be true that, as Brodmann stated, the claustrum (and the extreme capsule as well) always lives "within" cortex, although as Ramón y Cajal (1900) and Landau (1919) maintained, its structure and connections are very different from cells of other cortical regions.

It will be very instructive to see if the population of cortical cells internal to the claustrum in puddling regions is different from those in other regions. Do the "hearts" of puddles have more, or fewer, interloping cortical cells than do the thin-laminar regions? And it will be fascinating to learn in detail not only the structure, connections, and functions of these cortical cells that live scattered within the great mass of claustral cells, but also their relationship, if any, to the surrounding claustral cells. Do they offer any excitatory or inhibitory effects on the claustral cells, and how do the local astrocytes, ordinarily the regulators of their domains, deal with these two very diverse sets of tenants within their domains?

The Essence of Claustrumhood

What is a claustrum anyway? We know it is a sandwich of cells enclosed between fiber bundles, as noted initially by Vicq d'Azyr. But is it also a collection of cells with common ancestry, architecture, connections and functions, in common with most defined nuclear regions? This latter question, after 227 years, can still not be answered without proper data regarding ancestry, architecture, connections and function. Proteomic identification can thus be a giant step forward in understanding all aspects of this structure.

We might even see the day when the distinct nature of endopiriform regions and claustrum may turn out to be not so distinct after all. Suppose the endopiriform cells are doing for the variegated olfactory cortices just what the claustral cells are doing for the more uniform isocortical regions. That would be ironic, in that the endopiriform cells may turn out to be the claustrum for the olfactory cortex.

The Promising Puddles

The requirement of most techniques, for a spatially large study area, as Mathur et al. (2009) pointed out, has discouraged extensive research on the claustrum, whose canonical morphology is a thin layer of cells. In many species, however, there are regions of claustrum that form "puddles" of three-dimensional expanse, clearly distinct from surrounding non-claustral regions (Buchanan and Johnson, 2011; Figures 1.4, 1.7–1.9).

MONKEY

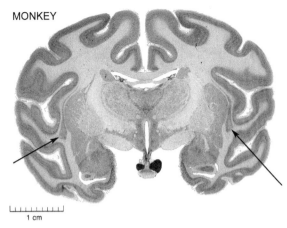

1 cm

FIGURE 1.7   The antero-inferior puddle (*arrows*) of a monkey claustrum. The cells of the enclosed wall of gray matter appear to spread out into a pool just external to the lower reaches of the putamen, where the putamen is about to fuse with the amygdaloid complex. This puddle is typical of anthropoid primates, including humans, and was recognized by Meynert (1885), who called it the ventral claustrum. This specimen is from a pig-tailed macaque, *Macaca nemestrina*, and the section was Nissl stained with thionine. *From the Welker Wisconsin Brain Collection at the National Museum of Health and Medicine in Silver Spring, MD.*

PIG

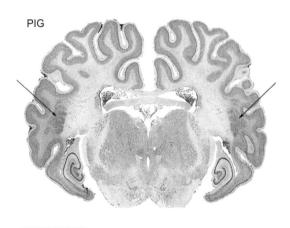

1 cm

FIGURE 1.8   The extraordinarily large posterior puddle of the pig claustrum (*arrows*). Claustral cells form into a massive structure consisting of a number of separate lobules, at the posterior end of the canonical narrow laminar claustrum, and reaching posteriorly to the level of the habernula and the red nucleus. This section is from a domestic pig, *Sus scrofa. From the Welker Wisconsin Brain Collection at the National Museum of Health and Medicine in Silver Spring, MD.*

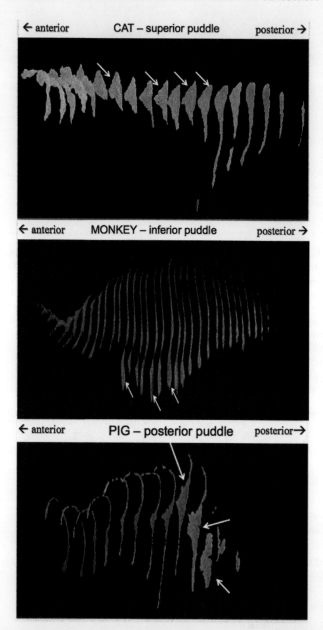

FIGURE 1.9 Three-dimensional reconstructions of diverse claustrums and puddles, made from tracings of coronal sections from domestic cat (*Felis catus*), vervet monkey (*Chlorocebus aethiops*), and domestic pig (*Sus scrofa*). From the canonical narrow lamina of claustrum, claustral cells pour out in different and distinctive directions and shapes in different mammalian taxa. The wider expanse of these puddle regions offer significant research opportunities. *Reconstructions created by Kenneth Buchanan.*

This leads to our recommendation that these large puddles be investigated (as the superior pyramidoid puddle in cats was used, possibly unwittingly, by LeVay and Sherk) to reveal complete claustral connectivity and much about function, and to generate reasonable hypotheses as to why claustrums exist and how they contribute to brain function.

Narkiewicz (1964) and Kowiański et al. (1999) also show this puddle in their diagrams of the morphology of the cat's claustrum: the superior pyramidoid puddle seen in many carnivores, as we have shown in Figure 1.4.

Puddles offer the wide expanse needed for mapping details of connectivity and some degree of electrophysiological activity, as seen in the retinotopic maps of LeVay and Sherk (1981b). They offer a similar advantage in studies using multi-contact electrodes that can simultaneously map activity of several individual entities up and down the shafts of a near-microscopic array of hundreds of recording points, to show population characteristics of tissue activity at rest as well as in response to stimulation of various kinds.

The availability of different shapes and sizes of puddles in different species strongly recommend a comparative approach to further claustral studies. The roomy posterior puddle of the pig (see Figure 1.8), in particular, invites attention, as pigs are now being produced and used as convenient and small laboratory subjects (e. g. Ellegaard Göttingen Minipigs at http://minipigs.dk/).

The Beckoning World of Astroglia

Another looming question, yet to be addressed in any way in claustrum research, is the role of astroglia in brain function. Recognized only in the past decade, early findings have given rise to a spate of speculations that isolated neuronal studies are revealing at best only a part of the functional systems involved in behavioral control. Early expectations, perhaps exaggerated, have taken popular book format, each by a recognized researcher in neuroscience (Fields, 2009; Koob, 2009).

Brain maps, based solely on the distinctive characteristics of populations of astroglial cells, look remarkably like those based on distinctive characteristics of neurons (Emsley and Macklis, 2006). Thus whole new programs of investigation will lie in identification of astroglia in the claustrum, and their function in such roles as neurogenesis, neuronal migration into adult locations, synaptic formation, and the permission or forbidding, or facilitation or inhibition, of transmission through these synapses, as dictated by current events and situations.

Ramón y Cajal (1900) recognized a variety of glial cells populating the extreme capsule, but mentioned none at all in the claustrum. It is time for another look.

There Will be Yet More

Astonishing new imaging using molecular constitution of various constituents may mean yet another fresh and revolutionary look at this still mysterious set of cells that are busy in our brain workings (Chung et al., 2013).

A coordinated and targeted approach, multidisciplinary and probably multi-institutional, is the optimal road ahead.

Acknowledgments

Kennneth Buchanan did much of the early stages of research and conceptualization for this review. Lawrence Edelstein was a constant source of historical references. Scans of the dolphin and human sections in Figure 1.1 were made with the cooperation of Archibald Fobbs at the National Museum of Health and Medicine. All of the other scans of stained sections were undertaken by Carol Dizack and the late Wally Welker at the University of Wisconsin. Use of all of this material is with the cooperation of Adrianne Noe, director, the National Museum of Health and Medicine, Silver Spring, MD.

References

Ariëns Kappers, C.U., 1920. Die vergleichende Anatomie des Zentralnervensystems des Wirbeltiere und des Menschen. De Erven F. Bohn, Haarlem.

Axer, H., Scheulen, K., Gerhards, C., Gräszel, D., Südfeld, D., v. Keyserlingk, D., 2003. Three-dimensional reconstruction of a Rhesus monkey brain from the Friedrich Sanides collection. Ann. Anat. 185, 315–323.

Arnold, F., 1838. Tabulae Anatomicae. Orelli, Fuesellini et Sociorum, Turici.

Behan, M., Haberly, L.B., 1999. Intrinsic and efferent connections of the endopiriform nucleus in rat. J. Comp. Neurol. 408 (4), 532–548.

Berke, J.J., 1960. The claustrum, the external capsule and the extreme capsule of Macaca mulatta. Neurology 115, 297–321.

Bianchi, L., 1922. The Mechanism of the Brain and the Function of the Frontal Lobes (J.H. MacDonald, Trans.). E & S. Livingstone, Edinburgh.

Brock, L.G., Coombs, J.S., Eccles, J.C., 1952. The recording of potentials from motoneurones with an intracellular microelectrode. J. Physiol. (London) 117, 431–460.

Brockhaus, H., 1940. Cytoarchitectural and myeloarchitectural study of claustral cortex and claustrum in man. J. Psychol. Neurol. 49, 249–348.

Brodmann K., 2006. J. Brodmann's localisation in the cerebral cortex. (L. Garey, Trans.). New York: Springer. (Originally published as Vergleichende Lokalisationslehre der Grosshirnrinde. Leipzig: Johann Ambrosius Bart, 1909).

Broesike, G., 1904. Lehrbuch der normalen Anatomie des menschlichen Körpers. Fischer's Medicin. Buchhandlung H. Kornfeld, Berlin.

Buchanan, K.J., Johnson, J.I., 2011. Diversity of spatial relationships of the claustrum and insula in branches of the mammalian radiation. Ann. N.Y. Acad. Sci. 1225 (Suppl. 1), E30–E63.

Burdach, K., 2012. Vom Baue und Leben des Gehirns, vol. 1. Nabu Press, New York, (Original work published 1819).

Burdach, K., 2012. Vom Baue und Leben des Gehirns, vol. 2. Nabu Press, New York, (Original work published 1822).

Catania, K.C., Lyon, D.C., Mock, O.B., Kaas, J.H., 1999. Cortical organization in shrews: evidence from five species. J. Comp. Neurol. 410 (1), 55–72.

Chung, K., Wallace, J., Kim, S.Y., Kalyanasundaram, S., Andalman, A.S., Davidson, T.J., et al., 2013. Structural and molecular interrogation of intact biological systems. Nature. Apr doi: 10.10.1038/nature12107. [Epub ahead of print]

Dejerine, J.J., Dejerine-Klumpke, A.M., 1895, 1901. Anatomie des centres nerveux. Rueff, Paris.

Druga, R., Chen, S., Bentivoglio, M., 1993. Parvalbumin and calbindin in the rat claustrum: an immunocytochemical study combined with retrograde tracing frontoparietal cortex. J. Chem. Neuroanat 6 (6), 399–406.

Edelstein, L.R., Denaro, F.J., 2004. The claustrum: A historical review of its anatomy, physiology, cytochemistry and functional significance. Cell. Mol. Biol. (Noisy le Grand) 50, 675–702.

Emsley, J.G., Macklis, J.D., 2006. Astroglial heterogeneity closely reflects the neuronal-defined anatomy of the adult murine CNS. Neuron Glia Biol. 2 (3), 175–186.

Fields, R.D., 2009. The other brain: The scientific and medical breakthroughs that will heal our brains and revolutionize our health. Simon and Schuster, New York.

Fink, R.P., Heimer, L., 1967. Two methods for selective silver impregnation of degenerating axons and their synaptic endings in the central nervous system. Brain Res. 4 (4), 369–374.

Golgi, C., 1873. Sulla struttura della sostanza grigia del cervello. Gazz. Med. Ital. 33, 244–246.

Gray, E.G., 1959. Axo-somatic and axo-dendritic synapses of the cerebral cortex: an electron microscope study. J. Anat. 93, 420–433.

Graybiel, A.M., 1978. A satellite system of the superior colliculus: the parabigeminal nucleus and its projections to the superficial collicular layers. Brain Res. 145 (2), 365–374.

Graybiel, A.M., Devor, M., 1974. A microelectrophoretic delivery technique for use with horseradish peroxidase. Brain Res. 68 (1), 167–173.

Hubel, D.H., 1957. Tungsten microelectrode for recording from single units. Science 125 (3247), 549–550.

Irvine, D.R., Brugge, J.F., 1980. Afferent and efferent connections between the claustrum and parietal association cortex in cat: a horseradish peroxidase and autoradiographic study. Neurosci. Lett. 20 (1), 5–10.

Knox, R., 1831. System of human anatomy: On the basis of the "Traité d'Anatomie Descriptive" of M.H. Cloquet, second ed. Maclachlan and Stewart, Edinburgh.

Koob, A., 2009. The root of thought: Unlocking glia – the brain cell that will help us sharpen our wits, heal injury, and treat brain disease. FT Press, Upper Saddle River, NJ.

Kowiański, P., Dziewiatkowski, J., Kowiańska, J., Moryś, J., 1999. Comparative anatomy of the claustrum in selected species: A morphometric analysis. Brain Behav. Evol. 53 (1), 44–54.

Kuypers, H.G., Catsman-Berrevoets, C.E., Padt, R.E., 1977. Retrograde axonal transport of fluorescent substances in the rat's forebrain. Neurosci. Lett. 6 (2–3), 127–133.

Landau, E., 1919. The comparative anatomy of the nucleus amygdalae, the claustrum, and the insular cortex. J. Anat. 53, 351–360.

LaVail, J.H., LaVail, M.M., 1972. Retrograde axonal transport in the central nervous system. Science 176 (4042), 1416–1417.

LeVay, S., 1986. Synaptic organization of claustral and geniculate afferents to the visual cortex of the cat. J. Neurosci. Dec. 6 (12), 3564–3575.

LeVay, S., Sherk, H., 1981a. The visual claustrum of the cat. I. Structure and connections. J. Neurosci. 1 (9), 956–980.

LeVay, S., Sherk, H., 1981b. The visual claustrum of the cat. II. The visual field map. J. Necrosci. 1 (9), 981–992.

Loo, Y.T., 1931. The forebrain of the opossum Didelphis virginiana. Part II. Histology. J. Comp. Neurol. 52 (1), 1–148.

Marchi, V., 1886. Des dégénérations consécutives à l'extirpation totale et partielle du cervelet. Arch. Ital. Biol. 7, 357–362.

Mathur, B.N., Caprioli, R.M., Deutch, A.Y., 2009. Proteomic analysis illuminates a novel structural definition of the claustrum and insula. Cereb. Cortex 19, 2372–2379.

Mesulam, M.M., Mufson, E.J., 1982. Insula of the old world monkey. I. Architectonics in the insulo-orbito-temporal component of the paralimbic brain. J. Comp. Neurol. 212 (1), 1–22.

Mettler, F.A., 1935. Corticofugal fiber connections of the cortex of Macaca mulatta. The frontal region. J. Comp. Neurol. 61, 509–542.

Meyer, A., 1971. Historical aspects of cerebral anatomy. Oxford University Press, London.

Meynert, T., 1885. Psychiatry. A Clinical Treatise on Diseases of the Forebrain Based upon a Study of its Structure, Functions, and Nutrition. Part 1. The Anatomy, Physiology, and Chemistry of the Brain. G.P. Putnam's Sons, New York, Reprinted by Ulan Press, Lexington KY, 2013.

Moreau, J.L., 1805. Oeuvres de Vicq-d'Azyr. Sciences Physiologiques et Médicales. Duprat-Duverger, Paris.

Narkiewicz, O., 1964. Degenerations in the claustrum after regional neocortical ablations in the cat. J. Comp. Neurol. 123, 335–356.

Nissl, F., 1894. Ueber eine neue Untersuchungsmethode des Zentralorgans speciell zur Feststellung der Localisation der Nervenzellen. Neurol. Zentralbl. 13, 507.

Nauta, W.J.H., Gygax, P.A., 1954. Silver impregnation of degenerating axons in the central nervous system. A modified technique. Stain Technol. 29, 91–93.

Olry, R., Haines, D.E., 2001. Claustrum: A sea wall between the island and the shell? J. Hist. Neurosci. 10 (3), 321–322.

Olson, C.R., Graybiel, A.M., 1980. Sensory maps in the claustrum of the cat. Nature 288, 479–481.

Parent, A., 2007. Félix Vicq d'Azyr: Anatomy, medicine and revolution. Can. J. Neurosci. 34, 30–37.

Paxinos, G., Watson, C., 1998. The Rat Brain in Stereotaxic Coordinates, fourth ed. Academic Press, San Diego.

Rae, A., 1954. The form of the human claustrum. J. Comp. Neurol. 1 (100), 15–39.

Ramón, y., Cajal, S., 1900. Estudios sobre la corteza cerebral humana. III. Estrutura de la corteza acustica. Rev. Trimmest. Micrográfica 5, 129–183.

Reil, J.C., 1809. Das Balken-System oder die Balken-Organisation im grossen Gehirn. Arch. Physiol. Halle 9, 172–195.

Schwab, M.E., Javoy-Agid, F., Agid, J., 1978. Labeled wheat germ agglutinin (WGA) as a new, highly sensitive retrograde tracer in the rat brain hippocampal system. Brain Res. 152, 145–150.

Sherk, H., 1979. Connections and visual-field mapping in cat's tectoparabigeminal circuit. J. Neurophysiol. Nov. 42 (6), 1656–1668.

Sherk, H., LeVay, S., 1981a. Visual claustrum: topography and receptive field properties in the cat. Science 212 (4490), 87–89.

Sherk, H., LeVay, S., 1981b. The visual claustrum of the cat. III. Receptive field properties. J. Neurosci. 1 (9), 993–1002.

Sherk, H., LeVay, S., 1983. Contribution of the cortico-claustral loop to receptive field properties in area 17 of the cat. J. Neurosci. 3 (11), 2121–2127.

Turner, W., 1866. The convolutions of the human cerebrum topographically considered. Edinb. Med. Surg. J. 11 (Part II), 1105–1122.

Vicq d'Azyr, F., 1786. Traité d'anatomie et de physiologie avec des planches coloriées représentant au naturel les divers organes de l'homme et des animaux, vol. 1. François, Paris.

Vicq d'Azyr, F., 1813a. Système Anatomique et de Physiologie, Tome 3, Mammifères et Oiseaux, Commence par Félix Vicq d'Azyr et continue par Hippolyte Cloquet. Agasse, Paris.

Vicq d'Azyr, F., 1813b. Traité de l'Anatomie du Cerveau. Louis Duprat-Duverger, Paris.

Weigert, C., 1885. Eine verbesserung der Haematoxylin Blutlaugensalzmethod fur das Centranlnervensystem. Fortschr. Deutsch. Med. 3, 236–239.

Wouterlood, F.G., Bol, J.G., Steinbusch, H.W., 1987. Double-label immunocytochemistry: combination of anterograde neuroanatomical tracing with Phaseolus vulgaris leuco-agglutinin and enzyme immunocytochemistry of target neurons. J. Histochem. Cytochem. 35 (8), 817–823.

Wundt, W., 1904. Principles of Physiological Psychology, vol. 1. Swan Sonnenschein, London, (E.B. Titchener, Transl. from fifth German ed., 1902).

The Structure and Connections of the Claustrum

Rastislav Druga

Institute of Anatomy, 2nd Medical Faculty, Charles University, Prague, Czech Republic

INTRODUCTION

The claustrum is a telencephalic, pallial subcortical nuclear mass which has been identified in all eutherian and metatherian mammals. Its function is still a matter of discussion (Ashwell et al., 2004; Edelstein and Denaro, 2004; Smythies et al., 2012). Differences in morphology and relationships to the cortex allow us to distinguish its two parts. The group of neurons located above the level of the rhinal fissure medial to the insular cortex has been called dorsal or insular claustrum (Buchanan and Johnson, 2011; Druga, 1966a, 1975; Kowiański et al., 1999; Narkiewicz, 1964, 1966; Narkiewicz and Mamos, 1990). The group of neurons situated below the rhinal fissure, medial to piriform cortex, has been designated ventral or piriform (prepiriform) claustrum, and also endopiriform nucleus (Druga, 1966a; Kowiański et al., 1999; Kretek and Price, 1977; Narkiewicz, 1964). Paxinos and Watson (1997) used the term endopiriform nucleus to describe the ventral claustrum in the rat, and they divided this into two parts – the dorsal endopiriform nucleus (DEN) and the ventral endopiriform nucleus (VEN). The DEN was localized rostrally to the pre-endopiriform nucleus (Behan and Haberly, 1999; Ekstrand et al., 2001). Recently the term endopiriform nucleus (END) has been frequently used for the ventral claustrum with the term claustrum (CL) being reserved for the dorsal (insular) claustrum.

An increasing body of evidence has been collected not only about the structure, ultrastructure, and hodology of the claustrum but also about its origin. On the basis of transcription factors expressed during development, the claustrum and the amygdaloid nuclei have been considered

J. Smythies, L. Edelstein and V. Ramachandran (Eds): The Claustrum.

DOI: http://dx.doi.org/10.1016/B978-0-12-404566-8.00002-7

to be derived from the lateral and ventral pallium.The dorsal part of the claustrum (the dorsal or insular claustrum, CL) was classified as being a derivative of the lateral pallium, whereas the endopiriform nucleus (the ventral claustrum, END) was considered to be a derivative of the ventral pallium. The CL, together with END and the pallial amygdala, may be regarded as a single entity – the claustroamygdaloid complex (Medina et al., 2004; Puelles et al., 2000).

The morphology and cytoarchitecture of this structure have been described in many mammals (Berke, 1960; Brockhaus, 1940; Druga, 1966a; Edelstein and Denaro, 2004; Filimonoff, 1966; Kowiański et al., 1999; Narkiewicz, 1964; Narkiewicz and Mamos, 1990; Rae, 1954; Wójcik et al., 2002). In small lissencephalic animals (mouse, bat, hedgehog, rat) the claustrum is small, and due to a weakly developed extreme capsule, is sometimes difficult to distinguish from the adjacent cortex. In some other rodents (beaver, guinea pigs), and in lagomorphs as well (as in carnivores and primates), the volume of the claustrum increases and the nucleus is easily differentiated from the adjoining structures.

STRUCTURE OF THE CLAUSTRUM

Gross Anatomy

Some authors have suggested that the claustrum is missing from the hemisphere of living monotremes (echidna and platypus) (Butler et al., 2002; Divac et al., 1987). This view was rejected by Ashwell et al. (2004), who described the dorsal and ventral claustrum in both representatives. The CL in the echidna was identified as a cluster of neurons with a circular outline lying ventrolateral to the putamen and separated from the cortex by the extreme capsule. The END is represented by an elongated sheet of neurons embedded in the white matter of the piriform lobe, dorsally intervening between the CL and insular cortex. The identification of the claustrum in the platypus is not as clear as in the echidna. It probably corresponds to a band of neurons extending dorsally from the core of the piriform lobe (the END) to the area between the striatum and sixth layer of the insular cortex (the CL). The CL was not clearly separated from the cortex by an extreme capsule. The claustrum of the North American opossum (*Didelphis marsupialis*) and kangaroo (*Macropus*) is described by Pilleri (1961) as a voluminous structure consisting of two parts, the claustrum dorsale and claustrum ventrale (see also Buchanan and Johnson, 2011).

In the hemisphere of hedgehog tenrec the neo-paleocortical transitional area contains several neuronal groups. Cyto- and chemoarchitectural findings suggest that the claustrum (neuronal group H) is not located subcortically but between the layers of neo-paleocortical transitional area (Künzle et al., 2001). Two groups of neurons were distinguished in the claustral area in the hedgehog and several other insectivores. The lateral group was interpreted either as a condensation of cells formed by merging of the fourth to the sixth layers (Druga, 1974) or as part of the claustrum itself (pars principalis). The medial group (lamina profunda claustri) is dorsally continuous with the deepest layer of the neocortex (lamina VIb), whereas ventrally it merges with the prepyriform claustrum (Dinopoulos et al., 1992; Narkiewicz and Mamos, 1990).

The claustrum of the Canadian beaver (*Castor canadensis*), and the claustra of eleven species of rodents, have been described by Pilleri (1961). In the beaver, as in other rodents, the claustrum consists of a cell-rich and more compact dorsal part, and a cell-poor ventral part. In addition to the description of cell types, Pilleri defined three cytoarchitectonic fields in the CL and two in the END.

The CL of the rat is located under the cortex adjacent to the rhinal fissure. Medially it is bounded by the external capsule, laterally it is in contact with insular cortex. The size of the nucleus decreases proceeding posteriorly.

The endopiriform nucleus was identified as a well-defined group of neurons that underlies the piriform cortex, lateral to the amygdala (Paxinos and Watson, 1997).

Detailed descriptions of the cytoarchitecture of the claustrum in the cat were provided by Narkiewicz (1964) and Druga (1966a). They described how the claustrum is embedded in the white matter, and elongated in the anteroposterior (AP) direction. The dorsal part is located deep to the insular cortex, whereas the ventral part adheres to the piriform cortex. At more caudal levels the nucleus has the form of a triangle, and occipitally the CL forms a dorsoventrally elongated structure. The CL, as seen in Nissl preparations, consists of densely packed large to medium-sized polymorphous cells (Chadzypanagiotis and Narkiewicz, 1971; Druga, 1966a; Kowiański et al., 1999; LeVay and Sherk, 1981a; Miodonski, 1975; Narkiewicz, 1964). The CL is continuous with the END via a transitional zone, which has the form of narrow cellular bridge and is situated at the level of the rhinal sulcus (Figure 2.1).

In non-human primates (*Macaca mulatta, Cercopithecus aethiops*) the claustrum forms a thin layer of gray matter that is vertically oriented and well demarcated from the surrounding structures by the external and extreme capsules (Berke, 1960; Kowiański et al., 1999).

The claustrum of monkey (*Macaca mulatta*) consists of two principal subdivisions i.e. CL and END. The anterior part of the END is in contact with prepiriform cortex (field Pp 2), medially and basally (Filimonoff, 1949) and has relations with the insular cortex laterally (Figure 2.2). At

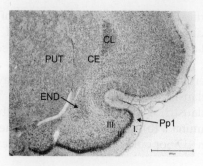

FIGURE 2.1   Claustrum of the guinea pig. Coronal section at intermediate third of the claustrum, stained with cresyl violet. CE, external capsule; CL, dorsal (insular) claustrum; END, endopiriform nucleus (ventral, piriform claustrum); Pp 1, prepiriform cortex; PUT, putamen; I–III, layers of prepiriform cortex. Bar = 1000 μm.

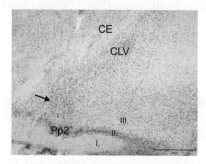

FIGURE 2.2   Claustrum of the monkey (*Macaca mulatta*). Coronal section stained with cresyl violet. An *arrow* indicates a contact of the ventral claustrum (CLV) with the prepiriform cortex (field Pp2). CE, external capsule; I–III, layers 1–3 of the prepiriform cortex. Bar = 1000 μm.

the level of the amygdala the medial extension of the END has relations to the amygdala and with the entorhinal cortex.

The human claustrum has been analyzed in several studies, but its parcellation is not uniform. Brockhaus (1940) distinguished two principal parts of the claustrum. The dorsally located claustrum insulare, which is subdivided into claustrum insulare dorsale and claustrum insulare ventrale, and ventral part of the claustrum, which is subdivided into four subdivisions – claustrum temporale laterale, claustrum praeamygdaleum, claustrum praepyriforme and claustrum reticulare. Filimonoff (1966, 1974) emphasized the complicated structure and relations of the ventral claustrum to adjoining structures, and described its relations to the human piriform cortex. The anterior part of the ventral claustrum (claustrum basale) has intimate relations with the prepiriform cortex (field Pp2a, Pp2b) according to Filimonoff (1966, 1974). The intermediate part of the ventral

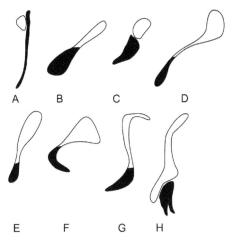

FIGURE 2.3   The schematic transverse sections of the claustrum representing various types. A, echidna; B, sorex; C, rat; D, Guinea pig; E, rabbit; F, cat; G, macaque; H, man. *Black* color indicates configuration of the ventral claustrum (END), *white* color represents the dorsal claustrum. *Adapted from Ashwell et al. (2004) (A), and Kowiański, Dziewiatkowski, and Kowianska (1999) (B–H).*

claustrum consists of two strips, the outer one adjoining the insular cortex, the inner one adjoining the amygdala and the prepiriform cortex. In its posterior part the ventral claustrum approaches the base of the insula.

On the basis of the volume and shape of the claustrum, Kowiański et al., (1999) differentiated five types of claustral shape. Type I is present in insectivores and rodents. In these species the claustrum is not well separated from the cortex, and its ventral part can be divided into the dorsal and ventral endopiriform nucleus (Dinopoulos et al., 1992; Druga, 1974; Narkiewicz and Mamos, 1990; Paxinos and Watson, 1997). Type II is found in guinea pig and rabbit, where the claustrum is distinctly separate from the insular cortex, its vertical dimension is larger, and the transitional zone between dorsal and ventral claustrum is longer than in type I. Type III is characteristic of carnivores (cat, dog), where the CL exhibits a triangular configuration (Druga, 1966a; Miodonski, 1975; Narkiewicz, 1964). The claustrum of non-human primates is type IV, and the most complex claustrum, designated as type V, is found in humans. The claustrum in primates is thin and elongated dorsally. Ventrally, it is expanded in the intermediate third of its anterior–posterior extent (Berke, 1960; Brockhaus, 1940; Filimonoff, 1966) (Figure 2.3).

The shape of the claustrum varies from species to species, but seems to increase in size in proportion to increasing neocortical volume. Volumetric data (Kowiański et al., 1999) indicates that the increase of the claustral log-volume follows the increase of the log-volume of the cerebral hemisphere. The ratio of the claustral volume to the volume of the cerebral hemisphere varies from 0.99 percent (mice) to 0.24 percent in humans. In all species analyzed, the volume of the dorsal part of the claustrum is larger than that of the ventral part (57.0–74.6%).

Neurons of the Claustrum

Basic Staining

In all mammals, cells of various sizes and shapes are found in the CL and END. The most common are medium-sized neurons with triangular, multipolar, oval or fusiform shapes (Ashwell et al., 2004; Dinopoulos et al., 1992; Druga, 1966a, 1974). In the rabbit claustrum, fusiform, pyramidal and multipolar neurons prevail in the anterior and posterior part of the nucleus, whereas in the central part oval are the commonest (Wójcik et al., 2002) (Figures 2.4 and 2.5).

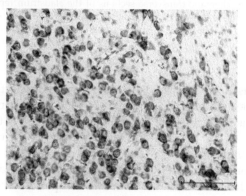

FIGURE 2.4   Microphotograph demonstrating structure and neurons of the dorsal claustrum (CL) in the rabbit. Coronal section stained with cresyl violet. Bar = 100 μm.

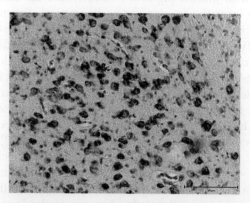

FIGURE 2.5   Microphotograph demonstrating structure and neurons of the ventral claustrum (END) in the rabbit. Coronal section stained with cresyl violet. Bar = 100 μm.

The CL in the cat, as seen in Nissl preparations, is characterized by its compact structure, a great density of cells, and the prevalence of multipolar cells. In Nissl preparations in the cat, Druga (1966a), and Chadzypanagiotis and Narkiewicz (1971) described three to five types of neurons. The commonest are multipolar cells, which are intermingled with a small number of pyramidal and spindle-shaped cells (Moryś et al., 1996). Norita and Hirata (1976) used semi-thin sections stained with toluidine blue to determine the basic typology of the claustral neurons in the cat. They described four types of neurons: large, fusiform, medium-sized and small. Myelinated fibers were sparse and most of them were less than 1 μm in diameter.

Marginal Zones

In the CL, it is possible to distinguish several marginal zones with structures that differ from that of the central zone (Druga, 1966a; Pilleri, 1961). The dorsal marginal zone fills the angle between the medial and the dorsolateral borders. Small, round and multipolar cells are characteristic in this zone. The lateral marginal zone, which forms the dorsolateral margin of the claustrum is characterized by spindle cells. The medial marginal zone is situated at the medial border of the claustrum facing the putamen. This zone exhibits a high density of large multipolar and triangular (pyramidal) cells (Druga, 1966a). The medial marginal zone together with the ventrally adhering narrow cell bridge of fusiform cells was designated an intermediate part of the claustrum on the basis of the organization of claustro-cortical projections (Witter et al., 1988). According to this concept, the claustrum is parcellated into dorsal, intermediate and ventral parts. Such parcellation appears to be useful for describing the limbic projections to the claustrum (Witter et al., 1988, see below) (Figures 2.6 and 2.7).

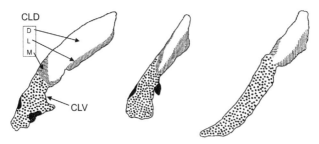

FIGURE 2.6 The schematic transverse sections of the anterior, middle and posterior third of the claustrum of the beaver. Three cytoarchitectonic fields can be defined in the dorsal claustrum (CLD): D, dorsal; L, lateral; M, medial, and two cytoarchitectonic fields in the ventral claustrum (CLV). *Adapted from Pilleri (1962).*

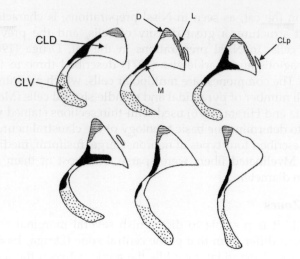

FIGURE 2.7   Cytoarchitectonic marginal zones in the dorsal claustrum of the cat. CLD, dorsal claustrum; CLV, ventral claustrum; D, dorsal marginal zone; L, lateral marginal zone; M, medial marginal zone; CLp, claustrum parvum. *Adapted from Druga (1966a).*

Golgi Stain Studies

Golgi impregnation studies of the claustrum in various species reveal two major classes of neurons: projection neurons with polymorphic perikarya and spiny dendrites, and local neurons (interneurons) characterized by round or oval perikarya and aspiny smooth dendrites. The stain also reveals different types of perikarya. In the hedgehog, Dinopoulos et al. (1992) described large spiny neurons (20–30 μm), that can be divided into polygonal, triangular and fusiform shapes. The somato-dendritic morphology is multipolar (20–30 μm), less frequently fusiform, ovoid or round. Dendrites are beaded, aspiny or sparsely spiny. Spines are numerous on distal dendrites. Aspiny neurons represent a small proportion of neurons and their axons have never been seen to leave the boundary of the claustrum (Dinopoulos et al., 1992).

The cat claustrum has large multipolar, fusiform neurons (30–40 μm) with extensively branched dendrites, as well as pyramidal cells (35–45 μm) with similarly arranged dendrites and spindle cells (25–40 μm). Their secondary and tertiary dendrites exhibit dendritic spines. Dendrites of these cells radiate out in all directions and span as much as 500 μm. Axons give off collaterals that ramifiy in the neighborhood of the cell body. Small polygonal or round aspiny cells are found less frequently (Druga, 1966a; LeVay and Sherk, 1981a) (Figures 2.8–2.11). In the cat, Mamos et al. (1986) distinguished three types of spiny neurons and two types of aspiny neurons. Spiny neurons included pyramidal cells (20–27 μm) with one main dendrite, fusiform cells (18–28 μm) with two

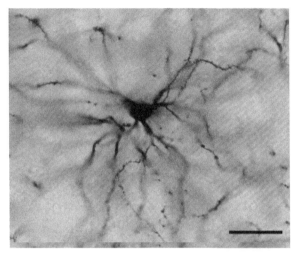

FIGURE 2.8   Multipolar spiny neuron in the dorsal claustrum of the cat. Golgi-Cox impregnation. Bar = 50 µm.

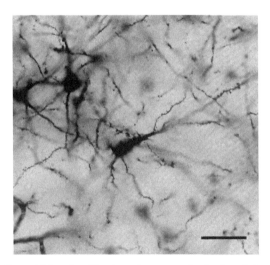

FIGURE 2.9   Pyramidal and multipolar spiny neuron in the dorsal claustrum of the cat. Golgi-Cox impregnation. Bar = 50 µm.

main dendrites and multipolar neurons (19–27 µm). The most frequent were pyramidal neurons. Aspiny or sparsely-spiny neurons were either large or small.

Three types of neurons have been described in the primate claustrum (Brand, 1981). Type I have pyramidal or spindle-shaped (17 × 8 µm) cell bodies with extensive dendritic arborization. Some cells gave off one

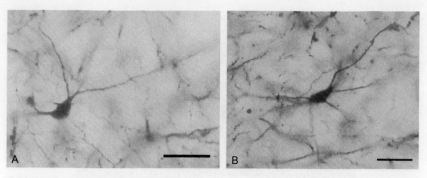

FIGURE 2.10   A, B: Medium-sized, oval aspiny neurons in the dorsal claustrum in the cat. Golgi-Cox impregnation. Bar = 50 μm.

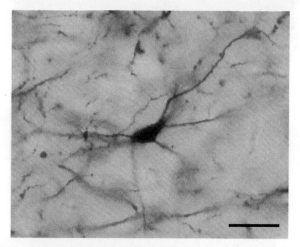

FIGURE 2.11   Medium-sized and small oval aspiny neurons in the dorsal claustrum of the cat. Golgi-Cox impregnation. Bar = 50 μm.

large, long dendrite, similar to the apical dendrites of pyramidal cells in the cerebral cortex, and such dendrites were laden with spines. The primary axon, or axon collaterals, left the claustrum for the external or extreme capsule. In addition to spiny neurons, Brand described two types of aspiny neurons: large ones (type II) with spherical cell bodies, and small ones (type III) with pear-shaped, smooth cell bodies. Axons and their collaterals formed plexuses within the dendritic domain.

Neuronal types in the claustrum of man have been investigated by means of a transparent Golgi technique combined with pigment (lipofuscin) staining (Braak and Braak, 1982). Five types of neurons have been described, four of which are aspiny varieties, the other one having spines. Two categories of aspiny neurons were described – large and small – and both differentiated pigmented and nonpigmented cells were found in each category. Most of the large aspiny cells have a spindle-shaped cell body with a small number of long dendrites spreading out radially. Small aspiny cells are frequently ovoid or globular, and give off a small number of dendrites radiating in all directions. Spiny neurons predominate, and usually their somata have a pyramidal-like shape. In addition, spiny neurons often have spindle-shaped and multipolar (polygonal) cell bodies. These cells contain fine and widely dispersed lipofuscin granules.

Electron Microscopy

In electron microscopic observations in the cat, Norita and Hirata (1976) identified four classes of neurons: Class I are small (<10 μm) and have large, dark nuclei with marked indentations. The narrow cytoplasm contains a few clearly defined organelles. Most claustral neurons belong to class II. Their perikarya are round, triangular or polygonal (multipolar) and medium-sized (15–20 μm). The nucleus is lighter than that of the previous type, and the cytoplasm contains more organelles. Class III neurons have elongated perikarya (15 × 25 μm). Other ultrastructural features of these neurons are similar to class II. Class IV neurons are large (24 × 25 μm) but scarce. Their cytoplasm is the most abundant among all described classes of claustral neurons and has well-developed organelles. The synaptic organization is complex and five types of presynaptic profiles were identified. Most common are terminals (boutons) with uniform, round vesicles forming asymmetrical axodendritic and axospinous contacts (type A). In addition there is another type of asymmetrical axodendritic or axospinous terminal that has round vesicles of various diameters (type B). The claustrum also contains two types of symmetrical axodendritic contacts with pleomorphic vesicles and one type of infrequent, asymmetrical axodendritic synaptic contact containing many granular vesicles (Hinova-Palova et al., 1979; Otellin, 1973).

In the posterior (visual) part of the cat claustrum there are three groups of neurons (Kubasik-Juraniec et al., 1998), medium-sized and large ones representing about 75 percent of the cell population. Their cell bodies are oval, round, fusiform or triangular; their ultrastructural characteristics are similar, and their dendrites exhibit spines evident under light (Golgi impregnation) and electron microscopy. These neurons accumulate retrogradely transported HRP and WGA-HRP following injections into the visual cortex. The third type of neurons are small cells that differ from the larger ones in several respects, and their dendrites do not have spines.

DISTRIBUTION OF NEUROCHEMICALS

Neurotransmitters

Most of the neurons in the CL and END, as in the cortex and subcortical structures, are glutamatergic projection neurons (excitatory). A small proportion of projecting neurons express nitric oxide synthase (NOS) (Kowiański et al., 2001).

GABA

It was found that CL neurons in the cat (Narkiewicz et al., 1988) and mouse (Feldblum et al., 1993), express GABA (γ-aminobutyric acid). The proportion of GABA-ergic interneurons in the CL of various mammals was estimated at between 7 percent and 12 percent (Braak and Braak, 1982; Gómez-Urquijo et al., 2000; Gutiérrez-Ibarluzea et al., 1998; Spahn and Braak, 1985). The CL contains a small number of glutamate decarboxylate (GAD)-positive neurons but has many immunostained puncta in the neuropil (Rahman and Baizer, 2007). GAD-positive cell bodies were oval or elongated and about 20 μm long.

Glutamate Transporters

The CL in the adult mouse has a vesicular glutamate transporter (VGLUT2)-positive neuropil, consisting of fine axons and puncta. Strong positivity was observed at the periphery of the CL, forming a shell around a core practically devoid of positive structures. VGLUT2-positive cell bodies were observed at postnatal day 0 but not in adult animals or at other embryonic ages (Real et al., 2006). The excitatory amino acid transporter EAAC1 is a glutamate transporter known to be expressed in glutamatergic cells of the cerebral cortex and is regarded as a marker for projection neurons. EAAC1 was reported in many neurons throughout the claustrum of the cat (Rahman and Baizer, 2007). Double-labeling experiments indicate that the majority of glutamate transporter expressing cells represent separate populations. Only a small number of cells were double-labeled for calcium-binding proteins (Rahman and Baizer, 2007).

Calcium-Binding Proteins

The calcium-binding proteins (CBPs) parvalbumin (PV), calbindin (CB), and calretinin (CR) are widely distributed in the central nervous system (Celio, 1990). It is generally accepted that the major function of these three CBPs is buffering of intracellular Ca^{2+}. In some neurons, CB and CR are bound to cellular structures, whereas PV is freely mobile in axons, neuronal somata and nuclei (Schmidt et al., 2007). In echidna, PV-immunoreactive (ir) neurons prevail in the central part of the CL, but are found only occasionally in the END. Both compartments of the claustrum contained a relatively high density of CB-ir neurons, the latter being found mainly at the dorsal and ventral edges of the CL, complementary to the distribution of the PV-ir neurons. In the END, CB-ir neurons were distributed throughout the structure. CR-ir neurons were not found in the CL but were present in the END (Ashwell et al., 2004).

Rat

The distribution of PV and CB have been examined in the dorsal and ventral claustrum of the rat by means of immunohistochemistry (Druga et al., 1993). PV and CB displayed different and largely complementary patterns of distribution. PV immunostaining was intense in the neuropil of the CL, but very weak in the neuropil of the END. Similarly, PV-ir neuronal cell bodies were relatively numerous in the CL, but were detected only occasionally in the END. The distribution of PV-ir cells did not show any clear topographical arrangement. The majority (approx. 80%) of the PV-ir neurons had a round or oval perikarya (12–20 × 8–13 μm). Some PV-ir cell bodies were smaller (up to 10 μm) or larger (longer than 20 μm). The perikarya of the smaller neurons were round or oval, whereas the subpopulation of larger neurons appeared polymorphous. The dendritic branches were frequently immunostained and in some neurons could be followed up to 200–250 μm from the cell body (Figure 2.12). A careful evaluation of the PV-positive dendritic arborization confirmed that dendrites in the periphery of the dendritic sphere were smooth, with no evidence of stained spines or thorns. PV immunostaining was hardly detectable in the neuropil of the END. Three to five neurons per section were detected in END tissue; their shape and size was similar to that of the PV-ir cells observed in the dorsal claustrum, and their dendrites were aspiny.

CB-positivity was weak in the CL whereas the neuropil of the END exhibited more intense staining. In addition, CB-ir cells were more numerous in the END than in the CL. The dendrites of the CB-ir neurons displayed a smooth surface.

Rat and guinea pig claustrum have been found to have different patterns of CR immunoreactivity (Edelstein et al., 2010). The CL of the rat consists of two regions, the first being a CR-negative zone in the core of the

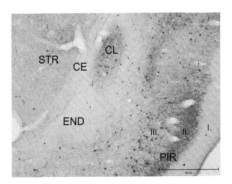

FIGURE 2.12   Strong and weak parvalbumin immunoreactivity in the dorsal claustrum (CL) and ventral claustrum (END) in the rat. CE, external capsule; PIR, piriform cortex; STR, striatum; I–III, layers 1–3 of the piriform cortex. Bar = 500 μm.

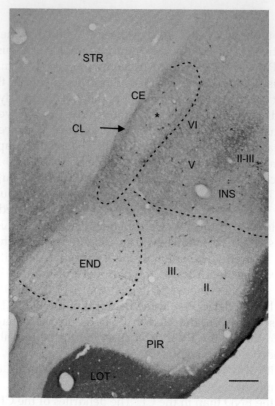

FIGURE 2.13   Calretinin-immunoreactivity in the dorsal claustrum (CL) and ventral claustrum (END) in the rat. In the core of the CL an oval area with weak CR-immunoreactivity (*asterisk*) is evident. CE, external capsule; INS, insular cortex; LOT, lateral olfactory tract; STR, striatum; PIR, piriform cortex; I, II, and III, layers 1–3 of piriform cortex; II-III, V and VI, layers 2, 3, 5, and 6 of insular cortex.

structure, largely devoid of CR-ir neurons. This region is dorsally, medially and basally surrounded by CR-positive neuropil containing a small number of CR-ir neurons. In the END, the neuropil contains few or no moderately staining neurons, but the density of CR-ir neurons is higher. In both subdivisions of the rat claustrum there are many multipolar and oval neurons with marked differences in immunostaining. Dendritic arborizations are smooth and devoid dendritic spines (Figure 2.13).

Mouse

The population of neurons containing CBPs in the mouse CL has been estimated at 12.3 percent of all claustral neurons. Among these, CB-ir neurons acount for 5.4 percent and PV-ir neurons 7.9 percent. PV-ir in the claustrum was characterized by the presence of a moderately stained

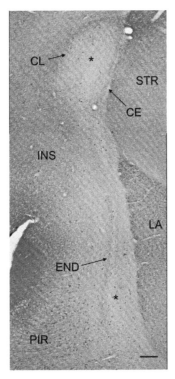

FIGURE 2.14 Calretinin-immunoreactivity in the dorsal claustrum (CL) and ventral claustrum (END) in the guinea pig. Areas of weak CR-immunoreactivity (*asterisks*) are evident in both the core of the CL and the END. CE, external capsule; INS, insular cortex; LA, lateral amygdaloid nucleus; PIR, piriform cortex; STR, striatum. Bar = 200 μm.

patch of neuropil which roughly corresponds to the CR-negative oval area. Most PV-ir neurons were medium in size and exhibited multipolar morphologies (Real et al., 2003).

The dorsal claustrum of the guinea pig contains a centrally situated, weakly CR-ir, ovoid area completely enveloped by a rim of strongly positive neuropil. In the ventral claustrum there is a moderately positive neuropil containing, in its lateral half, a vertically oriented elongated area of weaker CR-ir. Both subdivisions of the claustrum, the dorsal and the ventral, contain CR-ir neurons; however, the density of these neurons is higher in the ventral claustrum. In both the dorsal and ventral claustrum the majority of neurons have oval and triangular perikarya with smooth dendrites (Edelstein et al., 2010) (Figure 2.14).

Colocalization between three neuropeptides – neuropeptide Y (NPY), somatostatin (SOM), and vasoactive intestinal polypeptide (VIP) – and three CBPs (PV, CB, CR) was investigated in the rat claustrum during a 4-month postnatal period (Kowiański et al., 2009). The most frequent colocalization was observed for VIP/CR (51%). Moderate values were found for NPY/CB (32%) and SOM/CB (31%). The lowest values were observed in NPY/CR (9.5%) and NPY/PV (9%). Only one type of colocalization was present at birth (VIP/CR), the other types were noted later in the postnatal period.

The distribution of CBPs, neuropeptides and NOS and their colocalization has been described in the END of the rat (Kowiański et al., 2004). CBP immunoreactive neurons were evident throughout the nucleus. CB-ir neurons represent the most numerous population, followed by CR-ir and PV-ir neurons. Like the CL, the END contains neurons that are immunoreactive for SOM, NPY and VIP. SOM-ir neurons represent the most numerous population followed by NPY-ir and VIP-ir. In the DEN, neuropeptides reveal colocalization with CBPs and NOS. SOM colocalizes with CB, PV and NOS, whereas VIP colocalizes only with CR, and NPY colocalizes only with NOS. The most frequent colocalization in the DEN is SOM/CB (24%), NPY/NOS (68%) and NOS/CR (24%).

Rabbit

Wójcik et al. (2004) analyzed CBP immunoreactivity in the CL and END of the rabbit. Intense and homogenous PV-ir was found throughout the whole claustrum with frequent basket-like plexuses surrounding unlabeled cells. The density of PV-ir cells in the CL was three times larger than in the END. PV-ir levels in the END were low and few positive neurons were observed. CB-ir numbers were low in both subdivisions of the claustrum. The distribution of CB-ir neurons was homogenous and their number per unit area in CL and END did not differ significantly. The number of CR-ir neurons was moderate in CL, where thin fibers and immunoreactive puncta were observed, while in the END they were scarce.

Cat

PV-ir neurons were analyzed in cat claustrum using both light and electron microscopy (Hinova-Palova et al., 2007). PV-ir neurons surrounded by a meshwork of PV-ir fibers and puncta were observed throughout the claustrum. More PV-ir neurons were identified in the superior part than in the intermediate or inferior parts.

PV-ir neurons in the claustrum of the cat are divided into several subtypes, with oval, triangular, multipolar and bipolar cell bodies having been identified. Most numerous are medium-sized and small neurons. PV-ir dendrites analyzed in serial sections exhibited a smooth or wavy morphology in 66 percent of the samples, and about 23 percent of dendrites showed irregular swellings. Thirteen percent of PV-ir dendrites showed spines of different shapes and sizes. The majority of axon terminals (70%), which formed synapses with PV-ir dendrites, were of the asymmetric type. The remaining boutons contained round as well as elongated or flattened vesicles. The majority of PV-ir boutons (70%) contained large round vesicles and displaying synaptic contacts with dendritic spines and dendrites of different sizes.

The distribution of CBPs, GAD, NOS and glutamate transporter EAAC1 in the cat claustrum has been described by Rahman and Baizer (2007). PV-ir cells are distributed throughout the claustrum, have large somas that are 20–30 μm in the long diameter, and pear, oval, bipolar and multipolar shapes. Dendrites emerge from the cell bodies, and some of them extend long distances. Many dendrites are beaded. CB-ir neurons are less frequent than the PV-ir cells, and their cell bodies are multipolar (20–25 μm). CR-ir cells were less numerous than those immunoreactive to PV.

Monkey

Immunoreactivity for CBPs in the claustrum of the monkey varied with different antibodies, and several morphologically distinct types of positive cells were seen (Reynhout and Baizer, 1999). PV-ir cells had large cell bodies from which 3–7 processes emerged. Their cell bodies were oval or elongated. Some cell bodies resembled pyramidal cells with one stout apical dendrite. Dendrites were smooth and aspiny. CB-ir cells were more lightly stained and three different types were described.

Both the dorsal and ventral claustrum in human brain contain numerous CR-ir neurons that are sometimes arranged in clusters. Most CR-ir cells are small and bipolar with round or elongated cell bodies (10–15 μm) or larger (20–30 μm). The human claustrum also contains multipolar CB-ir neurons, a few of which are larger in size (20–30 μm). Dendrites of all labeled neurons were beaded and aspiny. Both subdivisions of the claustrum displayed intense and uniform LAMP (limbic system associated membrane protein) staining (Prensa et al., 2003).

Nitric Oxide Synthase

Guirado et al. (2003) demonstrated two types of nNOS-positive neurons in mouse claustrum. Densely stained neurons were GABAergic and aspiny. These cells were found throughout CL and END and represent the population of local interneurons. Lightly stained cells were more numerous and non-GABAergic. The authors suggested that these neurons were projecting neurons.

NOS-ir neurons in the rat dorsal claustrum (CL) were scattered throughout the entire nucleus and were intensively stained. The commonest were medium sized neurons showing oval, fusiform or triangular cell bodies with three or four poorly branched aspiny dendrites. In the population of claustro-cortical neurons (FluoroGold-labeled) 2 percent were NOS-ir. NOS was found to colocalize with two

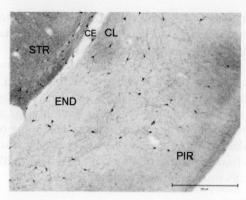

FIGURE 2.15 NADPH-diaphorase histochemistry. Difference in staining between the dorsal (CL) and ventral (END) claustrum and striatum in the rat. CE, external capsule; PIR, piriform cortex; STR, striatum.

neuropeptides, SOM and NPY. The colocalizing neurons were local interneurons (Kowiański et al., 2001) (Figure 2.15).

Colocalization of NOS with the CBPs PV, CB, and CR in the rat claustrum have been described (Kowiański et al., 2003). The percentage values of colocalizing neurons were low and constituted 18 percent for NOS/CB and 12 percent for NOS/PV. The approximate colocalization values for NOS-ir neurons with PV and CB were 20 percent and 30 percent, respectively. NOS/CR double-labeled neurons were not observed. Double-labeled neurons exhibited oval and fusiform cell bodies.

In cat CL, Hinova-Palova et al. (1997) described three types of neurons that were NADPH-d positive: large (22–25 μm), medium-sized (18–20 μm) and small (12–16 μm). These neurons exhibited various somatodendritic morphologies and long, sparsely branching dendrites, which, as a rule, were aspiny. This type of neuron prevailed in the anterior two-thirds of the claustrum and they were distributed in both subdivisions (CL, END) of the nucleus.

Similar NOS-positive neurons in the cat were described by Rahman and Baizer (2007).

Neuropeptides

Rat

Neuropeptide-ir neurons (VIP-, SOM-, and NPY-ir neurons) were randomly distributed in the CL. Their dendrites were aspiny (SOM-ir) or contained sparse spines (VIP-ir). Only NPY-ir neurons exhibited dendritic spines. The neuropeptide immunoreactivity was not observed in the FG-labeled cortico-claustral neurons. The majority of NPY-ir and SOM-ir neurons colocalized NOS (Kowiański et al., 2001).

In addition to the above-mentioned neuropeptides. isolated reports have described calcitonin gene-related peptide (CGRP) and cholecysto-kinin (CCK) immunoreactivity within the CL. The source of the dense CGRP-ir innervation to the claustrum and insular cortex is the ventro-lateral parabrachial subnucleus, with a smaller contribution from lateral subparafascicular nucleus (Yasui et al., 1989). A moderate density of CCK-ir fibers and cell bodies was present in the CL in the guinea pig, where labeled perikarya prevail in the posterior claustrum. Stronger CCK-positivity was evident in the END (Ciofi and Tramu, 1990).

Claustrum-Specific Proteins

Cadherins are a large family of cell–cell adhesion molecules that are expressed in specific cell groups and tracts of the vertebrate brain. Cadherin expression distinguishes three subdivisions of the CL (superior, intermediate and inferior). The superior and intermediate divisions contain perikarya that express Rcad (retinal cadherin). The END shows generally uniform and moderate expression of Rcad and Ncad (neural cadherin)(Obst-Pernberg et al., 2001).

In situ hybridization studies report a strong expression of netrin G2, a gene encoding a glycosyl phosphatidyl–inositol-linked membrane protein in monkey claustrum (Miyashita et al., 2005). Netrin G2 is expressed throughout the entire claustrum, but more so in the ventral claustrum. In addition to strong positivity within the main cellular mass of the ventral claustrum, strong positivity was also evident in the ventral and lateral cellular extensions adjoining the insular cortex, the prepiriform cortex and the amygdala. Netrin G2-positive neurons are not GABAergic, but correspond to claustro-cortical projection neurons. Netrin G2 immunostaining in the human material demonstrated positivity in the neurons distributed in the claustrum and the insular cortex but not in the putamen.

Latexin, an endogenous inhibitor of the metallocarboxypeptidases has been described as a robust and specific marker of the claustrum and selected areas of the cortex in the cat (Arimatsu et al., 2009). Latexin-ir neurons were densely distributed in the lower layers (5–6) of the neocortex and in both subdivisions of the claustrum (CL, END) in the cat, and the distribution was found to be similar in the rat and mouse. Latexin-ir neurons were absent in medial cortical areas and in the majority of allocortical fields. Double labeling for latexin, and either glutamate- or GABA-ir, demonstrated that neurons that were immunoreactive for latexin were also immunoreactive for glutamate, but not for GABA within selected cortical areas, CL and END (Arimatsu et al., 1999; Jin et al., 2006). In contrast, no latexin immunoreactivity was observed in the human claustrum and insular cortex (Pirone et al., 2012). Some data indicate that Latexin-positive neurons may represent a subpopulation of the glutamate-ir neurons.

The G-protein gamma 2 subunit (Gng2) was identified as a specific rat claustrum marker, and has been used to better delineate its boundaries and relations to surrounding structures (Mathur et al., 2009). Recently Gng2 immunoreactivity was detected in the human claustrum and in the insular cortex but not in the putamen. Within the claustrum the density of immunostaining was lower in the dorsal part of the claustrum and higher in the ventral but the boundary between the subdivisions was not distinct. Double staining showed that Gng2 and glial fibrillary acidic protein (GFAP) were colocalized in the same elements that were characterized morphologically as astrocytes (Pirone et al., 2012). In addition, Edelstein and Denaro (2004) included a list of more than 30 various neuroactive substances (neurotransmitters, neuropeptides, receptors, enzymes) demonstrated in the claustrum and reported in many disparate papers.

AFFERENT CONNECTIONS OF THE CLAUSTRUM

Cortical Afferents

Using Marchi's method, cortico-claustral connections have been studied in cat, goat and monkey. Early work by several groups that studied descending cortical projections from the frontal cortex in cats and monkeys failed to observe any degenerations in the claustrum or described only degenerated fibers passing through it (see Druga, 1966b, 1968). In contrast, projections from the temporal area (area 22) in monkeys were described by Hirasawa et al. (1938), and Berke (1960). Extensive cortical lesions in the goat resulted in sparse degeneration in the claustrum (Igarasi, 1940; Kamio, 1938).

These findings have been supplemented using methods visualizing anterograde and retrograde intra-axonal transport. In several studies it has been demonstrated that the most significant afferent connections of the claustrum arise in the entire neocortex (see also Edelstein and Denaro, 2004), as described in the next sections.

Rat

Descriptions of cortico-claustral projections for the rat are much less extensive than those for the cat and monkey. Efferent projections from the infralimbic cortex terminate in END and projections from the prelimbic cortex in the CL in the rat (Hurley et al., 1991; Sesack et al., 1989; Vertes, 2004). Lévesque and Parent (1998) reconstructed axonal trajectories of neurons located in prelimbic cortex injected with biotin-dextran or biocytin. These authors demonstrated the existence of cortico-striato/claustral axons projecting to both the striatum and claustrum. The majority of these axons do not cross the midline, and arborize within the striatum and the claustrum ipsilaterally. Subpopulations of axons project to the ipsilateral striatum and contralateral claustrum, or project to the striatum bilaterally and to the claustrum contralaterally. The terminal area of the cortico-striato/claustral fibers was at the rostral and mid-dorsoventral levels of the claustrum. As expected, injection of leucoagglutinin (PHA-L) into infralimbic cortex resulted in bilateral labeling in the END. In contrast to this, the injection of the same anterograde tracer into prelimbic cortex resulted in bilateral labeling of the CL.

Zhang et al. (2001) described cortical projections to the anterior CL with the use of intraclaustral ejections of the FluoroGold. Retrogradely labeled neurons were distributed in many neocortical areas but prevailed in medial prefrontal areas, in the cingulate, retrosplenial and perirhinal cortices. Moderate labeling was also evident in the lateral entorhinal cortex. Primary motor, sensory, auditory and visual cortices, together with their association fields, exhibited modest labeling.The rat claustrum receives strong bilateral projections from the whisker region of the primary motor (M1) cortex, but contralateral projections are more prominent. In contrast to this the M1 forepaw region has few efferent and afferent connections with the CL in either hemisphere (Alloway et al., 2009; Smith and Alloway, 2010).

Rabbit

Carman et al. (1964) were the first to offer evidence of topographically organized neocortical projections to the CL in the rabbit using the Nauta method (Nauta, 1964).

Buchanan et al. (1994), using PHA-L and horseradish peroxidase (HRP), demonstrated projections from the medial prefrontal cortex of the rabbit (areas 24, 25, and 32) to the CL.

Cat

Topographically organized cortico-claustral projections in the cat were described by Druga (1966b, 1968). They reported finding degenerated axons and their preterminal fragments in the frontal quarter of the nucleus following small lesions in the frontal cortex (areas 4, 6, 8, and 10). Thermocoagulations of the parietal, temporal and occipital cortex resulted in terminal degenerations in the posterior two-thirds of the claustrum. Although the Nauta method is probably incapable of depicting all terminal degenerations, the results of Carman et al. (1964) were confirmed in this study. A mediolateral as well as an AP organization of cortico-claustral projections was demonstrated (Figure 2.16).

Chadzypanagiotis and Narkiewicz (1971), Jayaraman and Updyke (1979), Squatrito et al. (1980a,b), and using different techniques, described projections from areas 17, 18, 19, and the Clare-Bishop area to the dorsocaudal sector of the claustrum. Projection to the contralateral claustrum was sparser and revealed after a longer survival time (8 days).

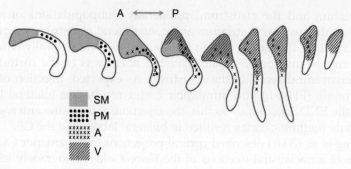

FIGURE 2.16   Coronal sections through cat claustrum, illustrating schematically termination fields of cortico-claustral projections from the sensorimotor (SM), premotor (PM), auditory (A) and visual (V) cortical areas. The most anterior section is on the left. A–P, anteroposterior axis.

Olson and Graybiel (1980) demonstrated that projections to and from areas 17 and 18 are organized retinotopically in the cat, and projections from somatosensory cortex are organized somatotopically. Projections from the primary somatosensory cortex (S1) terminate in the dorsolateral part of the anterior half of the CL. LeVay and Sherk (1981a) summarized several important findings relating to the organization of the cortico-claustral loop associated with the visual sector of the claustrum. Following injections of HRP into the visual sector of the claustrum, they demonstrated retrogradely labeled cells in ipsilateral visual cortical areas. Retrogradely labeled spiny neurons were found in layer 6, but smaller numbers were seen at the border of layers 3 and 4. Labeled cells prevailed in the representation of the peripheral visual fields. In area 17 about 3.5 percent of the cells in the layer 6 were labeled, whereas in area 19 only 4 percent were labeled. Cortical neurons projecting to the claustrum and into the LGN at corresponding retinotopic loci overlapped extensively, but no double-labeled cells were found. This finding suggests that the corticogeniculate and cortico-claustral pathways arise from two distinct populations of neurons. The termination of cortico-claustral fibers, studied by the autoradiographic tracing technique, forms a focus in the ipsilateral clustrum and indicates the retinotopic organization of this projection.

The projections of the associative auditory areas AII and Ep areas to the CL were established in several studies (Druga, 1968; Otellin and Makarov, 1972; Otellin, 1972, 1969; Olson and Graybiel, 1980; Hinova-Palova et al., 1988). However, the cortico-claustral connection arising in the AI area was a matter of controversy, and in several studies was denied (Clarey and Irvine, 1990; LeVay and Sherk, 1981a; Reale and Imig, 1983). Nevertheless, Neal et al. (1986), following the making of small lesions in the primary auditory cortex, found a small focus of terminal degeneration posteriorly in the inferior half of the claustrum. The distribution of terminal degeneneration corresponded to the position of retrogradely labeled cells after the injection of HRP into the same cortical area. Hinova-Palova et al. (1988), in an extensive study, demonstrated degenerating cortico-claustral axons and their terminations following lesions of the AI, AII and Ep areas. Degeneration was observed in all three areas but was greatest following AI lesions. An AP organization of the cortical projections was also seen, together with a weak contralateral component.

Witter et al. (1988) described reciprocal and bilateral projections with limbic and paralimbic cortical regions in the cat. Following injections of tritiated amino acids into the cingulate, agranular insular, perirhinal and anterior limbic areas, anterograde labeling was evident in the dorsal, intermediate and ventral claustrum. Following injections into the entorhinal cortex, strong labelling was evident in the ventral claustrum, whereas weak projection from the hippocampal formation terminated within the intermediate claustrum.

Electron microscopic studies show that, after ablation of visual and auditory cortical areas in cats, many synaptic boutons in the dorsal claustrum degenerate. Cortico-claustral connections are mostly axodendritic of the asymmetrical type. Degenerated boutons ended preferentially on dendritic spines, less frequently on the shafts of beaded dendrites (Juraniec et al., 1971; LeVay and Sherk, 1981a). It appears that the bulk of the cortico-claustral projection terminates on principal projecting neurons of the nucleus (spiny cells), whereas the weaker component terminates on the dendrites of aspiny neurons, which are considered to be local interneurons (Hinova-Palova and Paloff, 1982; Hinova-Palova et al., 1988).

Monkey

The somatosensory and motor cortices project to the dorsal region of the monkey's CL, and a significant overlapping of both projections within the CL is evident. Projections from the SI terminate in the dorsocaudal half of the CL (Kemp and Powell, 1970; Pearson et al., 1982), whereas projections from the motor cortex terminate in the dorsal part of the whole AP extent of the CL (Kemp and Powell, 1970; Künzle, 1975; Pearson et al., 1982; Leichnetz, 1986). Künzle (1975) reported some level of somatotopy in the motor projections with the face cortical area projecting in the anterior part of the motor zone, whereas the cortical areas for the limbs project more posteriorly. Efferent projections of cortical area 8 terminate in the AP extent of the CL in two anterograde patches, one in the central region, the other in the caudoventral part (Künzle and Akert, 1977; Stanton et al., 1988).

Projections from the visual cortical areas (areas 17 and 18) into the posteroinferior part of the CL were reported by Riche and Lanoir (1978) and Pearson et al. (1982) in baboon and macaque monkey. Data about projections from the primary and associative visual areas to the CL were summarized by Sherk (1986). Nevertheless, there is uncertainty as to whether area 17 projects to the CL. No evidence of direct projections from area 17 to the CL was found in the owl monkey by Graham et al. (1979) but such projections were reported in galago and tree shrew (Carey et al.,1980; Carey et al., 1979a,b). Recently Day-Brown et al. (2010) confirmed bidirectional excitatory connections between the claustrum and area VI in tree shrew. Ungerleider et al. (1984) using autoradiography, described weak projections from the area MT, which is specialized for the analysis of visual motion, to the small, limited caudoventral zone of the claustrum. Leichnetz (2001) described heavy reciprocal connections of the medial posterior parietal cortex (area 7m) with the middle region of the claustrum. Anterograde labeling was evident throughout the AP extent of the nucleus. It is not known whether the primary auditory cortex (area 41) projects to the CL, but projections from the associative area 22 terminating in the ventral part of the nucleus were described by Pearson et al. (1982). The cingulate gyrus has strong reciprocal connections with the CL, terminating in the ventral part of the CL (Baleydier and Mauguiere, 1980). The entorhinal cortex and prepiriform cortex project to the anteroventral part of the CL (Mufson and Mesulam, 1982).

Subcortical Afferents

Narkiewicz et al. (1983), Narkiewicz et al. (1985), Sloniewski (1983), and Sloniewski et al. (1985, 1986a) described labeled neurons in the anterior pretectal nucleus (pars compacta) and in the lamina medullaris pretectothalamica in the cat following HRP injections involving the central part of the claustrum. Following injections of the tritiated amino acids into the lamina medullaris pretectothalamica and into the anterior pretectal nucleus in the cat, anterograde transport of the tracer was found in the dorsomedial part of the claustrum (somatosensory part according Olson and Graybiel, 1980). Another study demonstrated that the visual sector of the claustrum receives subcortical projections from the lateral hypothalamus and from the thalamic nucleus centralis medialis. Projections from the lateral hypothalamus are ipsilateral and terminate in both visual and nonvisual sectors of the dorsal claustrum. Weaker projections from the intralaminar thalamic nucleus terminate in the dorsal claustrum bilaterally (Kaufman and Rosenquist, 1985; LeVay and Sherk, 1981a). Reciprocal connections were demonstrated between the claustrum and visual thalamus (lateral intermediate nucleus) in the tree shrew. Terminations of anterogradely labeled fibers were restricted to the most ventral part of the CL (Carey and Neal, 1985). The CL receives projections from the supramammillary nucleus in the rat (Vertes, 2002, 2004). The CL and the END in the rat also receive projections from the amygdaloid complex, namely from the magnocellular division of the basal nucleus, the accessory basal nucleus, and the lateral division of the amygdalohippocampal area, the latter being the source of the most substantial projections to the END. Smaller projections originate in the cortical amygdaloid nuclei, periamygdaloid cortex and the accessory basal nucleus (Majak et al., 2002). Projections between the claustrum and basal nucleus are reciprocal (Zhang et al., 2001). The subcortical projections to the anterior CL were described by Zhang et al. (2001), who noted retrogradely labeled neurons in the midline and intralaminar thalamic nuclei and in the basolateral amygdala.

In the brainstem, labeled neurons were distributed in the medial part of the substantia nigra pars compacta (SNc), in the ventral tegmental area (VTA) and in the dorsal raphe nucleus. Dense projections from the dorsal raphe nucleus terminate in the END and in the CL Serotoninergic innervation of rodent and primate CL was also described by Baizer (2001).

EFFERENT CONNECTIONS OF THE CLAUSTRUM

Cortical Efferents

In 1964 Narkiewicz reported that cortical ablations in cats led to topographically organized areas of neuronal degeneration in the claustrum (also Narkiewicz, 1966, personal communication). He concluded that degenerative changes within the claustrum are the manifestation of retrograde neuronal degeneration and that the claustrum is a structure comparable to the complex of ventroposterior thalamic nuclei in its dependency on the neocortex (see also Sloniewski et al., 1985, 1986b). Chadzypanagiotis and Narkiewicz (1971) confirmed this general scheme of the claustro-cortical projections, and collected new data about the dynamics of the degeneration process within the claustrum. Narkiewicz's view was later criticized by Cowan (1970), who interpreted the neuronal changes in the claustrum as a result of transneuronal anterograde process rather than retrograde degeneration. However, the introduction of retrograde intra-axonal transport methods conclusively confirmed the existence of claustro-cortical projections and the original Narkiewicz concept (Kievit and Kuypers, 1975; Norita, 1977; and see Edelstein and Denaro, 2004).

Insectivores

Following injections of HRP-WGA (wheat germ agglutinin) into various neocortical areas of the hedgehog tenrec, retrogradely labeled neurons were found in the claustrum (group Hr). The connection of the claustrum with the neocortex is bilateral and reciprocal. Projections from the ventral claustrum (group Hp) terminate in paleocortical region and in the subiculum (Künzle and Radtke-Schuller, 2001).

In the hedgehog, claustro-neocortical projections originate in the dorsal claustrum, are distributed to the entire neocortex, and show a rough topographic organization with a substantial degree of overlap. These projections are mainly ipsilateral (Dinopoulos et al., 1992).

Rat

For their study in the rat, Sloniewski and Pilgrim (1984) and Minciacchi et al. (1985) used HRP and injected large amounts of a fluorescent tracer into frontal and occipital cortex. The majority of labeled neurons were found in the ipsilateral claustrum. Their distribution showed a topographic organization with some degree of overlap of neurons projecting to the sensorimotor and occipital cortex (areas 17 and 18). Neurons labeled from the frontal injections prevailed in the anterior part of the CL, while occipital injections resulted in a labeled population predominantly located in the posterior part of the CL. Double-labeled neurons were observed after unilateral injections of tracer into frontal and occipital areas and after bilateral symmetrical injections (2–3% of double-labeled neurons).

Sloniewski et al. (1986b) injected fluorescent tracers to the prefrontal, motor, somatosensory auditory and visual cortical fields in the rat. They demonstrated prominent ipsilateral claustro-cortical projections and weaker contralateral ones. The projections to the prefrontal and auditory cortex were modest, those to the motor, somatosensory and visual cortex more prominent. Symmetrical bilateral injections into the motor and visual cortex, as well as unilateral simultaneous injections of fluorescent markers in the motor and visual cortex, revealed no double-labeled neurons. Claustro-cortical projections exhibited coarse AP topographic distribution with marked overlapping. Most anterior sections of the claustrum contained neurons projecting to the prefrontal and motor cortex, whereas, in the most posterior sections, neurons projecting to the somatosensory, auditory and visual cortical areas were found. Condé, Maire-Lepoivre, Audinat, and Crépel (1995) reported retrogradely labeled neurons in both the dorsal claustrum and the endopiriform nucleus in rat, following injections of fluorescent tracers into the medial frontal cortex. Subsequently, Sadowski et al. (1997) distinguished two principal zones within the claustrum. The sensorimotor zone occupies the anterodorsal part of the CL, whereas the visuoauditory zone occupies the posteroventral part of the CL. Between these zones, only scanty overlap was observed, an example of which is seen in the sensorimotor and visuoauditory areas. In contrast to this, a large overlap exists between neurons projecting to the motor and sensory cortical areas from the visual and auditory cortical areas.

Neurons projecting to the limbic cortex (cingulate and retrosplenial areas) were found only in the CL, and were organized over the whole rostrocaudal extent of the structure. Experiments with double tracers revealed that neurons projecting to both limbic areas were intermingled. Double-labeled neurons were rare (Majak et al., 2000). Massive efferent projections of the anterior CL were distributed to the prefrontal cortex (cingulate area), insular and entorhinal cortex (Condé et al., 1995; Zhang et al., 2001).

Retrograde tracer injections in the rat M1 (whisker and forepaw region) produced many labeled neurons in the ipsilateral claustrum projecting to the whisker area. Projection to the forepaw region was weaker (Colechio and Alloway, 2009; Smith and Alloway, 2010). Hoover and Vertes (2007) described ipsi- and contralateral projections from the claustrum to the medial prefrontal cortex and anterior cingulate cortex. Both subdivisions of the claustrum (CL and END) were shown to project to the prelimbic cortex, but the majority of claustral neurons projecting to the infralimbic cortex were located in the CL. Injections of different retrograde tracers into whisker area of the M1 and S1 revealed intermingled populations of labeled neurons in the CL, as well as many double-labeled neurons. This finding indicates that the same part of the CL projects to the whisker representations in both S1 and M1 (Smith et al. (2012).

Claustral projections to area Oc. 2.2 of the occipital cortex (Druga, 1989) and to area 6 of the visual cortex (Jakubowska-Sadowska et al., 1998) arise from the caudal half of the claustrum in the guinea pig.

In the rabbit, FluoroGold injections into a visual area (mostly into area 17) resulted in retrogradely labeled neurons throughout almost the whole AP extent of the claustrum. The distribution of labeled neurons was similar to that in the rat and the majority were located in the ventral part of the claustrum (Jakubowska-Sadowska et al., 1998). Kowiański et al. (1998) demonstrated ipsilateral projections from the CL to several neocortical areas. Neurons projecting to the cortex were present along the whole AP extent of the nucleus (Figure 2.17). Neurons projecting to somatosensory and auditory cortex, to the motor (trunk area) and somatosensory cortex and to the auditory and visual cortex overlap. Similarly, as in the rat, neurons in the CL projecting to the cingular or retrosplenial cortex were present in the whole AP extent of the CL and slightly prevailed in the medial part of the nucleus. The highest density of neurons projecting to the cingular cortex was found in the intermedial third of the CL, and the highest density of neurons projecting to the retrosplenial cortex was found in the caudal half of the nucleus. Small numbers of labeled neurons were found in the contralateral CL (Majak et al., 2000).

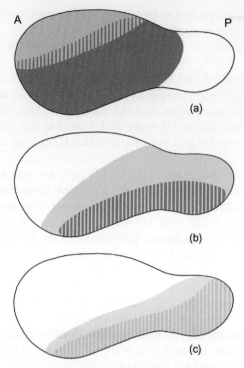

FIGURE 2.17 Schematic picture showing overlap of several claustrocortical projections systems within the rabbit claustrum: (a) *pink* – neurons projecting to the area Prc 2,3 (projection zone of the motor representation of the head), *purple* – neurons projecting to the area Prc 1 (projection zone of the motor representation of the trunk); (b) *light blue* – neurons projecting to the primary somatosensory cortex, *dark blue* – neurons projecting to the secondary somatosensory cortex; (c) *yellow* – neurons projecting to the visual cortex, *green* – neurons projecting to the auditory cortex. A, anterior pole of the claustrum; P, posterior pole of the claustrum. *Adapted from Kowiański et al. (1998).*

Cat

Norita (1977), and Irvine and Brugge (1980) demonstrated that neurons in the bilateral dorsal claustrum project to different neocortical areas with a preponderance of them projecting to the ipsilateral site. Distribution of labeled neurons exhibited a rostrocaudal as well as dorsoventral topographical arrangement. Norita (1977) reported weaker projections from the claustrum to the sensorimotor and to the AII areas. Riche and Lanoir (1978) described claustro-cortical projections to the gyrus proreus (frontal eye field). HRP-positive neurons prevailed in the ipsilateral CL, and the majority of them were in that part of the nucleus, which was specified as a "medial marginal zone" by Druga (1966a). A small number of labeled neurons were also discernible in the END and in the rostral half of the contralateral CL. Neurons projecting to area 6 were widely distributed along the AP axis and prevailed in the dorsal and intermediate part of the CL (Olson and Jeffers, 1987).

Macchi et al. (1981, 1983) examined the projections from the claustrum to the cerebral cortex using HRP and fluorescent markers injected into various areas of the medial and lateral surfaces of the neocortex. Retrogradely labeled cells prevailed in the ipsilateral dorsal claustrum with the minority of cells located contralaterally. The number of labeled cells was higher following HRP injections in motor, visual, insular and cingulate cortex. The patterns of distribution of labeled cells following the different HRP injections provide evidence of an AP and dorsoventral topographical arrangement. This study also indicates some degree of overlapping in the distribution of HRP-positive cells. Similarly, as in the previous study, a retrograde transport of cortically injected HRP was demonstrated bilaterally in the CL (Druga, 1982). Following an HRP injection into the sensorimotor cortical area, labeled neurons were found in the anterior two-thirds of the nucleus. Following a similar injection into the auditory area, labeled neurons were located in the claustrum more occipitally, and a considerable extent of these neurons along the AP axis was conspicuous. In the contralateral claustrum the number of labeled cells was smaller. Our results indicate an AP arrangement of the claustro-cortical projections with distinct overlapping in the distribution of cells labeled from injections in different cortical fields. After injecting fluorescent tracers into pericruciate cortex, Minciacchi et al. (1985) demonstrated that the majority of labeled neurons were in the anterior half of the dorsal claustrum. Labeled neurons significantly prevailed in the dorsolateral part of the ipsilateral claustrum but also extended more ventrally to the narrow isthmic part. Anteroposterior organization of the claustro-cortical projections were also confirmed by Hughes (1980), Guldin et al. (1986) and Jakubowska-Sadowska et al. (1998).

Projections from the claustrum to the motor and frontal association areas in the cat were analyzed in detail by Clascá et al. (1992). In the CL, the neurons projecting to each cytoarchitectonic area of the frontal cortex were found to be distibuted according to specific patterns. Neurons projecting to area 4 were situated only in the dorsal third of the claustrum within the anterior half of the nucleus. Following the injection of marker into area 6, labeled neurons in the CL were situated more ventrally. HRP injections in the prefrontal cortex elicited labeling in more ventral parts of the CL, and in the END. Following injections into more ventral parts of prefrontal cortex, the number of labeled neurons within the END increased (see also Guldin et al., 1986). Injections that involved the frontal pole of the hemisphere produced relatively scarce labeling in the CL. Clascá et al. (1992) support the notion that claustral projections terminating in the various frontal areas are organized according to specific patterns. First, there is an AP organization because the majority of labeled neurons projecting to the frontal areas are located in the anterior half of the claustrum. Second, neurons projecting to areas 4, 6, and prefrontal cortex are distributed along the dorsoventral axis of claustrum. The medial zone of area 4γ (motor representation of hindlimb muscles) receives projections from most medial and dorsal part of the claustrum. The more lateral part of area 4γ (motor representation of face and neck muscles) receives projections from the more lateral portion of the claustrum. The precruciate part of area 6αβ, containing units whose stimulation elicits movements of axial musculature, receives projections from a wide area in the central portion of the CL. The results of the Clascá et al. (1992) study also indicate that the claustral neurons projecting to area 4 are intermingled with those projecting to the somatosensory areas. A small number of claustral neurons give off axonal collaterals to closely related points of area 4γ and the somatosensory area (areas 3b and 1).

Projections from the claustrum to the somatosensory forepaw digit cortex in the raccoon was examined by Standage and Doetsch (1988). The HRP-positive cells were concentrated in the dorsolateral part of the CL in the middle third of the AP extent of the nucleus. Neurons projecting to the digit 2 cortical area were located anterior to neurons projecting to the digit 5 area. Injections of the tritiated amino acids into the visual claustrum demonstrated strong projection from the nucleus to the ipsilateral visual areas 17, 18, and 19. Weaker projections terminated in several associative areas (20a, 20b, 21a, PLLS, VLS and DLS). Contralaterally, the main projection was to a strip of the cortex situated along the area 17–18 border (LeVay and Sherk, 1981b). Projection from the CL was heaviest in the cortical representation of the periphery of the visual field and decreased toward more central locations. The highest densities of labeled terminals were in layers 4 and 6. Layer 5 and supragranular layers were labeled more weakly, and labeling in layer 1 was weakest.

Afferents from the claustrum to the cortical pupillo-constrictor area (posterior suprasylvian gyrus, area 20) of the cat were demonstrated by S. Kuchiiwa et al. (1984, 1985). Neurons projecting to this area were observed ipsilaterally in the caudal half of the CL. Ultrastructural studies indicated that claustral afferents to layers 1 and 6 predominantly contacted dendritic spines and formed asymmetric (excitatory) synapses (LeVay, 1986). Claustral afferents to layer 4, on the other hand, terminated about equally on spines and dendritic shafts. This finding suggests the possibility that the claustral afferents to layer 4 have a substantial proportion of terminals on inhibitory cortical interneurons.

Neal et al. (1986) demonstrated labeled neurons in a restricted part of the claustrum after injections of HRP or HRP-WGA in the primary auditory cortex. Labeled neurons were demonstrated in the claustrum bilaterally, but their number was small. The claustro-cortical projections to the auditory cortex/area AI, AII and Ep) were similar to the cortico-claustal system with an AP organization (Hinova-Palova et al., 1988). The central claustral third projects heavily upon the AII area and weakly upon the AI area. Neurons located in the posterior claustral third project heavily to the Ep area. Each major subdivision of the auditory cortex is reciprocally connected with the claustrum, but connections with the AI area are weaker.

Several studies elucidated projections from the claustrum to the limbic cortex. Markowitsch et al. (1984) analyzed the claustral projections to the cat's limbic cortex with retrograde and anterograde tracing techniques. Both methods revealed claustral efferents to the cingulate, retrosplenial, subicular, and prepiriform cortex; ipsilateral projections prevailed. The END projects to the cingulate, entorhinal and prefrontal cortex. Room and Witter (1984), and Room and Groenewegen (1986) described strong projections from the ventral (END) and intermediate claustrum to the entorhinal cortex (ventrolateral, VLEA, and dorsolateral, DLEA, fields). Following injections of retrogradely transported tracer into the medial subdivision (MEA) of the entorhinal cortex, labeled neurons were also discernible in the CL. Injection of the tracer into the subiculum resulted in labeled neurons being found only in the END.

The dorsal, intermediate and ventral claustrum project to subdivisions of the entorhinal cortex (MEA, presubiculum) whereas the END projects to the perirhinal cortex (Room and Groenewegen, 1986). Witter et al. (1988) analyzed the connections of the claustrum with the limbic and paralimbic cortex in the cat. Injections of anterograde tracer (tritiated amino acids) and retrograde tracer (WGA-HRP) into the anterior cingulate, anterior limbic, agranular insular and perirhinal cortex resulted in reciprocal labeling in all subdivisons of the claustrum (dorsal, intermediate and ventral). Injections into the entorhinal cortex resulted in anterograde labeling in the END and in an intermediate transitional part of the nucleus. Significantly weaker projections from the caudal part of the hippocampal formation terminated within the intermediate part of the nucleus. Connections between the injected cortical areas and the claustrum were in all cases reciprocal.

Minciacchi et al. (1985) studied projections from the claustrum to different representations within the primary somatosensory and visual areas by using the multiple retrograde fluorescent tracing technique. Retrograde labeling in claustrum reveals two different projection patterns, the first being more compact, the second one rather diffuse. The somatosensory part of the claustrum shows a dorsoventral organization of cells projecting to the hindpaw, forepaw, and face representations of S1 (Figure 2.18). The visual part of the claustrum has cells projecting to the vertical meridian representation of V1 surrounded dorsally (and laterally and medially) by neurons projecting to the representations of the retinal periphery. A second pattern of claustral projections consists of

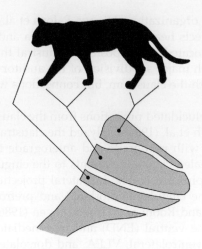

FIGURE 2.18   Dorsoventral organization of the somatosensory claustrum in the cat. Schematic picture shows localization of claustro-cortical neurons projecting to the hindpaw, forepaw, and face representations of S1 area. *Adapted from Minciacchi et al. (1985).*

neurons that are distributed diffusely throughout relatively large sectors of nucleus. Another important feature of this study is the comparison of claustral projections with the size of cortical representations of various parts of the body and central and peripheral part of the visual field. In other words, the S1 projection preference is for the forepaw representation in claustrum, whereas the V1 projection preference is for the retinal periphery representation.

Monkey

Kievit and Kuypers (1975) were the first to report the existence of direct projections from the anterior half of the claustrum to the cortex of the frontal lobe in the monkey. Following HRP injections into the frontal area 8 of the baboon, Riche and Lanoir (1978) found labeled neurons in the whole AP extent of the ipsilateral nucleus with a distinct condensation in its anterodorsal portion. Later, the projections from the claustrum to the frontal lobe were studied in detail. It was shown that the claustrum projects to all frontal lobe functional areas (M1, premotor (PM), supplementary motor area (SMA), frontal eye field (FEF), orbitofrontal cortex, cingulate cortex, and frontopolar cortex).

Following HRP injections into different cortical areas on the lateral surface of the cerebral hemisphere, Pearson et al. (1982) demonstrated that the anterior half of the claustrum projects to the cortex of the frontal lobe and the posterior half projects to the cortex of the parietal lobe. The posterior and inferior part of the claustrum projects to the occipital and temporal cortex. In addition to retrogradely labeled neurons, there were

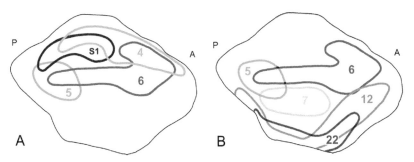

FIGURE 2.19   A, B: Flattened map of of the monkey claustrum. The pattern of overlap between the parts of the claustrum projecting (and receiving axons) to the primary motor area (4), premotor area (6), somatosensory area (S1), parietal areas (5, 7), prefrontal area (12) and associative temporal area (22). A, anterior pole of the claustrum; P, posterior pole of the claustrum. *Adapted from Pearson et al. (1982).*

fine extracellular granules (puncta) co-extensive with the cells. This indicated anterograde transport from the cortex, and showed that projections between the claustrum and the cortex are precisely reciprocal. Injections in the areas 4, 6, 12, 22 and areas 8 and 9 resulted in labeling that had the form of elongated bands of considerable AP extension. These bands partially overlapped in the anterior third of the claustrum. Following injection into areas S1, S5, and S7, the bands of labeled neurons were significantly shorter. The areas of cortex, which are interconnected by means of associative fibers, are related to bands within the claustrum that overlap. Thus, the parts of the claustrum related to area 4 and S1, to 5 and 6, and to 7 and 9, all overlap anteroposteriorly. The authors suggested that one cortical area may influence another not only through associative cortical fibers but also through the claustrum (Figure 2.19).

Projections of the claustrum to the M1, PM and prefrontal cortices were analyzed using several retrograde tracers (Druga et al., 1990; Leichnetz, 1986; Tanné-Gariépy et al., 2002). The distribution of retrogradely labeled neurons showed their distribution along the entire AP extent of the CL. Labeled neurons projecting to all motor areas occupied the dorsal and intermediate parts of the CL along the dorsoventral axis. Subpopulations of neurons projecting to area 8 (FEF) and area 46 were localized differently in the middle, ventral and most ventral part of the CL. The projections from the CL to the multiple subareas of the motor cortex and to area 46 arise from largely overlapping territories, with some degree of local segregation. Connections between the claustrum and frontal cortex were reciprocal. Although bilateral claustral connections exist between the claustrum and prefrontal and posterior parietal regions (Leichnetz and Goldberg (1988), this was not observed with area 4 (Leichnetz, 1986).

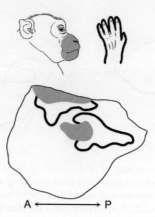

FIGURE 2.20   Flattened map of the monkey claustrum. *Gray* and *white* areas indicate schematically the regions where neurons projecting to the face and hand representations of the somatosensory cortex (S1) were concentrated. Claustro-cortical neurons are present in two distinct regions, both somatotopically organized. The dorsal region is organized dorso-ventrally, while the ventral region rostrocaudally. *Adapted from Minciacchi et al. (1991). A–P anteroposterior axis.*

In the macaque monkey, Minciacchi et al. (1991) described two populations of neurons projecting to the somatosensory area. Following injections of fluorescent retrograde tracers into the face and the hand representation of S1, retrogradely labeled cells were found throughout most of the rostro-caudal extent of the ipsilateral claustrum in the dorsal half of the nucleus. These were concentrated in two regions of the claustrum. The first region was more dorsal, the second region was ventral in the caudal half of the claustrum. Neurons labeled from the face and hand cortical representations were segregated in both the dorsal and more ventral populations. In the dorsal population, neurons projecting to the face area were dorsal to neurons projecting to the hand area. In the ventral neuronal population, S1 face cells were located rostrally and S1 hand ones caudally. Claustral cells double labeled from both representations were rare. These data suggest that claustrum contains two separate somatosensory representations, with a dorsoventral segregation (Figure 2.20).

In a metareview, Weller et al. (2002) collected a list of studies of projections from the claustrum to the various visual areas. Perkel et al. (1986) and Lysakowski et al. (1988) found, after injections of the fluorescent tracers into visual areas V1 and V4, retrogradely labeled neurons in the ventral caudal claustrum. Cells projecting to areas V1 and V4 formed two intermingled subpopulations, even though there were very few double-labeled cells (<1%) projecting to both areas. Projections from the ventral caudal claustrum to the visual areas V1 and V2 with small colateralizations were reported by Doty (1983); Kaske et al. (1991); Kennedy and Bullier (1985); Riche and Lenoir, 1978; Tigges et al. (1982); and Tigges et al. (1983).

In an investigation of projections of the claustrum to the object vision and spatial vision cortical areas and pathways, the inferior temporal and the posterior parietal cortices were injected with retrograde or antero-grade tracers (Baizer et al., 1993). The temporal injections (areas TE and TEO) resulted in labeling concentrated in the most ventral and ventro-medial part of the claustrum. The parietal injections (intraparietal sulcus, areas 7, 5) resulted in labeling extending to the more dorsal and poste-rior part of the nucleus. Neurons projecting to the posterior parietal cor-tex were situated lateral to cells projecting to the temporal areas. The two populations of labeled cells occupied contiguous but separate territories and no double-labeled cells were seen. If anterograde tracer were used (WGA-HRP, tritiated amino acids), the preterminal and terminal label-ing occupied the same parts of the claustrum in which the retrogradely labeled cells were found. Following injections of the fluorescent retro-grade tracer into visual areas V4 and PM, the distribution of retrogradely labeled cells was analyzed. The area V4 injections resulted in labeled neurons in the ventromedial claustrum. The area PM injection resulted in label that was in an adjacent and more dorsal part of the claustrum. Neurons projecting to both injected areas were interspersed, but no double-labeled cells were reported (Baizer et al., 1997).

In three primates (owl monkey, squirrel monkey, macaque monkey) the dorsolateral and the middle temporal area had reciprocal connections with the claustrum. The retrogradely labeled neurons and anterograde projec-tions were found in the ventromedial portion of the CL (Weller et al., 2002).

Large numbers of retrogradely labeled neurons (choleratoxin labeled cells) were found in the claustrum after injections into prefrontal cor-tex in marmoset. Labeled neurons were observed in both parts of the claustrum, in the CL as well as in the ventral claustrum (endopiriform nucleus). The most abundant labeling in both parts of the nucleus was found after the injections into the lateral prefrontal cortex. Positive cells were present throughout the whole AP extent of the CL (Roberts et al., 2007).

Strong input to the frontal pole (rostral area 10) in the marmoset mon-key originated in the ipsilateral claustrum. Labeled cells were observed rostrocaudally distributed in the middle and basal part of the CL. An additional population of claustro-cortical neurons occupied the most ventral part of the caudal half of the claustrum. Claustral neurons pro-jecting to the the lateral, rostral and dorsomedial region of area 10 rep-resented as many as 50 percent of all neurons labeled in subcortical structures (including thalamic nuclei). Projection from the CL to the medial region of area 10 was much sparser and represented only 4.4% of the total subcortical input (Burman et al., 2011).

Neurons projecting to the medial posterior parietal cortex (area 7m) were located in the middle region of claustrum rostrocaudally. These connections were reciprocal, and the distribution of anterogradely labeled cortico-claustral terminals corresponds with the distribution of retrogradely labeled neurons (Leichnetz, 2001). The temporopolar cortex was innervated by neurons located in the ventral claustrum and in the END. Retrogradely labeled cells were observed in the middle third of the claustrum (Morán et al., 1987).

Moryś et al. (1993) demonstrated several cases of ulegyria in human brain where severe neuronal loss in the CL confirmed claustro-cortical connections. Neuronal loss in the anterior and central part of the CL was observed after pathological lesions that involved frontal cortex. Pathological changes in the parietal and occipital cortices caused neuronal loss in the central and posterior parts of the CL. In both groups, the dorsal half of the dorsal claustrum was more affected. These findings suggest that the human claustrum is connected widely with the cerebral cortex, and that these connections, as in experimental animals, are topographically organized.

Somatodendritic Morphology of the Claustro-Cortical Neurons

Introduction of retrogradely transported markers (HRP, WGA-HRP), which fill the perikaryon and extend to the main dendrites, in some cases as far as their first bifurcation, enable the diameters and shape of labeled neurons to be evaluated. Our results indicated that intensively labeled neurons exhibited easily discernible stem dendrites reaching up 30–80 μm and, in some cases, the initial segment of the axon. Considerable perikaryal polymorphism and differences in the size of perikarya are characteristic features of retrogradely labeled claustro-neocortical neurons in rat, cat and monkey (Druga, 1982; Druga et al., 1990). Neurons with multipolar or pyramidal (triangular) perikarya outnumber the others and represent about 60–70 percent of all labeled neurons. Remaining neurons are oval or spindle shaped. In 75 percent of the labeled neurons the longer perikaryal dimension ranged from 19 μm to 25 μm. Neurons that were smaller or larger were found less frequently. Retrogradely labeled neurons within the dorsal claustrum projecting to the cortex were similarly described in other studies using HRP as a marker (Macchi et al., 1978,1981; Riche and Lanoir, 1978) (Figure 2.21).

Subcortical Efferents

Although the majority of hodological data relating to the claustrum involves cortical connections, the claustrum was also found to project to several subcortical structures. Claustrothalamic projections in the hedgehog were demonstrated after large WGA-HRP or HRP

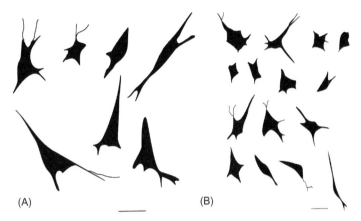

(A)   (B)

FIGURE 2.21   Various types of claustro-cortical (HRP-positive) neurons in the dorsal claustrum of the cat (A) and rhesus monkey (B). *Redrawn from the darkfield and brightfield photomicrographs (Druga, 1982; Druga et al., 1990). Bar = 20 µm.*

injections in the thalamus. A small number of retrogradely labeled cells (5–10 per section) were found throughout the ipsilateral dorsal claustrum. Thalamoclaustral projections were not demonstrated in this study (Dinopoulos et al., 1992). Due to the large thalamic injections of the tracer, it was impossible to state in which thalamic nucleus or nuclei the claustral projections terminated.

Subcortical projections of the anterior CL in the rat terminate in the shell and core of the nucleus accumbens, the amygdala and midline thalamic nuclei, where a prominent focus is evident in the submedial nucleus (Behan and Haberly, 1999; Zhang et al., 2001). Projections from the CL to the thalamus in the rat were studied using the retrograde tracer FluoroGold method (McKenna and Vertes, 2004). Following injections of the tracer into the rostral and caudal part of the reuniens nucleus, retrogradely labeled neurons were described in many limbic structures including the claustrum. Projections from the CL to the rhomboid/centromedial nucleus, to the submedial nucleus and to the ventromedial nucleus of the thalamus were also confirmed.

Druga (1972) found, using the anterograde axonal degeneration method in the cat, that only the sensorimotor part the CL projects to the ipsilateral putamen. Flindt-Egebak and Olsen (1978) injected tritiated proline into cat CL and described label in area 19 (Clare-Bishop area), in the parahippocampal gyrus, putamen, mediodorsal thalamic nucleus, supragenic ulate thalamic nucleus and zona incerta. Projections from the claustrum to the thalamus of the cat were described by Sloniewski et al. (1985), Carey et al. (1979a,b), and Carey and Neal (1985). Following tracer injections into the CL, labeled cells and terminals were found throughout the lateral intermediate thalamic nucleus in the tree shrew (Carey and Neal, 1985). Injection of the fluorescent retrograde tracer into the posterior thalamus and anterior pretectal region of rats resulted in retrograde neuronal labeling of the ipsilateral claustrum. The majority of labeled cells were in the central third of the claustrum (Sloniewski et al., 1985).

Arikuni and Kubota (1985), following injection of HRP into the head of the caudate nucleus in monkeys, found retrogradely labeled cells in the rostral part of the ipsilateral CL and in the amygdala. In contrast to above mentioned results, LeVay and Sherk (1981a) rejected the idea that there might be claustral projections to the thalamus in the cat and concluded that the cortex is the sole target of visual claustral axons.

Erickson et al. (2004) described strong projections from the CL to the caudoventral and ventrolateral division of the mediodorsal thalamic nucleus and to the magnocellular division of the ventral anterior thalamic nucleus in monkey. Following thalamic injections of the cholera toxin b subunit, retrogradely labeled neurons were seen distributed throughout the entire rostrocaudal and dorsoventral extent of the claustrum. The highest number of cells was found in the rostral and dorsal half of the nucleus. Large numbers of labeled neurons were found also in the external and extreme capsules, which authors termed "periclaustrum."

PROJECTIONS OF THE VENTRAL CLAUSTRUM (ENDOPIRIFORM NUCLEUS)

Afferent Projections

Cortical projections originate in the pyriform cortex (Druga, 1971; Krettek and Price, 1978; Schwabe et al., 2004), in the entorhinal cortex (Witter et al., 1988) and in the infralimbic area (Hurley et al., 1991; Insausti et al.,1997; Vertes, 2004). Shi and Cassell (1999) described projections from the perirhinal and insular cortex (area 35). A strong projection from the amygdala to the END of the rat originates in the lateral division of the amygdalohippocampal area. Lighter projections originate in the cortical nucleus, periamygdaloid cortex and the accessory basal nucleus (Majak et al., 2002).

Efferent Projections

Several papers describe the projection of the ventral claustrum (END) to neocortical areas. Macchi et al. (1983) demonstrated projections from the END to the prefrontal cortex in the cat. Projections from the DEN to the infralimbic cortex in the rat were described by Condé et al. (1995). Efferent projections from the END to the anterior cingulate cortex have been reported in the cat by Guldin et al. (1986). Efferent projections from the END to the piriform and entorhinal cortex in rat were described by Haberly and Price (1978a), Hoffman and Haberly (1991), Kerr et al. (2007), and Krettek and Price (1977) using anterograde and retrograde transport methods.

Projections from the most rostral part of the END are seen most in the rostral half of the piriform cortex. Projections from the middle and caudal part of the END terminate mainly in the middle and posterior part of the piriform cortex. These projections were seen to terminate in all layers of the piriform cortex including layer 1 (Behan and Haberly, 1999).

The END projects massively to all subdivisions of the entorhinal cortex, where the majority of terminals were found in layers 1b, 2, and 3. Weaker projections terminated in perirhinal cortex (areas 35, 36) (Room et al., 1985). Although the majority of neurons projecting to the MEA and DLEA were located within the ventral claustrum, some neurons were also found above the level of the rhinal fissure in the dorsal claustrum (compare to Witter et al., 1988). Markowitsch et al. (1984) reported efferent projections from the CL, as well as from the END, to both the entorhinal and subicular cortex The findings of Behan and Haberly (1999) indicate that efferent projections of the END in rat are oriented to the cortex of the basal forebrain rather than to subcortical nuclear structures. Massive projections from the END terminate in several basal cortical areas including the piriform, entorhinal, insular, perirhinal, postrhinal, orbital, and all cortical amygdaloid areas.

Extensive projections to the entorhinal cortex originate in all parts of the END. The rostral part of the END projects more laterally, the caudal part more medially. Deeper layers of the entorhinal cortex are supplied largely from the middle and posterior parts of the END. In both lateral and medial entorhinal areas (LEA, MEA) the densest terminal labeling was present in layers 4 and 5. Behan and Haberly (1999) also collected new data about the organization of the moderate projections from the END to the perirhinal, insular and orbital cortices.

As in the cat, the ventral portion of the claustrum in the monkey gives rise to projections to the hippocampal formation and subiculum (Amaral and Cowan, 1980). Insausti et al. (1987) showed, in macaque monkey, a strong projection from the claustrum to the entorhinal cortex. The retrograde tracer (WGA-HRP) injected to subdivisions of the entorhinal cortex resulted in strong claustral labeling, indicating that the claustrum is a major source of subcortical projections to this cortical area. The majority of labeled neurons were observed in the ventral half of the nucleus, but in some cases they were also located above it in the intermediate part. The ventral portion of the claustrum projects to the rostral part of the entorhinal cortex, whereas the intermediate third of the claustrum projects to the caudal part of the entorhinal cortex.

Intrinsic Connections in the Dorsal and Ventral Claustrum

Behan and Haberly (1999) and Zhang et al. (2001) confirmed the presence of heavy intrinsic projections in all parts of the END in rat, where labeled axons extended anteriorly as well as posteriorly. A few isolated axons derived from the END were also observed in the CL. In addition to END, Zhang et al. (2001) confirmed the presence of heavy intrinsic connections within the CL. Smith and Alloway (2010), following tracer injection in the CL, revealed many labeled neurons throughout its rostrocaudal extent. This finding indicates the presence of long-range intrinsic claustral connections and confirms the finding of Behan and Haberly (1999) and Zhang et al. (2001). The latter also reported commissural fibers interconnecting the ipsi- and contralateral anterior claustri.

DISCUSSION AND CONCLUSIONS

The claustrum is a telencephalic, subcortical structure traditionally divided into two subdivisions, the dorsal (insular) claustrum connected with the neocortex, and the ventral (piriform) claustrum or endopiriform nucleus connected with the allocortex.

In the primitive insectivore (hedgehog tenrec) the claustrum is located between the layers of the rhinal/insular cortex (Künzle and Radtke-Schuller, 2001). Narkiewicz and Mamos (1990) and Dinopoulos et al. (1992), in several representatives of insectivora, described two subdivisions in the dorsal claustrum – lateral (principal) and a medially situated lamina profunda claustri. It is interesting that, in echidna, a cellular extension of the END lateral to the neuronal condensation has been described, and designated as a dorsal claustrum (Ashwell et al., 2004). In contrast, in Lagomorpha, Carnivora and Primates, using basic staining methods, mediolateral differences in the cellular structure of CL have not been described, and the CL appears more homogenous (Kowiański et al., 1999). The intraperitoneal injection of methylazoxymethanol acetate (MAM, an inhibitor of neuronal migration influences) to pregnant rats alters the configuration of the CL (Moryś et al., 1994). High MAM doses (30 mg/kg) resulted in the lack of superficial layers (2–4) of the cortex while the deep layers were preserved. The claustrum was present but composed of loosely packed neurons. After administration of low MAM doses (14 mg/kg) two different parts (medial and lateral) of the CL were observed in newborn pups. Dorsal claustrum consisting of two different subpopulations of neurons was observed in primitive insectivores (hedgehog tenrec), but in development these two parts fuse into a unitary structure (Narkiewicz and Mamos, 1990; Künzle and Radtke-Schuller, 2000). The existence of several marginal zones in the CL of the cat and beaver is not associated with phylogenetically related fusion of CL components and the creation of a unitary structure. The marginal zone M, situated on the medial margin of the dorsal claustrum in the cat, together with a narrow cellular bridge between the dorsal and ventral claustrum (Druga, 1966a), was later classified as an intermediate claustrum (Witter et al., 1988). On the basis of the volume and configuration, Kowiański et al. (1999) differentiated five types of claustrum. The shape of the claustrum varies from species to species but seems to increase in size in proportion to increasing neocortical volume.

Although the description of neuronal types within the claustrum varies to a certain degree, multipolar, triangular, oval and spindle perikarya of different sizes are described most frequently (Chadzipanagiotis and Narkiewicz, 1971; Druga, 1966a; Norita and Hirata, 1976; Pilleri, 1961). This has been confirmed by the electron microscopic observations of Hinova-Palova et al. (1979), Kubasik-Juraniec et al. (1998), and Norita and Hirata (1976).

There is a general agreement that claustrum is constituted of two subpopulations of neurons: (1) projecting neurons – with medium-sized and large polymorphic perikarya and spiny dendrites; and (2) medium-sized and small interneurons with round and oval perikarya and aspiny dendrites (Braak and Braak, 1982; Brand, 1981; Druga et al., 1993; Kubasik-Juraniec et al, 1998; Mamos, 1984; Wójcik et al., 2004). In carnivores and primates different CBPs, neuropeptides and NOS are distributed somewhat uniformly with no functional segregation or inhomogeneity (Hinova-Palova et al., 2007; Rahman and Baizer, 2007; Reynhout and Baizer, 1999). A similarly uniform distribution of the CB-ir and CR-ir neurons has been reported in the dorsal and ventral claustrum in human brain (Prensa et al., 2003). All mammal neurons expressing CBPs have smooth, non-spiny dendrites. However, in an electronmicroscopic study Hinova-Palova et al. (2007) reported that 13 percent of PV-ir dendrites in the cat claustrum have spines of different shapes and sizes.

Colocalization of NOS with CB and PV in rat were described by Kowiański et al. (2003). Further colocalizations have been reported between three neuropeptides (NPY, SOM, VIP) and three CBPs (PV, CB, CR). Abundant colocalizations have been observed for VIP/CR, NPY/CB and SOM/CB (Kowiański et al., 2009).

In addition to CBPs and neuropeptides, several other molecules are expressed in the CL and END. Cadherins were used for parcelation of the CL into three subdivisions (Obst-Pernberg et al., 2001). Netrin G2 and latexin have been described as specific markers of the claustrum and several cortical areas (Miyashita et al., 2005; Arimatsu et al., 2009). Latexin-ir is a marker for glutamate-positive neurons within the CL and END, and indicates a relationship between the claustrum and the infragranular neocortical layers (Arimatsu et al., 1999, 2009).

Afferent Projections

The CL receives its major input from all regions of the cerebral cortex. Cortico-claustral projections were demonstrated by using several methodological approaches (see Edelstein and Denaro, 2004). Each cortical region sends fibers terminating in a distinct zone within the CL. Cortico-claustral projections are bilateral, but the ipsilateral system is more prominent. AP and dorsoventral organization of the cortico-claustral projections was repeatedly confirmed (Carman et al., 1964; Druga, 1966b, 1968; Leichnetz, 2001; LeVay and Sherk, 1981a; Pearson et al., 1982; and others). Sherk (1986) divided CL, on the basis of cortical connections and physiological mapping, into three zones – somatosensory, visual and ventral. The ventral zone receives input from the cingulate cortex and from the auditory cortex (Hinova-Palova et al., 1988). The overlapping of terminations of cortico-claustral projections was repeatedly confirmed. Although the claustrum receives projections from all the cortical areas so far analyzed, data about intensity of these projections are scarce. Projections from the whisker area of the M1 are strong and terminate mainly in the contralateral CL, whereas projections from the forelimb area of the M1 are weak. The whisker area of the S1 does not project to the CL in either hemisphere (Alloway et al., 2009; Smith and Alloway, 2010; Smith et al., 2012).

Weak projections were also reported between primary auditory cortex and the CL (Beneyto and Prieto, 2001; Clarey and Irvine, 1990; Hinova-Palova et al., 1980a,b).

On the basis of cortico-claustral projections, several anteroposteriorly elongated zones have been distinguished in the CL of rat, rabbit, cat and monkey (somatomotor, auditory, visual, limbic) (Carman et al., 1964; Druga, 1968; Leichnetz, 1986; Pearson et al., 1982; Sherk, 1986). The main source of cortico-claustral projections are neurons in layer 6 (Carey et al., 1980).

At least part of the cortico-claustral projections arise from a specific category of cortical neurons projecting to both the striatum and claustrum, ipsi-, contra- or bilaterally (Lévesque and Parent, 1998). This finding suggests that prefrontal cortex may send efferent copies of the same information to both the claustrum and the striatum. Introduction of the HRP, and other retrograde intra-axonal tracers, have confirmed basic data about the projections from the cortex to the claustrum and have also led to the discovery that cortical neurons of the sixth layer provide the axons for this projection.

Subcortical Afferents

Although the majority of claustral afferentation is of cortical origin, the claustrum is also targeted by fibers originating in the thalamus and hypothalamus. Neurons projecting to the CL are distributed to midline, intralaminar, and visual thalamic nuclei, and to the pretectal area (Carey and Neal, 1986; Narkiewicz et al., 1983, 1985; Sloniewski, 1983; Zhang et al., 2001). Projections from the thalamic lateral intermediate nucleus (Li) to the CL indicate another parallel route via which visual thalamic nuclei may influence the cortical mechanism. Given that Li receives ascending inputs from the pretectum as well as from the superior colliculus, perhaps connections from Li to CL to cortex provide a route for relay of visuomotor information to the respective cortical regions (see Carey and Neal, 1986). In the cat and rat hypothalamus these projections are distributed in the lateral hypothalamus and in the supramammillary nucleus (LeVay and Sherk, 1981a; Vertes, 1991). Projections from the amygdala to the CL and END may serve as a route by which the claustral complex is supplied by signals from the amygdaloid complex in order to process emotionally significant information. In addition, the claustral complex may spread epileptiform activities from the amygdala to other brain regions, especially to the neocortical and hippocampal formation (Majak and Morys, 2007; Majak et al., 2002; Zhang et al., 2001). In the brainstem, neurons projecting to the CL are distributed in the medial part of the SNc, in the VTA and in the dorsal raphe nucleus (Baizer, 2001; Vertes, 1991; Zhang et al., 2001).

Efferent Projections

The main output of the claustrum is the return pathway to the cortex. However the claustrum also has significant non-cortical projections. Claustro-cortical projections belong among several "nonthalamic" subcortical systems. Apart from those arising from the claustrum, nonthalamic projections to the neocortex also originate from the basal forebrain, the basolateral amygdala and several brainstem structures.

The existence of claustro-cortical projections was first suggested by Narkiewicz (1964), who demonstrated that lesions of various parts of the neocortex led to retrograde neuronal degenerations within the claustrum. The claustro-cortical projection was later confirmed using retrograde intra-axonal transport in all animals investigated. The data reviewed in this chapter indicate that the claustral projections are distributed to the entire neocortex in all the mammalian representatives so far analyzed. Such diffuse projections indicate that the functional role of the claustrum cannot be interpreted only in relation to the function of a given cortical area. The pattern of the projections strongly suggests that the inputs ascending from the claustrum may be integrated in many functional circuits. The fact that the claustrum has bidirectional connections with all sensory and motor cortical areas, as well as to the associative and limbic areas, suggests participation of this area in more general and more complex functions.

Primary visual areas are reciprocally connected with the most dorsal part of caudal claustrum. These cortico-claustral projections reproduce the cortical retinotopic map. Association visual areas (20, 21, PLLS and DLS), whose retinotopic organization is more complex, are connected with a wider and more ventral part of the claustrum (LeVay and Sherk, 1981a,b; Macchi et al., 1981,1983; Norita, 1977; Olson and Graybiel, 1980). In addition to the retinotopic map located within the dorsal and posterior part of the claustrum, motor and somatosensory hemibody maps are located in the anterior part of the claustrum (Clascá et al. (1992). Neurons projecting to area 4 have been found in the same part of the claustrum that projects to area S1, and overlapping between these two subpopulations of neurons has been demostrated (see also Macchi, 1981; Minciacchi et al., 1985; Norita, 1977).

The most ventral part of the claustrum is interconnected with the inferior temporal and other cortical visual areas (Baizer et al., 1997, and see also Pearson et al., 1982; Riche and Lanoir, 1978; Turner et al., 1980; Ungerleider et al., 1984). These data also suggest that monkey claustrum, like that of cat, probably contains a single map of the visual field, with the central vision represented ventrally and peripheral vision represented more dorsally.

An increasing body of data supports the concept that the striate cortex (area 17) in primates is the source of two major visual processing streams (Ungerleider and Mishkin, 1982). One of these is directed to the inferior temporal cortex and is concerned with pattern recognition, the other is directed to the posterior parietal cortex and is involved with visuospatial functions (Felleman and Van Essen, 1991). End stations of these streams project to different parts of the claustrum (see Pearson et al., 1982).

After initial disagreements, projections from the primary auditory cortex to the claustrum have been confirmed (Olson and Graybiel 1980, Reale and Imig, 1983). The primary auditory cortex is reciprocally connected with the claustrum, but these projections are probably weak (Hinova-Palova et al., 1988; Neal et al., 1986).

The limbic zone in the claustrum in the rat and rabbit contains neurons projecting to the cingulate and retrosplenial cortex. This zone is localized throughout the whole AP extent and prevails in the medial part of the dorsal claustrum. Only small numbers of projecting neurons were localized in the contralateral nucleus and only a few neurons were double-labeled (Condé et al., 1995; Majak et al., 2000).

Thus the present results, in conjunction with earlier reports, indicate that most (if not all) of the neocortex receives a projection from the dorsal claustrum. The ipsilateral projection is always much stronger than the contralateral one. Detection of the contralateral projection seems to depend largely on the size of injection and efficient uptake of marker in the terminals of claustral axons. The claustrum receives convergent tactile, acoustic and visual inputs and also inputs from the M1, PM and limbic cortical areas, and thus may be responsible for transmitting signals between different areas.

In contrast to the large number of reports concerning the dorsal claustrum, only a few studies have been dedicated to the connections of the ventral claustrum. From several studies it has emerged that the ventral claustrum projects to several allocortical regions including the piriform cortex, the subiculum and the entorhinal cortex. These cortical formations are also reached by fibers arising from the most ventral part of the dorsal claustrum, and from the transition between both subdivisions (Amaral and Cowan, 1980; Markowitsch et al., 1984; Witter et al., 1988). Projections from the ventral claustrum to the ventral zones of the prefrontal cortex in the cat have been described by Clascá et al. (1992).

Although parcellation of the claustrum into the CL and END reflects to some degree differences in the structure and chemoarchitecture of the claustrum, it does not reflect the actual organization of its cortical connections. Witter et al. (1988) favour a more complex view: that the ventral claustrum is reciprocally connected with the olfactory related cortices and the entorhinal cortex; the intermediate claustrum is connected with limbic and paralimbic areas; and the most dorsal part is associated with the sensory, parasensory and motor cortical areas. The claustrum also projects to several subcortical structures, including the intralaminar and anterior thalamic nuclei, the mediodorsal thalamic nucleus, the lateral hypothalamus and amygdala (Behan and Haberly, 1999; Ericson et al., 2004; Zhang et al., 2001).

The Function of the Claustrum

In spite of many structural, neurochemical and hodological studies, the function of the claustrum is still unclear. An initial concept indicates a "relay" function for the claustrum. Experiments in cat and monkey using retrograde tracers have suggested that claustrum receives glutamatergic input from cortical layer 6, similar to some thalamic nuclei (somatosensory, auditory, visual). The principal (or projection) neurons that receive input from the cortex provide reciprocal projection back to cortex terminating mainly in layer 4 (Clascá et al., 1992; LeVay, 1986; LeVay and

Sherk, 1981a; Macchi et al., 1981). Such cortico-claustro-cortical circuits are established in all functionally determined parts of the claustrum.

It may be that somatosensory, auditory or visual signals are sent to the appropriate part of the claustrum, processed and then sent back to the relevant cortical area. However, this relatively simple arrangement does not take into account eventual integration processes within the nucleus. In all animals so far analyzed, there is a distinct overlapping of cortico-claustral projections, as well as overlapping of claustral neurons projecting to different cortical areas. Such overlapping indicates the possibility of integrative processes within the claustrum. It is also possible that the claustrum can influence the cortex, in addition to this direct link, by a link via the thalamus. The existence of claustrothalamic projections and consecutive thalamocortical projections may represent a parallel circuit. Thus there is a possibility that the claustrum influences the cortex directly via claustro-cortical projections as well as indirectly via a claustro-thalamo-cortical link or thalamo-claustro-cortical link (Carey and Neal, 1985; Ericson et al., 2004; Zhang et al., 2001).

The data demonstrating the convergence of information from different sensory cortical areas and subcortical structures within the claustrum indicate that the claustrum may be the site at which modality-specific sensory inputs are bound into a multimodal representation. The structural substrate of such binding at the cellular level is not clear, especially in view of a recent finding indicating that the visual and auditory zones within the primate CL consist of unimodal neurons (Remedios et al., 2010). On the other hand visual-tactile integration in the human claustrum/insular area has been detected by positron emission tomography (Banati et al., 2000; Hadjikhani and Roland, 1998). Intrinsic claustral interconnections could play some role in the integrative processes, but axons interconnecting different parts of the CL and END have so far only been demonstrated in the rat. Clearly more research is needed to ascertain whether such interconnections also play some integrative role in the claustrum of carnivores and non-human primates. Another possible mechanism of intraclaustral integration could be networks of GABAergic neurons, especially PV-ir and CR-ir GABAergic interneurons.

References

Alloway, K.D., Smith, J.B., Beauchemin, K.J., Olson, M.L., 2009. Bilateral projections from rat MI whisker cortex to the neostriatum, thalamus, and claustrum: forebrain circuits for modulating whisking behavior. J. Comp. Neurol. 515, 548–564.

Amaral, D.G., Cowan, W.M., 1980. Subcortical afferents to the hippocampal formation in the monkey. J. Comp. Neurol. 189 (4), 573–591.

Arikuni, T., Kubota, K., 1985. Claustral and amygdaloid afferents to the head of the caudate nucleus in macaque monkeys. Neurosci. Res. 2 (4), 239–254.

Arimatsu, Y., Kojima, M., Ishida, M., 1999. Area- and lamina-specific organization of a neuronal subpopulation defined by expression of latexin in the rat cerebral cortex. Neuroscience 88, 93–105.

Arimatsu, Y., Nihomantsu, I., Hatanaka, Y., 2009. Localization of latexin-immunoreactive neurons in the adult cat cerebral cortex and claustrum/endopirioform formation. Neuroscience 162, 1398–1410.

Ashwell, K.W.S., Hardman, C., Paxinos, G., 2004. The claustrum is not missing from all monotreme brains. Brain Behav. Evol. 64, 223–241.

Baizer, J.S., 2001. Serotoninergic innervation of primate claustrum. Brain Res. Bull. 55, 431–434.

Baizer, J.S., Desimone, R., Ungerleider, L.G., 1993. Comparison of subcortical connections of inferior temporal and posterior parietal cortex in monkeys. Vis. Neurosci. 10, 59–72.

Baizer, J.S., Lock, T.M., Youakim, M., 1997. Projections from the claustrum to the prelunate gyrus in the monkey. Exp. Brain Res. 113, 564–568.

Baleydier, C., Mauguiere, R.F., 1980. The duality of the cingulate gyrus in monkey. Neuroanatomical study and functional hypothesis. Brain 103, 525–554.

Banati, R.B., Goerres, G.W., Tjoa, C., Aggleton, J.P., et al., 2000. The functional anatomy of visual-tactile integration in man: a study using positron emission tomography. Neuropsychology 38, 115–124.

Behan, M., Haberly, L.B., 1999. Intrinsic and efferent connections of the endopiriform nucleus in rat. J. Comp. Neurol. 408 (4), 532–548.

Beneyto, M., Prieto, J.J., 2001. Connections of the auditory cortex with the claustrum and the endopiriform nucleus in the cat. Brain Res. Bull. 54, 485–489.

Berke, J.J., 1960. The claustrum, the external capsule and the extreme capsule of Macaca mulatta. Neurology 115, 297–321.

Braak, H., Braak, E., 1982. Neuronal types in the claustrum of man. Anat. Embryol. 163 (4), 447–460.

Brand, S.A., 1981. A serial section Golgi analysis of the primate claustrum. Anat. Embryol. 162 (4), 475–488.

Brockhaus, H., 1940. Cytoarchitectural and myeloarchitectural study of claustral cortex and claustrum in man. J. Psychol. Neurol. 49, 249–348.

Buchanan, K.J., Johnson, J.I., 2011. Diversity of spatial relationships of the claustrum and insula in branches of the mammalian radiation. Ann. NY Acad. Sci. 1225, 30–63.

Buchanan, S.L., Thompson, R.H., Maxwell, B.L., Powell, D.A., 1994. Efferent connections of the medial prefrontal cortex in the rabbit. Exp. Brain Res. 100, 469–483.

Burman, K.J., Reser, D.H., Richardson, K.E., Worthy, K.H., Rosa, M.G., 2011. Subcortical projections to the frontal pole in the marmoset monkey. Eur. J. Neurosci. 34, 303–319.

Butler, A.B., Molnár, Z., Manger, P.R., 2002. Apparent absence of the claustrum in monotremes: implication for evolution of amniotes. Brain Behav. Evol. 60, 230–240.

Carey, R.G., Neal, T.L., 1985. The rat claustrum: afferent and efferent connections with visual cortex. Brain Res. 329 (1-2), 185–193.

Carey, R.G., Neal, T.L., 1986. Reciprocal connections between the claustrum and visual thalamus in the tree shrew (Tupaia glis). Brain Res. 386 (1-2), 155–168.

Carey, R.G., Fitzpatrick, D., Diamond, I.T., 1979a. Layer I of striate cortex of Tupaia glis and Galago senegalensis: projections from thalamus and claustrum revealed by retrograde transport of horseradish peroxidase. J. Comp Neurol. 186 (3), 348–393.

Carey, R.G., Fitzpatrick, D., Diamond, I.T., 1979b. Subcortical projections to layer I of striate cortex shown by retrograde transport of horseradish peroxidase. Anat. Rec. 193, 497–498.

Carey, R.G., Bear, M.F., Diamond, I.T., 1980. The laminar organization of the reciprocal projections between the claustrum and striate cortex in the tree shrew, Tupaia glis. Brain. Res. 184 (1), 193–198.

Carman, J.B., Cowan, W.M., Powell, T.P.S., 1964. The cortical projection upon the claustrum. J. Neurol. Neurosurg. Psychiat. 27, 46–51.

Celio, M.R., 1990. Calbindin D-28k and parvalbumin in the rat nervous system. Neuroscience 35, 375–475.

Chadzypanagiotis, D., Narkiewicz, O., 1971. Connections of the visual cortex with the claustrum. Acta Neurobiol. Exp. 31 (3), 291–311.

Ciofi, P., Tramu, G., 1990. Distribution of cholecystokinin-like-immunoreactive neurons in the Guinea pig forebrain. J. Comp. Neurol. 300, 82–112.

Clarey, J.C., Irvine, D.R., 1990. The anterior ectosylvian sulcal auditory field in the cat: II. A horseradish peroxidase study of its thalamic and cortical connections. J. Comp. Neurol. 30, 304–324.

Clascá, F., Avendaño, C., Román-Guindo, A., et al., 1992. Innervation from the claustrum of the frontal association and motor areas: axonal transport studies in the cat. J. Comp. Neurol. 326 (3), 402–422.

Colechio, E.W.M., Alloway, K.D., 2009. Differential topography of the bilateral cortical projections to the whisker and forepaw regions in rat motor cortex. Brain Struct. Funct. 213, 423–439.

Condé, F., Maire-Lepoivre, E., Audinat, E., Crépel, F., 1995. Afferent connections of the medial frontal cortex of the rat. J. Comp. Neurol. 352, 567–593.

Cowan, W.M., 1970. Anterograde and retrograde transneuronal degeneration in the central and peripheral nervous system. In: Nauta, W.J.H., Ebbesson, S.O.E. (Eds.), Contemporary Research Methods in Neuroanatomy Springer Verlag, Berlin–Heidelberg–New York, pp. 217–251.

Day-Brown, J.D., Wei, H., Chornsung, R.D., et al., 2010. Pulvinar projections to the striatum and amygdala in the tree shrew. Front. Neuroanat. 15, 143–153.

Dinopoulos, A., Papadopoulos, G.C., Michaloudi, H., et al., 1992. Claustrum in the hedgehog (Erinaceus europaeus) brain: cytoarchitecture and connections with cortical and subcortical structures. J. Comp. Neurol. 316 (2), 187–205.

Divac, I., Holst, M.C., McKenzie, J.S., 1987. Afferents of the frontal cortex in the echidna (Tachyglossus aculeatus). Indication of an outstandingly large prefrontal area. Brain Behav. Evol. 30, 303–320.

Doty, R.W., 1983. Nongeniculate afferents to striate cortex in Hques. J. Comp. Neurol. 218 (2), 159–173.

Druga, R., 1966a. The claustrum of the cat (Felis domestica). Folia Morphol. (Prague) 14 (1), 7–16.

Druga, R., 1966b. Cortico-claustral connections. I. Fronto-claustral connections. Folia Morphol. (Prague) 14 (4), 391–399.

Druga, R., 1968. Cortico-claustral connections. II. Connections from the parietal, temporal and occipital cortex to the claustrum. Folia Morphol. (Prague) 16 (2), 142–149.

Druga, R., 1971. Projection of prepyriform cortex into claustrum. Folia Morphol. (Prague) 19 (4), 405–410.

Druga, R., 1972. Efferent projections from the claustrum (an experimental study using Nauta's method). Folia Morphol. (Prague) 20 (2), 163–165.

Druga, R., 1974. The claustrum and the transitional neo-paleocortical area of the hedgehog (Erinaceus europaeus). Anat. Anz. 135 (5), 442–454.

Druga, R., 1975. Claustrum (Struktura, Ontogenese a Spoje). Prague. PhD Dissertation. Charles University, pp. 193.

Druga, R., 1982. Claustro-neocortical connections in the cat and rat demonstrated by HRP tracing technique. J. Hirnforsch. 23 (2), 191–202.

Druga, R., 1989. Reciprocal connections between the claustrum and the gyrus sigmoideus posterior in the cat. An experimental study using the antegrade degeneration methods and the HRP retrograde axonal transport. Anat. Anz. 156 (2), 109–118.

Druga, R., Rokyta, R., Benes Jr., V., 1990. Claustro-neocortical projections in the rhesus monkey (projections to area 6). J. Hirnforsch. 31 (4), 487–494.

Druga, R., Chen, S., Bentivoglio, M., 1993. Parvalbumin and calbindin in the rat claustrum: an immunocytochemical study combined with retrograde tracing frontoparietal cortex. J. Chem. Neuroanat. 6 (6), 399–406.

Edelstein, L.R., Denaro, F.J., 2004. The claustrum: a historical review of its anatomy, physiology, cytochemistry and functional significance. Cell Molec. Biol. 50, 675–702.

Edelstein, L.R., Druga, R., Stamm, J.S., Cerman, J., Salaj, M., 2010. Calretinin immunoreactivity in the dorsal and ventral claustrum of the rat and guinea pig. Soc. Neurosci. 36 [Abstr.].

Ekstrand, J.J., Domroese, M.E., Johnson, D.M., Feig, S.L., Knodel, S.M., Behan, M., et al., 2001. A new subdivision of anterior piriform cortex and associated deep nucleus with novel features of interest for olfaction and epilepsy. J. Comp Neurol. 434, 289–307.

Erickson, S.L., Melchitzky, D.S., Lewis, D.A., 2004. Subcortical afferents to the lateral mediodorsal thalamus in cynomolgus monkeys. Neuroscience 129, 675–690.

Feldblum, S., Erlander, M.G., Tobin, A.J., 1993. Different distributions of GAD65 and GAD67 mRNAs suggest that the two glutamate decarboxylases play distinctive functional roles. J. Neurosci. Res. 34, 689–706.

Felleman, D.J., Van Essen, D.C., 1991. Distributed hierarchical processing in the primate cerebral cortex. Cereb. Cortex 1, 1–47.

Filimonoff, I.N., 1966. The claustrum, its origin and development. J. Hirnforsch. 8 (5), 503–528.

Filimonov, I.N., 1949. Comparative anatomy of mammalian cerebral cortex. 158 p. Ed. Academy of Medical Sciences, Moscow (in Russian).

Filimonov, I.N., 1974. Selected works. Ed. Meditsina, Moscow (in Russian).

Flindt-Egebak, P., Olsen, R.B., 1978. Some efferent connections of the feline claustrum. Neurosci. Lett. 1 (Suppl), S150–S159.

Gomez-Urquijo, S.M., Gutierrez-Ibarluzea, I., Bueno-Lopez, J.L., et al., 2000. Percentage incidence of gamma-aminobutyric acid neurons in the claustrum of the rabbit and comparison with the cortex and putamen. Neurosci. Lett. 282 (3), 177–180.

Graham, J., Lin, C.S., Kaas, J.H., 1979. Subcortical projections of six visual cortical areas in the owl monkey, Aotus trivirgatus. J. Comp. Neurol. 187 (3), 557–580.

Guirado, S., Real, M.A., Olmos, J.L., et al., 2003. Distinct types of nitric oxide-producing neurons in the developing and adult mouse claustrum. J. Chem. Neuroanat. 465 (3), 431–444.

Guldin, W.O., Markowitsch, H.J., Lampe, R., et al., 1986. Cortical projections originating from the cat's insular area and remarks on claustrocortical connections. J. Comp. Neurol. 243 (4), 468–487.

Gutiérrez-Ibarluzea, I., Gomez-Urquijo, S.M., Arana-Arri, E., Bueno-López, J.L., Reblet, C., 1998. Quantitative distribution of GABAergic neurons in the rabbit claustrum. Eur. J. Neurosci. 10, 187–195.

Haberly, L.B., Price, J.L., 1978a. Association and commissural fiber systems of the olfactory cortex in the rat. I. Systems originating in the piriform cortex and adjacent areas. J. Comp. Neurol. 178, 711–740.

Hadjikhani, N., Roland, P.E., 1998. Cross-modal transfer of information between the tactile and the visual representations in the human brain: a positron emission tomographic study. J. Neurosci. 18, 1072–1084.

Hinova-Palova, D.V., Paloff, A.M., 1982. Corticoclaustar connections. An electron-microscopic study. Verh. Anat. Ges. 76, 503–504.

Hinova-Palova, D.V., Paloff, A.M., Penev, D.I., 1979. Synaptic organization of the claustrum in the cat. Dokl. Akad. Nauk. Bulg. 32, 831–834.

Hinova-Palova, D.V., Paloff, A.M., Usunoff, K.G., 1980a. Identification of three types of degenerated boutons in claustrum dorsale in the cat after lesion of auditory cortex. Dokl. Akad. Nauk. Bulg. 33, 129–132.

Hinova-Palova, D.V., Paloff, A.M., Usunoff, K.G., 1980b. Identification of three types of degenerated boutons in claustrum dorsale of the cat after lesion of the temporal cortex. CR Acad. Bulg. Sci. 33, 125–128.

Hinova-Palova, D.V., Paloff, A.M., Usunoff, K.G., et al., 1988. Reciprocal connections between the claustrum and the auditory cortical fields in the cat. An experimental study using light- and electron microscopic anterograde degeneration methods, and the horseradish peroxidase retrograde axonal transport. J. Hirnforsch. 29, 255–278.

Hinova-Palova, D.V., Paloff, A.M., Christova, T., et al., 1997. Topographical distribution of NADPH-diaphorase-positive neurons in the cat's claustrum. Eur. J. Morphol. 35 (2), 105–116.

Hinova-Palova, D.V., Edelstein, L.R., Paloff, A.M., et al., 2007. Parvalbumin in the cat claustrum: ultrastructure, distribution and functional implications. Acta Histochemica. 109 (1), 61–77.

Hirasawa, K., Okano, S., Kamio, S., 1938. Beitrag zur kenntnis uber die corticalen extrapyramidalen fasern aus der area temporalis superior (area 22) beim affen. Z. Mikr. Anat. Forsch. 44, 74–84.

Hoffman, W.H., Haberly, L.B., 1991. Bursting-induced epileptiform EPSPs in slices of piriform cortex are generated by deep cells. J. Neurosci. 11, 2021–2031.

Hoover, W.B., Vertes, R.P., 2007. Anatomical analysis of afferent projections to the medial prefrontal cortex in the rat. Brain Struct. Funct. 212, 149–179.

Hughes, H.C., 1980. Efferent organization of the cat pulvinar complex, with a note on bilateral claustrocortical and reticulocortical connections. J. Comp. Neurol. 193 (4), 937–963.

Hurley, K.M., Herbert, H., Moga, M.M., et al., 1991. Efferent projections of the infralimbic cortex of the rat. J. Comp. Neurol. 308 (2), 249–276.

Igarasi, Y., 1940. Uber die fasernverbindung des vorderer teiles der lateralen grosshirn-hemisphärenfläche bei der Ziege. Morph. Jb. 84, 108–153.

Insausti, R., Amaral, D.G., Cowan, W.M., 1987. The entorhinal cortex of the monkey. III. Subcortical afferents. J. Comp. Neurol. 264, 396–408. 1987.

Insausti, R., Herrero, M.T., Witter, M.P., 1997. Entorhinal cortex of the rat: cytoarchitectonic subdivisions and the origin and distribution of cortical efferents. Hippocampus 7 (2), 146–183.

Irvine, D.R.F., Brugge, J.F., 1980. Afferent and efferent connections between the claustrum and parietal association cortex in cat: a horseradish peroxidase and autoradiographic study. Neurosci. Lett. 20 (1), 5–10.

Jakubowska-Sadowska, K., Morys, J., Sadowski, M., et al., 1998. Visual zone of the claustrum shows localizational and organizational differences among rat, guinea pig, rabbit and cat. Anat. Embryol. 198 (1), 63–72.

Jayaraman, A., Updyke, B.V., 1979. Organization of visual cortical projections to the claustrum in the cat. Brain Res. 178 (1), 107–115.

Jin, M.-h., Ishida, M., Katoh-Fukui, Y., et al., 2006. Reduced pain sensitivity in mice lacking latexin, an inhibitor of metallocarboperoxidases. Brain Res. 1075, 117–121.

Juraniec, J., Narkiewicz, O., Wrzolkowa, T., 1971. Axon terminals in the claustrum of the cat: an electron microscope study. Brain Res. 35, 277–282.

Kamio, S., 1938. Uber die Fasernverbindung des hinteren Teiles der lateralen Grosshirnhemisphärenfläche der Ziege unter besondere Berücksichtigung der corticalen extrapyramidalen Bahnen. Folia Anat. Jap. 16, 1–37.

Kaske, A., Dick, A., Creutzfeldt, O.D., 1991. The local domain for divergence of subcortical afferents to the striate and extrastriate visual cortex in the common marmoset (Callithrix jacchus): a multiple labelling study. Exp. Brain Res. 84, 254–265.

Kaufman, E.F., Rosenquist, A.C., 1985. Efferent projections of the thalamic intralaminar nuclei in the cat. Brain Res. 335, 257–279.

Kemp, J.M., Powell, T.P.S., 1970. The cortico-striate projection in the monkey. Brain 93 (3), 525–546.

Kennedy, H., Bullier, J., 1985. A double-labelling investigation of the afferent connectivity to cortical areas V1 and V2 of the macaque monkey. J. Neurosci. 5, 2815–2830.

Kerr, K.M., Agster, K.L., Furtak, S.C., Burwell, R.D., 2007. Functional neuroanatomy of the parahippocampal region: the lateral and medial entorhinal areas. Hippocampus 17 (9), 697–708.

Kievet, J., Kuypers, H.G.J.M., 1975. Subcortical afferents to the frontal lobe in the rhesus monkey studied by means of retrograde horseradish peroxidase transport. Brain Res. 85 (2), 261–266.

Kowiański, P., Moryś, J.M., Karwacki, Z., et al., 1998. The cortico-related zones of the rabbit claustrum-study of the claustrocortical connections based on the retrograde axonal transport of fluorescent tracers. Brain Res. 784 (1-2), 199–209.

Kowiański, P., Lipowska, M., Moryś, J., 1999. The piriform cortex and the endopiriform nucleus in the rat reveal generally similar pattern of connections. Folia Morphol. (Warsaw) 58 (1), 9–19.

Kowiański, P., Dziewiatkowski, J., Kowiańska, J., 1999. Comparative anatomy of the claustrum in selected species: a morphometric anapysis. Brain Behav. Evol. 53, 44–54.

Kowiański, P., Timmermans, J.P., Moryś, J., 2001. Differentiation in the immunocytochemical features of intrinsic and cortically projecting neurons in the rat claustrum : combined immunocytochemical and axonal transport study. Brain Res. 905, 63–71.

Kowiański, P., Moryś, J.M., Wojcik, S., et al., 2003. Co-localization of NOS with calcium-binding proteins during the postnatal development of the rat claustrum. Folia. Morphol. (Warsaw) 62 (3), 211–214.

Kowiański, P., Moryś, J., Wojcik, S., et al., 2004. Neuropeptide-containing neurons in the endopiriform region of the rat: morphology and colocalization with calcium-binding proteins and nitric oxide synthase. Brain Res. 996 (1), 97–110.

Kowiański, P., Dziewiatkowski, J., Moryś, J.M., et al., 2009. Colocalization of neuropeptides with calcium-binding proteins in the claustral interneurons during postnatal development of the rat. Brain Res. Bull. 80 (3), 100–106.

Krettek, J.E., Price, J.L., 1977. Projections from the amygdaloid complex to the cerebral cortex and thalamus in the rat and cat. J. Comp. Neurol. 172 (4), 687–722.

Krettek, J.E., Price, J.L., 1978. A description of the amygdaloid complex in the rat and cat with observations on intra-amygdaloid axonal connections. J. Comp. Neurol. 178 (2), 255–280.

Kubasik–Juraniec, J., Dziewiatkowski, J., Morys, J., Narkiewicz, O., 1998. Ultrastructural organization of the visual zone in the claustrum of the cat. Folia Morphol. (Warsaw) 57, 287–289.

Kuchiiwa, S., Shoumura, K., Kuchiiwa, T., et al., 1984. Afferents to the cortical pupillo-constrictor areas of the cat, traced with HRP. Exp. Brain Res. 54 (2), 377–381.

Kuchiiwa, S., Kuchiiwa, T., Matsue, H., et al., 1985. Efferent connections of area 20 in the cat: HRP-WGA and autoradiographic studies. Exp. Brain Res. 60 (1), 179–183.

Künzle, H., 1975. Bilateral projections from precentral motor cortex to the putamen and other parts of the basal ganglia. An autoradiographic study in Macaca fascicularis. Brain Res. 88 (2), 195–209.

Künzle, H., Akert, K., 1977. Efferent connections of cortical, area 8 (frontal eye field) in Macaca fascicularis. A reinvestigation using the autoradiographic technique. J. Comp. Neurol. 173, 147–164.

Künzle, H., Radtke-Schuller, S., 2000. Multiarchitectonic characterization of insular, perirhinal and related regions in a basal mammal, Echinops telfairi. Anat. Embryol. 202 (6), 507–522.

Künzle, H., Radtke-Schuller, S., 2001. Cortical connections of the claustrum and subjacent cell groups in the hedgehog tenrec. Anat. Embryol. (Berlin) 203, 403–415.

LeVay, S., 1986. Synaptic organization of claustral and geniculate afferents to the visual cortex of the cat. J Neuroscience 6, 3564–3575.

LeVay, S., Sherk, H., 1981a. The visual claustrum of the cat. I. Structure and connections. J. Neurosci. 1 (9), 956–980.

LeVay, S., Sherk, H., 1981b. The visual claustrum of the cat. II. The visual field map. J. Neurosci. 1 (9), 981–992.

Leichnetz, G.R., 1986. Afferent and efferent connections of the dorsolateral precentral gyrus (area 4,hand/arm region) in the macaque monkey, with comparisons to area 8. J. Comp. Neurol 254, 460–492.

Leichnetz, G.R., 2001. Connections of the medial posterior parietal cortex (area 7m) in the monkey. Anat. Rec. 263, 215–236.

Leichnetz, G.R., Goldberg, M.E., 1988. Higher centers concerned with eye movement and visual attention: cerebral cortex and thalamus. Rev. Oculomot. Res. 2, 365–429.

Levesque, M., Parent, A., 1998. Axonal arborization of corticostriatal and corticothalamic fibers arising from prelimbic cortex in the rat. Cerebral Cortex 8, 602–613.

Lysakowski, A., Standage, G.P., Benevento, L.A., 1988. An investigation of collateral projections of the dorsal lateral geniculate nucleus and other subcortical structures to cortical areas V1 and V4 in the macaque monkey: a double label retrograde tracer study. Exp. Brain Res. 69, 651–661.

Macchi, G., Bentivoglio, M., Minciacchi, D., et al., 1978. The claustro-cortical projections: a HRP study in the cat. Neurosci. Lett. 1 (Suppl), S166.

Macchi, G., Bentivoglio, M., Minciacchi, D., et al., 1981. The organization of the claustroneo-cortical projections in the cat studied by means of the HRP retrograde axonal transport. J. Comp. Neurol. 195 (4), 681–695.

Macchi, G., Bentivoglio, M., Minciacchi, D., et al., 1983. Claustroneocortical projections studied in the cat by means of multiple retrograde fluorescent tracing. J. Comp. Neurol. 215 (2), 121–134.

Majak, K., Moryś, J., 2007. Endopiriform nucleus connectivities: the implications for epileptogenesis and epilepsy. Folia Morphologica. Warszawa 66 (4), 267–271.

Majak, K., Kowiański, P., Morys, J., Spodnik, J., Karwacki, Z., et al., 2000. The limbic zone of the rabbit and rat claustrum: a study of the claustrocingulate connections based on the retrograde axonal transport of fluorescent tracers. Anat. Embryol. (Berlin) 201, 15–25.

Majak, K., Pikkarainen, M., Kemppainen, S., et al., 2002. Projections from the amygdaloid complex to the claustrum and the endopiriform nucleus: a Phaseolus vulgaris leucoagglutinin study in the rat. J. Comp. Neurol. 451 (3), 236–249.

Mamos, L., 1984. Morphology of claustral neurons in the rat. Folia Morphol. (Warsaw) 43 (2), 73–78.

Mamos, L., Narkiewicz, O., Moryś, J., 1986. Neurons of the claustrum in the cat: a golgi study. Acta Neurobiol. Exp. 46, 171–178.

Markowitsch, H.J., Irle, E., Bang-Olson, R., et al., 1984. Claustral efferents to the cat's limbic cortex studied with retrograde and anterograde tracing techniques. Neuroscience 12 (2), 409–425.

Mathur, B.N., Caprioli, R.M., Deutch, A.Y., 2009. Proteomic analysis illuminates a novel structural definition of the claustrum and insula. Cerebr. Cortex 19, 2372–2379.

McKenna, J.T., Vertes, R.P., 2004. Afferent projections to nucleus reuniens of the thalamus. J. Comp. Neurol. 480, 115–142.

Medina, L., Legaz, I., Gonzalez, G., De Castro, F., Rubenstein, J.L., Puelles, L., 2004. Expression of Dbx1, Neurogenin 2, Semaphorin 5A, Cadherin 8, and Emx1 distinguish ventral and lateral pallial histogenetic divisions in the developing mouse claustroamygdaloid complex. J. Comp. Neurol. 474 (4), 504–523.

Minciacchi, D., Molinari, M., Bentivoglio, M., et al., 1985. The organization of the ipsi- and contralateral claustrocortical system in rat with notes on the bilateral claustrocortical projections in cat. Neuroscience 16 (6), 557–576.

Minciacchi, D., Granato, A., Barbaresi, P., 1991. Organization of claustro-cortical projections to the primary somatosensory area of primates. Brain Res. 553, 309–312.

Miodonski, R., 1975. The claustrum of the dog brain. Acta Anat.(Basel) 91 (3), 409–422.

Miyashita, T., Nishimura-Akyoshi, S., Itohara, S., et al., 2005. Strong expression of Netrin-G2 in the monkey claustrum. Neuroscience 136, 487–496.

Morán, M.A., Mufson, E.J., Mesulam, M.M., 1987. Neural inputs into the temporopolar cortex of the rhesus monkey. J. Comp. Neurol. 256, 88–103.

Moryś, J., Narkiewicz, O., Wisniewski, H.M., 1993. Neuronal loss in the human claustrum following ulegyri. Brain Res. 616 (1-2), 176–180.

Moryś, J., Bobinski, M., Maciejewska, B., Berdel, B., et al., 1994. The insular claustrum in the methylazoxymethanol acetate (MAM) treated rats shows two different population of neurons. Folia Neuropathol. 32 (2), 107–112.

Moryś, J., Berdel, B., Maciejewska, B., et al., 1996. Division of the human claustrum according to its architectonics and morphometric parameters. Folia Morphol. (Warsaw) 55 (2), 69–82.

Mufson, E.J., Mesulam, M.M., 1982. Insula of the old world monkey. II: afferent cortical input and comments on the claustrum. J. Comp. Neurol. 212 (1), 23–37.

Narkiewicz, O., 1964. Degenerations in the claustrum after regional neocortical ablations in the cat. J. Comp. Neurol. 123, 335–356.

Narkiewicz, O., 1966. Connections of the claustrum with the cerebral cortex. Folia Morphol. (Warsaw) 25 (4), 555–561.

Narkiewicz, O., Mamos, L., 1990. Relation of the insular claustrum to the neocortex in insectivora. J. Hirnforsch. 31 (5), 623–633.

Narkiewicz, O., Juraniec, J., Wrzolkova, T., 1983. Early degeneration of the axon terminals in the central nervous system. Ann. Med. Sect. Pol. Acad. Sci. 18, 17–19.

Narkiewicz, O., Moryś, J. Sloniewski, P. 1985. Neurons of the lamina medullaris pretectothalamica and their projections to the insulaclaustral area in the cat. Paper presented at the 79th Versammlung der Anatomischen Gesellschaft, Bochum: FRG, pp. 427–428.

Narkiewicz, O., Nitecka, L., Mamos, L., et al., 1988. The pattern of the GABA-like immunoreactivity in the claustrum. Folia Morphol. 47, 21–30.

Nauta, W.J.H., 1964. Some efferent connections of the prefrontal cortex in the monkey. In: Warren, J.M., Akert, K. (Eds.), The Frontal Granular Cortex and Behavior McGraw-Hill, New York, pp. 397–409.

Neal, J.W., Pearson, R.C.A., Powell, T.P.S., 1986. The relationship between the auditory cortex and the claustrum in the cat. Brain Res. 366 (1-2), 145–151.

Norita, M., 1977. Demonstration of bilateral claustro-cortical connections in the cat with the method of retrograde axonal transport of horseradish peroxidase. Arch. Histol. Japon 40 (1), 1–10.

Norita, M., Hirata, I., 1976. Some electron microscope findings of the claustrum of the cat. Arch. Histol. Japon 39 (1), 33–49.

Obst-Pernberg, K., Medina, L., Redies, C., 2001. Expression of R-cadherin and N-cadherin by cell groups and fiber tracts in the developing mouse forebrain : relation to the formation of functional circuits. Neuroscience 106, 505–533.

Olson, C.R., Graybiel, A.M., 1980. Sensory maps in the claustrum of the cat. Nature 288, 479–481.

Olson, C.R., Jeffers, I., 1987. Organization of cortical and subcortical projections to area 6m of the cat. J. Comp. Neurol. 266, 73–94.

Otellin, V.A., 1969. Connections of the auditory cortex (zone Ep) with the claustrum. Dokl. Akad. Nauk. SSSR (Ser. Biol.) 187 (5), 1198–1200.

Otellin, V.A., 1972. Connections of the auditory and vestibular cortex with the claustrum in cats (microscopic and electron-microscopic investigation. Neurosci. Behav. Physiol. 5 (3), 275–281.

Otellin, V.A., 1973. Ultrastructural organization of the claustrum. Arkh. Anat. Gistol. Embriol. 64, 62–69.

Otellin, V.A., Makarov, F.N., 1972. Descending connections of cat auditory cortex with contralateral neostriatal complex and claustrum. Dokl. Akad. Nauk. SSSR (Ser. Biol.) 202 (3), 723–725.

Paxinos, G., Watson, C., 1997. The Rat Brain In Stereotaxic Coordinates. Academic Press, London.

Pearson, R.C.A., Brodal, P., Gatter, K.C., et al., 1982. The organization of the connections between the cortex and the claustrum in the monkey. Brain Res. 234 (2), 435–441.

Perkel, D.J., Bullier, J., Kennedy, H., 1986. Topography of the afferent connectivity of area 17 in the macaque monkey: a double – labelling study. J. Comp. Neurol. 253, 374–402.

Pilleri, G., 1961. On the structure of the claustrum in Didelphis marsupialis Lin. (Marsupialia, Didelphoidea). Acta Anat. 45, 310–314.

Pilleri, G., 1962. The claustrum of the Canadian beaver (Castor canadensis Kuhl)-structure and comparative anatomy. J. Hirnforsch. 5, 59–81.

Pirone, A., Cozzi, B., Edelstein, L., Peruffo, A., et al., 2012. Topography of Gng2 – and NetrinG2- expression suggests an isular origin of the human claustrum. PLoS One 7, e44745.

Prensa, L., Richard, S., Parent, A., 2003. Chemical anatomy of the human ventral striatum and adjacent basal forebrain structures. J. Comp. Neurol. 460, 345–367.

Puelles, L., Kuwana, E., Puelles, E., Bulfone, A., Shimamura, K., Rubenstein, J.L., 2000. Pallial and subpallial derivatives in the embryonic chick and mouse telencephalon, traced by the expression of the genes Dlx-2, Emx-1, Nkx-2.1, Pax-6, and Tbr-1. J. Comp. Neurol. 424 (3), 409–438.

Rae, A.S.L., 1954. The connections of the claustrum. Confin. Neurol. 14 (4), 211–219.

Rahman, F.E., Baizer, J.S., 2007. Neurochemically defined cell types in the claustrum of the cat. Brain Res. 1159, 94–111.

Real, M.A., Davila, J.C., Guirado, S., 2003. Expression of calcium-binding proteins in the mouse claustrum. J. Chem. Neuroanat. 25 (3), 151–160.

Real, M.A., Dávila, J.C., Guirado, S., 2006. Immunohistochemical localization of the vesicular glutamate transporter VGLUT2 in the developing and adult mouse claustrum. J. Chem. Neuroanat. 31, 169–177.

Reale, R.A., Imig, T.J., 1983. Auditory cortical field projections to the basal ganglia of the cat. Neuroscience 8, 67–86.

Remedios, R., Logothetis, N.K., Kayser, C.H., 2010. Unimodal responses prevail within the multisensory claustrum. J. Neurosci. 30, 12902–12907.

Reynhout, K., Baizer, J.S., 1999. Immunoractivity for calcium-binding proteins in the claustrum of the monkey. Anat. Embryol. 199, 75–83.

Riche, D., Lanoir, J., 1978. Some claustro-cortical connections in the cat and baboon as studied by retrograde horseradish peroxidase transport. J. Comp. Neurol. 177 (3), 435–444.

Roberts, A.C., Tomic, D.L., Parkinson, C.H., Roeling, T.A., Cutter, D.J., Robbins, T.W., et al., 2007. Forebrain connectivity of the prefrontal cortex in the Marmoset monkey (Callitrix jacchus): an anterograde and retrograde tract-tracing study. *J. Comp. Neurol.* 502, 86–112.

Room, P., Groenewegen, H.J., 1986. Connections of the parahippocampal cortex. II. Subcortical afferents. J. Comp. Neurol. 251, 451–473.

Room, P., Witter, M.P., 1984. The connections between claustrum and entorhinal and perirhinal cortices in the cat: topography and reciprocity. Acta Morphol. Neerl. Scand. 22, 150.

Room, P., Russchen, F.T., Groenewegen, H.J., et al., 1985. Efferent connections of the prelimbic (area 32) and the infralimbic (area 25) cortices: an anterograde tracing study in the cat. J. Comp. Neurol. 242 (1), 40–55.

Sadowski, M., Morys, J., Wisniewski, K., et al., 1997. Rat's claustrum shows two main cortico-related zones. Brain Res. 756 (1-2), 147–152.

Schmidt, H., Arendt, O., Brown, E.B., et al., 2007. Parvalbumin is freely mobile in axons, somata and nuclei of cerebellar Purkinje neurons. J. Neurochem. 100, 727–735.

Schwabe, K., Ebert, U., Löscher, W., 2004. The central piriform cortex : anatomical connections and anticonvulsant effect of GABA elevation in the kindling model. Neuroscience 126 (3), 727–741.

Sesack, S.R., Deutch, A.Y., Roth, R.H., Bunney, B.S., 1989. Topographical organization of the efferent projections of the medial prefrontal cortex in the rat: an anterograde tract-tracing study with *Phaseolus vulgaris* leucoagglutinin. J. Comp. Neurol. 290, 213–242.

Sherk, H., 1986. The claustrum and the cerebral cortex. In: Jones, E.G., Peters, A. (Eds.), Cerebral Cortex, vol. 5, Senzori-Motor Areas and Aspects of Cortical Connectivity Plenum Press, New York, pp. 467–499.

Sherk, H., LeVay, S., 1981a. The visual claustrum of the cat. III. Receptive field properties. J. Neurosci. 1 (9), 993–1002.

Sherk, H., LeVay, S., 1981b. Visual claustrum: topography and receptive field properties in the cat. Science 212 (4490), 87–89.

Sherk, H., Le Vay, S., 1983. Contribution of the cortico-claustral loop to receptive field properties in area 17 of the cat. J. Neurosci 3 (11), 2121–2127.

Shi, C.J., Cassell, M.D., 1999. Perirhinal cortex projections to the amygdaloid complex and hippocampal formation in the rat. J. Comp. Neurol. 406 (3), 299–328.

Sloniewski, P., 1983. Pretectal connections to the claustrum: an HRP retrograde transport study in cats. Acta Neurobiol. Exp. 43 (3), 165–182.

Sloniewski, P., Pilgrim, C., 1984. Claustro-neocortical connections in the rat as demonstrated by retrograde tracing with Lucifer yellow. Neurosci. Lett. 49 (1-2), 29–32.

Sloniewski, P., Usunoff, K.G., Pilgrim, C., 1985. Efferent connections of the claustrum to the posterior thalamic and pretectal region in the rat. Neurosci. Lett. 60 (2), 195–199.

Sloniewski, P., Usunoff, K.G., Pilgrim, C., 1986a. Diencephalic and mesencephalic afferents of the rat claustrum. Anat. Embryol. 173 (3), 401–411.

Sloniewski, P., Usunoff, K.G., Pilgrim, C., 1986b. Retrograde transport of fluorescent tracers reveals extensive ipsi- and contralateral claustrocortical connections in the rat. J. Comp. Neurol. 246 (4), 467–477.

Smith, J.B., Alloway, K.D., 2010. Functional specifity of claustrum connections in the rat: interhemispheric communication between specific parts of motor cortex. J. Neurosci. 30, 16832–16844.

Smith, J.B., Radhakrishnan, H., Alloway, K.D., 2012. Rat claustrum coordinates but does not integrate somatosensory and motor cortical information. J. Neurosci. 32 (25), 8583–8588.

Smythies, J., Edelstein, L., Ramachandran, V., 2012. Hypotheses relating to the function of the claustrum. Front. Integr. Neurosci. 6, 1–16.

Spahn, B., Braak, H., 1985. Percentage of projection neurons and various types of interneurons in the human claustrum. Acta Anat. 122 (4), 245–248.

Squatrito, S., Battaglini, P.P., Galletti, C., Riva Sanseverino, E., 1980a. Autoradiographic evidence for projections from cortical visual areas 17, 18, 19 and the Clare-Bishop area to the ipsilateral claustrum in the cat. Neurosci. Lett. 19 (3), 265–269.

Squatrito, S., Battaglini, P.P., Galletti, C., Riva Sanseverino, E., 1980b. Projections from the visual cortex to the contralateral claustrum of the cat revealed by an anterograde axonal transport method. Neurosci. Lett. 19 (3), 271–275.

Standage, G.P., Doetsch, G.S., 1988. Projections from cortical area SmII and claustrum to two functional subdivisions of SmI forepaw digit cortex of the raccoon. Brain Res. Bull. 21, 207–213.

Stanton, G.B., Goldberg, M.E., Bruce, C.J., 1988. Frontal eye field efferents in the macaque monkey:I. Subcortical pathways and topography of striatal and thalamic terminal fields. J. Comp. Neurol. 271, 473–492.

Tanné-Gariépy, J., Boussaoud, D., Rouiller, E.M., 2002. Projections of the claustrum to the primary motor, premotor, and prefrontal cortices in the macaque monkey. J. Comp. Neurol. 454 (2), 140–157.

Tigges, J., Tigges, M., Gross, N.A., et al., 1982. Subcortical structures projecting to visual cortical areas in squirrel monkey. J. Comp. Neurol. 209 (1), 29–40.

Tigges, J., Walker, L.C., Tigges, M., 1983. Subcortical projections to the occipital and parietal lobes of the chimpanzee brain. J. Comp. Neurol. 220 (1), 106–115.

Turner, B.H., Mishkin, M., Knapp, M., 1980. Organization of the amygdalopetal projections from modality-specific cortical association areas in the monkey. J. Comp. Neurol. 191 (4), 515–543.

Ungerleider, L.G., Desimone, R., Galkin, T.W., et al., 1984. Subcortical projections of area MT in the macaque. J. Comp. Neurol. 223 (3), 368–386.

Vertes, R.P., 1991. A PHA-L analysis of ascending projections of the dorsal raphe nucleus in the rat. J. Comp. Neurol. 313 (4), 643–668.

Vertes, R.P., 2002. Analysis of projections from the medial prefrontal cortex in the rat with emphasis on nukleus reuniens. J. Comp. Neurol. 442, 163–187.

Vertes, R.P., 2004. Differential projections of the infralimbic and prelimbic cortex in the rat. Synapse 51, 32–58.

Weller, R.E., Steele, G.E., Kaas, J.H., 2002. Pulvinar and other subcortical connections of dorsolatersl visual cortex in monkeys. J. Comp. Neurol. 450, 215–240.

Witter, M.P., Room, P., Groenewegen, H.J., et al., 1988. Reciprocal connections of the insular and piriform claustrum with limbic cortex: an anatomical study in the cat. Neuroscience 24 (2), 519–539.

Wójcik, S., Dziewiatkowski, J., Kowiański, P., Ludkiewicz, B., et al., 2002. Qualitative and quantitative study of the postnatal development of the rabbit insular claustrum. Int. J. Dev. Neurosci. 20, 113–123.

Wójcik, S., Dziewiątkowski, J., Spodnik, E., Ludkiewicz, B., Domaradzka-Pytel, B., Kowiański, P., et al., 2004. Analysis of calcium binding protein immunoreactivity in the claustrum and the endopiriform nucleus of the rabbit. Acta Neurobiol. Exp. 64 (4), 449–460.

Yasui, Y., Saper, C.B., Cechetto, D.F., et al., 1989. Calcitonin gene-related peptide immunoreactivity in the visceral sensory cortex, thalamus, and related pathways in the rat. J. Comp. Neurol. 290 (4), 487–501.

Zhang, X., Hannesson, D.K., Saucier, D.M., Wallace, A.E., et al., 2001. Susceptibility to kindling and neuronal connections of the anterior claustrum. J Neurosci. 21, 3674–3687.

The Neurochemical Organization of the Claustrum

Joan Susan Baizer

Department of Physiology and Biophysics,
University of Buffalo, NY, USA

OVERVIEW

The claustrum is a forebrain structure that has been extensively studied in multiple species (review in Edelstein and Denaro, 2004); however, its function remains mysterious. The purpose of this review is to consider how the data about the neurochemical organization of the claustrum may help in the evaluation of different hypotheses about its function.

THE CLAUSTRUM: BASIC ORGANIZATION

The claustrum is a subcortical structure found in all mammals that have a cerebral cortex. It is well-established that the major input to the claustrum is from the cerebral cortex, and its major output is back to the cortex (Baizer et al., 1993; Chachich and Powell, 2004; Clascá et al., 1992; Crescimanno et al., 1989; Druga et al., 1990; Guldin et al., 1986; Hinova-Palova et al., 1988; Kennedy and Bullier, 1985; LeVay and Sherk, 1981a; Minciacchi et al., 1991; Morecraft et al., 1992; Perkel et al., 1986; review and older references in Sherk, 1986; Sloniewski et al., 1986). These connections are overall topographically organized, and electrophysiological data reinforce anatomical data showing that the claustrum, like the cerebral cortex, is divided into multiple sensory and motor areas (review in Baizer et al., 1993, 1997; Clascá et al., 1992; Druga, 1989; Minciacchi et al., 1995; Pérez-Cerdá et al., 1996; Sherk, 1986; Shima et al., 1996; Tokuno and Tanji,

J. Smythies, L. Edelstein and V. Ramachandran (Eds):
The Claustrum.

DOI: http://dx.doi.org/10.1016/B978-0-12-404566-8.00003-9

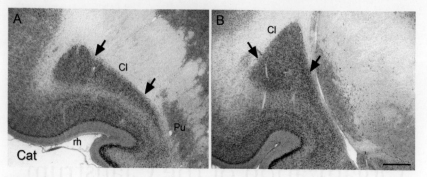

FIGURE 3.1   Cresyl-violet stained sections through the claustrum (*arrows*) of the cat showing the different shape at different rostro-caudal levels. The section in A is the more rostral. Note the dorsal enlargement of the claustrum in B. Cl, claustrum; Pu, putamen; rh, rhinal sulcus. Scale bar = 1 mm.

1993). In the cerebral cortex, different functional areas are cytoarchitectonically different, as described in detail by the early anatomists Brodmann (Brodmann and Gary, 2006), and Bonin and Bailey (1947). Despite the anatomical and electrophysiological evidence for functional subdivisions in the claustrum, unlike the cerebral cortex the claustrum appears cytoarchitectonically uniform. Figures 3.1 and 3.2 show cresyl-violet stained sections through the claustrum of the cat and the monkey revealing the uniform structure at different dorsoventral and rostrocaudal levels; Figure 3.3 shows Weil-stained sections from the macaque monkey.

Species Differences in Organization: What is the "Claustrum"?

The relationship between the claustrum and the cerebral cortex is well-established. However, the size and complexity of the cerebral cortex change dramatically over evolution, with its highest development seen in humans (discussion and references in Preuss, 2011; Striedter, 2005). The size and complexity of the claustrum show some parallel changes. A comparative study of the claustrum (Buchanan and Johnson, 2011) in 26 mammalian species illustrates the different forms it can take. The claustrum has been extensively studied in the cat and in the macaque monkey, animals in which the borders of the claustrum are clear but whose complexity of cortical organization is different. This difference in cortical organization introduces the possibility that the organization of the claustrum might show parallel variations. There are certainly differences in the gross appearance of the claustrum between the cat and the monkey. In both species the claustrum has an enlarged "bulb-like" portion from which extends a long, thin stem. However, in the monkey the enlarged portion is ventral and in the cat it is dorsal (see Figures 3.1 and 3.2).

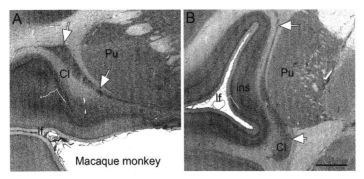

FIGURE 3.2 The shape of the claustrum in the monkey also varies with rostro-caudal level. Cresyl-violet stained sections through the claustrum (*arrows*) of the macaque monkey. A is the more rostral. Note the ventral enlargement of the claustrum in B. Cl, claustrum; lf, lateral fissure; ins, insula; Pu, putamen. Scale bar = 2mm.

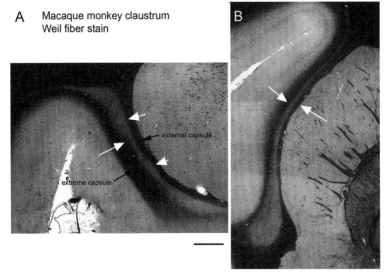

FIGURE 3.3 Weil-stained sections through the claustrum (*arrows*) of the macaque monkey. Note the fiber bundles medial (external capsule) and lateral (extreme capsule) to the claustrum. A is the more rostral section.

The organization of the claustrum is qualitatively different between lissencephalic animals and those species with a folded cerebral cortex. While most studies in the cat and monkey recognize a single structure called the "claustrum," this is not the case in rodents. Almost all rodent studies have recognized at least two subdivisions of the claustrum, but the number and nomenclature of these subdivisions has varied with the study. Two main subdivisions have been suggested: the *dorsal claustrum* and the *endopiriform nucleus* (Bayer and Altman, 1991; Dávila et al., 2005). These have been considered to be subdivisions of a single structure or distinct structures. In the mouse, one study used the term "claustral complex" to describe dorsal claustrum, which is probably homologous to the claustrum in primates and cats, and "ventral claustrum" was used for the endopiriform nucleus (Dávila et al., 2005). Yet another scheme recognized the terms *insular claustrum* for the dorsal claustrum and *prepiriform claustrum* as synonymous with the endopiriform nucleus (Zhang et al., 2001). The dorsal claustrum has been in turn further subdivided into dorsolateral and ventromedial regions (Dávila et al., 2005). For rabbit, the "endopiriform claustrum" and "insular claustrum" have been distinguished, suggesting that both structures are subdivisions of a single claustrum (Reblet et al., 2002). These different parcellation schemes and terminology can make the interpretation and comparison of different papers very difficult.

One approach to determining the correct parcellation of the claustrum is to look at the developmental origins of its putative components, comparing the expression of developmental markers or the time and site of neurogenesis (or both) for cell groups in the different structures included as "claustrum." It has been suggested that the pallial amygdala, which includes the lateral basolateral and basomedial amygdalar complexes represents an extension of the claustrum complex, itself consisting of both claustrum proper and endopiriform nuclei (references cited in Medina et al., 2004a). Another study suggests that the endopiriform nucleus develops earlier than the dorsal claustrum (claustrum proper) and argues for treating them as separate structures (Bayer and Altman, 1991). The endopiriform nucleus is also divided into dorsal and ventral subdivisions (Dávila et al., 2005). Medina et al. (2004b), who looked at the development of the claustrum in the mouse, used the general term *claustroamygdaloid complex*. They studied the developmental patterns of expression of several markers including Tbr1, Dbx1, and Neurogenin 2, Emx1, Cadherin 8 and Semaphorin 5A. The data suggested that a region they classed as the *ventromedial claustrum* was separate from, but associated with, the endopiriform nucleus, and that both were derivative of the ventral pallium. Another structure that they called the *dorsolateral claustrum* was instead a derivative of the lateral pallium. The weight of the evidence suggests that the embryological origins of the dorsal claustrum and endopiriform nucleus are different, with the claustrum of cortical origin and the endopiriform nucleus of basal ganglia origin (references and discussion in Dávila et al., 2005; Majak et al., 2002; Zhang et al., 2001). The connections are also different, with the claustrum reciprocally connected with all of the neocortex and the endopiriform nucleus connected preferentially with piriform, entorhinal, insular, and orbital cortices, again arguing that they are separate and distinct structures

(Zhang et al., 2001). Recently, Mathur et al. (2009), using a very different approach to defining the claustrum in the rat, also suggested that the claustrum and endopiriform nucleus are different entities. They identified a protein called G-protein gamma2 subunit (Gng2) which was enriched in the rat claustrum but not in adjacent structures including the endopiriform nucleus. The claustrum, defined as the region with high levels of this protein, projected only to cortical and not to subcortical structures.

In summary, the organization of the rodent claustrum seems both qualitatively and quantitatively different from the organization seen in nonlissencephalic mammals. In the present discussion, we will focus on data from cats and primates and only consider data from rodents that are from the dorsal or insular claustrum and not from the endopiriform nucleus.

MULTIPLE CORTICAL MAPS AND THE ORGANIZATION OF THE CLAUSTRUM

The early cytoarchitectonic studies of the cortex showed multiple distinct regions (Bonin and Bailey, 1947; Brodmann and Gary, 2006). However, the organization of cortex is considerably more complex than shown by cytoarchitectonics, with additional functional subdivisions found within cytoarchitectonically uniform regions. Visual, auditory and motor cortex all consist of multiple functional areas (Felleman and Van Essen, 1991; Kaas et al., 1999; Picard and Strick, 2001; Strick, 1986; Wiesendanger and Wise, 1992). These results raise the question of how multiple cortical maps are represented in the claustrum; for example, are there 30+ distinct visual areas in the claustrum as there are in cerebral cortex? The data suggest that there are many fewer areas in claustrum than in cortex, and that cortical topography is not mirrored. For the visual claustrum, data from retrograde tracing experiments suggest that single neurons in the claustrum project to more than one cortical visual area (Bullier et al., 1984). There is considerable divergence in the projections from the claustrum to the cortex (Salin et al., 1989), which suggests that there is divergence of signals from the claustrum across different visual cortical areas or maps. In turn, these data imply extensive integration of information across different visual field maps as well as intraclaustral interactions beyond precise topography within maps. Macchi et al. (1981) calculated that the part of the claustrum projecting to a particular cortical field is larger than the part receiving fibers from that same field, again suggesting an integrative or modulatory role of the claustral input to cortex rather than a simple shaping of receptive field properties.

Similar complexity exists in the connections between motor claustrum and motor cortex, with bilateral connections between claustrum and motor cortex (Smith and Alloway, 2010). For auditory cortex, Beneyto and Prieto (2001) showed that the eleven cortical auditory areas described in the cat each had reciprocal connections with the claustrum, but that there was a high degree of convergence of cortico-claustral projections from different cortical areas.

Connections of the Monkey Claustrum with Multiple Prestriate Visual Areas

In the monkey, electrophysiological and anatomical studies established that there are multiple visual areas (review in Felleman and Van Essen, 1991). These areas have been conceptualized as belonging to two different processing streams, a dorsal and a ventral. We asked whether the end stations of the two different visual processing streams, the inferotemporal cortex and the posterior parietal cortex, interconnect with the same or different regions of the claustrum (Baizer et al., 1993). We made injections of two different retrograde tracers in the two areas (fast blue, FB, in posterior parietal cortex, and diamidino yellow, DY in inferotemporal cortex) and compared the distribution of labeled cells in the claustrum.

Figure 3.4 illustrates the data from one case. Figure 3.4A shows a lateral view of the monkey brain and the locations of the injections. Figure 3.4B shows three coronal cross-sections through the claustrum showing the distribution of labeled cells. The filled symbols represent cells labeled by the DY injections and the open symbols represent cells labeled by the FB injection. While there is some overlap of the two populations, it is clear that the FB-labeled cells are found more dorsally to the majority of the DY-labeled cells. In a second study (Baizer et al., 1997) we looked at the distribution of retrogradely labeled cells in the claustrum after injections of tracer into two visual areas on the prelunate gyrus, V4, and a more dorsal area we named the posteromedial prelunate area, PM. Figure 3.5 shows the injection sites on a lateral view of the brain and the levels of the four parasagittal sections shown below.

Again, retrogradely labeled cells from the two different injections form geographically separate populations. The more ventral V4 injections resulted in label in the claustrum in the same general territory as is labeled after injections of tracers into several visual areas including V1, V2, MT, FST, MST, TEO and TE (Boussaoud et al., 1992; Kennedy and Bullier, 1985; Ungerleider et al., 1984; Webster et al., 1993). These data show that in the monkey there is a region of the claustrum which is interconnected with multiple cortical visual areas with dramatically different response characteristics. However, the pattern of connections between these many areas and single cells in the claustrum is unknown. A single cell in the claustrum might be connected with most or all of the areas.

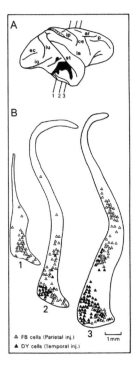

FIGURE 3.4   Claustrum projections to inferotemporal and posterior parietal cortex. A: Lateral view of the brain of a macaque monkey. The *shading* and *cross-hatching* show the sites of injection of fast blue (FB) and diamidino yellow (DY), respectively. B: The claustrum on three sections at the levels shown in A showing the distribution of cells retrogradely labeled by the cortical tracer injections. ar, arcuate sulcus; ce, central sulcus; ec, external calcarine sulcus; io, inferior occipital sulcus; ip, intraparietal sulcus; la, lateral sulcus; lu, lunate sulcus; p, principal sulcus; st, superior temporal sulcus. *Redrawn from Baizer et al., 1993.*

Alternatively, there may be segregated processing modules within a single visuotopically defined area of the claustrum that are related to different areas or subsets of areas. The concept of mosaic or compartmental organization is well-established for many structures at several levels of the CNS including the superior colliculus, striatum, visual cortex, medial vestibular nucleus, and the nucleus prepositus hypoglossi (Baizer and Baker, 2005, 2006; Graybiel, 1983; Illing and Graybiel, 1986; Nakamura et al., 1993; Sandell et al., 1986). There are anatomical data that suggest that a single neuron in the claustrum can project to both V1 and V2 (Kennedy and Bullier, 1985), but receptive field properties are not dramatically different in those two cortical areas (Baizer et al., 1977; Bullier et al., 1984; Hubel and Wiesel, 1968). These two possible patterns of connections of single cells in the claustrum with cortex impose very different constraints on the hypotheses about the function of the claustrum.

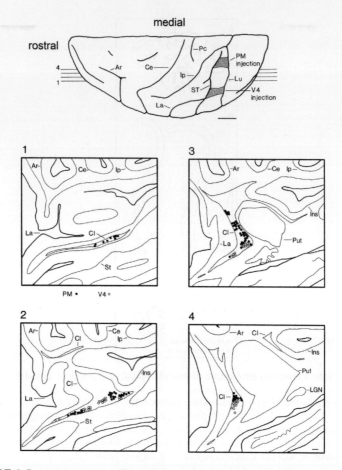

FIGURE 3.5 Projections from the claustrum to two visual areas of the prelunate gyrus. The diagram at the top shows a dorsal view of the left half of a monkey brain with the injection sites on the prelunate gyrus shaded. The lines 1–4 show the levels of the parasagittal sections illustrated below. Sections 1–4 show parasagittal sections through the claustrum with cells labeled by the V4 injection as *open circles* and cells labeled by the more dorsal posteromedial prelunate (PM) injection as *solid circles*. Ar, arcuate sulcus; Ce, central sulcus; Cl, claustrum; Ins, insula; Ip, intraparietal sulcus; La, lateral sulcus; lu, lunate sulcus; Put, putamen; ST, superior temporal sulcus. *Redrawn from Baizer et al., 1997.*

It is difficult to conceptualize a simple contribution to receptive field organization that would be the same for V1, inferotemporal cortex and MT since the response properties of neurons in those areas are so very different (Gross et al., 1969,1972; Hubel and Wiesel, 1968; Zeki, 1974). On the other hand, segregated modules could well mediate a simple contribution to receptive field properties. If there is extensive convergence at a single cell level there may be additional response characteristics of cells in the claustrum that have not been examined and therefore not found, for example the elaboration of more global visual properties as in visual cortex (Duffy and Wurtz, 1991a,b, 1995, 1997a,b; Wurtz and Duffy, 1992) or superior colliculus (Bender and Davidson, 1986; Davidson and Bender, 1991; Davidson et al., 1992).

What does the pattern of connections of the claustrum and cortex and the condensation of multiple cortical maps suggest about function? The claustrum is not the only subcortical structure with input from widespread cortical territories. The same is true of the basal ganglia, the cerebellum (via the pontine nuclei), the thalamus, and, for visual areas, the superior colliculus (Glickstein et al., 1980; Kemp and Powell, 1970; Lock et al., 2003; Mower et al., 1980; Saint-Cyr et al., 1990; Walker, 1938). However, the claustrum is unique in that its outflow is not into motor structures, as for the superior colliculus and the cerebellum, nor restricted to certain cortical domains, as for basal ganglia (Alexander and Crutcher, 1990; Hoover and Strick, 1999; Wurtz, 1996). Like the thalamus, the claustrum receives input from cortical layer 6 and projects back to layer 4 (Grieve and Sillito, 1995; LeVay and Sherk, 1981a; McCourt et al., 1986). Unlike the thalamus, the claustrum does not have a major source of ascending input. While there are some parallels in the organization of inputs and outputs between thalamus and claustrum, the feedback from cortex clearly has different functions for the two since different populations of layer 6 cells carrying different information project to the two (Grieve and Sillito, 1995; McCourt et al., 1986).

Further, the design of the claustrum allows compression of cortical topography. In the cat, the entire brain extends about 35 mm in the anteroposterior direction whereas the AP extent of the claustrum is only about 11 mm (Snider and Niemer, 1961). This means that the distance between functional regions and the time necessary for signals to travel between them are much shorter in the claustrum than in the cortex, allowing for much more rapid transmission of information between or among domains.

WHAT DOES THE CLAUSTRUM CONTRIBUTE TO INFORMATION PROCESSING IN THE CEREBRAL CORTEX? STUDIES OF THE VISUAL CLAUSTRUM

The anatomical data suggest that the claustrum receives selected cortical input, performs some analysis of that input and provides signals back to cortex. A major and unanswered question is what does that feedback do? How has that question been approached experimentally? One approach to understanding the function of the cortico-claustral-cortical loop has been to study the organization of individual functional subdivisions using both anatomical and electrophysiological techniques. The best studied and best understood functional subdivision of the claustrum is the visual claustrum of the cat (Boyapati and Henry, 1985; LeVay, 1986; LeVay and Sherk, 1981a,b; Macchi et al., 1981; Minciacchi et al., 1995; Olson and Graybiel, 1980; Pérez-Cerdá et al., 1996; Sherk and LeVay, 1981; Updyke, 1993). Electrophysiological analysis showed a visuotopic map in the part of the claustrum connected with visual cortex. Cells had exclusively visual responses and receptive field properties similar to those of cells in V1 (LeVay and Sherk, 1981b; Sherk and LeVay, 1981). Two studies addressed the function of the input from visual claustrum to visual cortex and suggested a rather simple contribution to visual processing. Sherk and LeVay (1983) suggested that input from the claustrum was critical for generating length sensitivity in V1 since lesions of visual claustrum result in a significant loss of end-stopping in visual receptive fields (hypercomplex cells) of V1. This interpretation was challenged by Boyapati and Henry (1985), who found little end-stopping in cells electrophysiologically determined to receive input from the claustrum. They suggested instead that the claustrum contributes to detection of stimulus motion independent of direction. Neither of these studies suggested a role for the function of the nonvisual subdivisions of the claustrum.

Implications for Function

The data of Sherk and LeVay (1983) prompted the theory that overall the claustrum subserves a simple "housekeeping" function for cortex, like adjustment of length sensitivity in visual receptive fields. This would imply a myriad of separate functions for different regions of the claustrum with the function of each subdivision tailored to a different cortical area or small subset of areas. This theory is incompatible with a number of observations. First, the general picture of convergence of projections from cortex suggests widespread integration of cortical input. While connections are overall topographic, there is some degree of topographic mismatch (Minciacchi et al., 1995). The existence of bilateral projections from claustrum to cortex is also at odds with a very specific function that depends on precise topographic matching. Together, these data support the idea that the claustrum has a more integrative role, extending across both sensory and motor domains, and that the function is unlikely to be simply the sculpting of cortical receptive field properties as suggested by Sherk and LeVay (1983).

NEUROCHEMISTRY OF THE CLAUSTRUM

The neurochemistry of the claustrum has been studied in the context of four different questions. As discussed above, the first question is whether the claustrum is better viewed as a derivative of the cortex or of the basal ganglia (review of the older literature in Bayer and Altman, 1991). Following the developmental patterns of expression of different markers has contributed to answering that question. A second, and related, question is whether the claustrum and endopiriform nucleus should be considered subdivisions of the same structure or as entirely different structures; again, neurochemical techniques have been used to see if these two regions have the same embryological origins. This question is more important for the rodent, in which the two structures are often considered subdivisions of a single claustral complex, than for other mammals with more developed cerebral cortex and a more clearly defined (dorsal) claustrum. Still other neurochemical studies have been concerned with two additional questions: first, whether there are neurochemically defined compartments in the claustrum and, second, whether there are neurochemically defined populations of neurons within the claustrum.

Are there Neurochemical Compartments in the Claustrum?

Cat and Monkey

The claustrum is functionally linked to the cerebral cortex, which is composed of many distinct cyto- and myeloarchitectonically distinct areas (Felleman and Van Essen, 1991; Lock et al., 2003). By contrast, despite evidence for similar functional subdivisions, the appearance of the claustrum is uniform in cell and fiber stains. We used immunohistochemistry and antibodies for calcium-binding proteins calretinin (CR), parvalbumin (PV) and calbindin (CB) to see if the expression of these proteins might allow us to visualize different functional subdivisions as has been seen in other structures, such as the pulvinar (Cusick et al., 1993) and the thalamus (Hashikawa et al., 1991; Kondo et al., 1999). Our results instead showed uniform label throughout the anatomically defined functional zones of the claustrum in both the cat and the monkey (Rahman and Baizer, 2007; Reynhout and Baizer, 1999). Analysis of the expression of a different marker, latexin, in both cortex and claustrum in the cat again reinforced the qualitatively different neurochemical organization between them (Arimatsu et al., 2009). Latexin is an endogenous inhibitor of the A/B family of metallocarboxypeptidases (references in Arimatsu et al., 2009). The expression of latexin in the rat cortex had been shown to be both region and layer-specific (Arimatsu et al., 1999). Similarly, in the cat, the cortical distribution of immunoreactive latexin (latexin-ir) neurons varied among cortical regions, whereas their distribution in the claustrum was homogeneous across functional zones (Arimatsu et al., 2009).

However, neurochemical inhomogeneity has been suggested in one study that looked at the localization of the netrin G2 gene in neurons of the monkey claustrum (Miyashita et al., 2005), finding stronger label in ventral than in dorsal claustrum.

Human

There are limited data on the neurochemical organization of the claustrum in the human. Prensa et al. (2003), in a study that focused on the human ventral striatum, also included some data on the adjacent ventral claustrum. They used immunohistochemistry and several markers, including the calcium-binding proteins CR and CB, substance P (SP), a marker unique to this study called limbic system associated protein (LAMP), and choline acetyltransferase (ChAT), as well as acetylcholinesterase (AChE) histochemistry. They found uniform staining for LAMP and CR-ir neurons in the claustrum, with minimal staining for SP, ChAT and AChE. A few CB-ir neurons were found. The extent of the claustrum studied was too restricted to answer questions of neurochemical compartments or to allow a comparison of the numbers and distribution of different cell types in the human claustrum with data from other species. Pirone et al. (2012) looked at the expression of netrin G2 and GNG2, (the G-protein gamma 2 subunit found in neurons with G-protein coupled receptors) and latexin in the human claustrum. The subjects studied had average postmortem intervals (PMIs) of 26 hours, a long enough time that protein preservation could be compromised. The results showed staining for GNG2 and netrin-G2 throughout the extent of the claustrum. There was no latexin labeling, surprising in light of results in rat and cat (Arimatsu et al., 2009; Miyasaka et al., 1999) but possibly a result of the very long PMIs and not a true species difference.

Rodent

While most data from the cat and the monkey show neurochemical uniformity in the claustrum, data from the rodent provides evidence for neurochemical compartments. In the mouse, immunoreactivity to CR and the vesicular glutamate transporter VGLUT2 suggested a division of the claustrum into an inner core and an outer shell region (Dávila et al., 2005; Real et al., 2006). The core region was further defined by an absence of CR immunoreactivity (Real et al., 2003). Again, it is not clear how the core and shell regions of the claustral complex relate to the claustrum as seen in the cat and primate. Eiden et al. (1990) looked at the distribution of cells expressing somatostatin (SOM), cholecystokinin (CCK) and vasoactive intestinal peptide (VIP) and did not see any evidence for neurochemical compartments with these proteins. The pattern of expression of cadherins (Ncad and Rcad, cell-cell adhesion molecules) in mouse led to the proposal of a Cl proper (dorsal claustrum) with three subdivisions, the Cl-i (inferior), Cl-int (intermediate) and Cl-s (superior) (Obst-Pernberg et al., 2001). More data about those proposed subdivisions (e.g. whether the connections differ), however, are necessary to see if this is a useful way of characterizing the adult claustrum or whether the different expression patterns are a legacy of developmental processes and not relevant to the adult function. The pattern of expression in the ventral claustrum/endopiriform nucleus did distinguish it from the dorsal claustrum, supporting them as separate entities.

Cell Types and Circuitry in the Claustrum: Evidence from Golgi Stains and Neurochemistry

Understanding the function of information processing in the claustrum depends on understanding the circuitry within the claustrum itself. Is it a simple relay, with information sent to the claustrum and then back to cortex relatively unchanged, or does the structure of the claustrum provide a substrate for information processing? Classic studies of the cellular organization of the claustrum used the Golgi staining method (Braak and Braak, 1982; Brand, 1981; LeVay and Sherk, 1981a) and different studies identified a number of different cell types in several species. For the cat, LeVay and Sherk (1981a) found only two major cell types, the most common being the large "spiny dendrite cell" with somata 15–29 μm in diameter. These cells are the putative projection neurons of the claustrum that make excitatory synapses on their cortical targets (LeVay, 1986). They are morphologically similar to the glutamatergic projection cells of the cerebral cortex (noted by Bayer and Altman, 1991). The second cell type was a smaller "Golgi type II" with somata 10–15 μm in diameter, thought to be local circuit interneurons. The presence of only two cell types suggests very limited intraclaustral processing. Brand (1981) identified three types of neurons in the monkey. Type 1 neurons were similar to the projection neuron of Sherk and LeVay; the Type 2 and Type 3 neurons seemed to contribute only to local circuitry. A total of five cell types were recognized in humans (Braak and Braak, 1982). In the rat, as in the monkey, three different cell types were proposed (Mamos, 1984).

Do the different estimates of cell types represent species differences or methodological issues? Using Golgi material poses major problems for determining the numbers of different cell types. How different do cells need to be in appearance to be reliably assigned to different classes rather than representing variations in the morphology of cells belonging to a single class? Further, a more serious limitation of this technique is that it does not allow an estimation of the total density or distribution of the different cell types that are visualized. Golgi staining is highly idiosyncratic, following rules that are still mysterious, and stains only a tiny subset of the cells present.

More recent studies have used immunohistochemistry (IHC) to identify cell types based on the expression of particular proteins and have suggested considerably more complex circuitry in the claustrum. IHC allows visualization of the numbers and densities of the different neurochemically identified types and can also reveal information about the internal circuitry in the claustrum by showing the transmitters used by claustral neurons.

What have these studies shown about the cellular organization of the claustrum? In reviewing the data there are two key questions to keep in mind: First, is there any evidence of differential distribution of particular cell types, receptors or input fibers suggesting differences in information processing in the various functional zones? Secondly, is there any evidence for species differences in the number and distribution of the various cell types that might reflect the species differences in cortical organization?

Expression of Calcium-Binding Proteins Defines Cell Classes in the Claustrum

Several studies have looked at the expression of the calcium-binding proteins CB, CR and PV in a number of species (Dávila et al., 2005; Hinova-Palova et al., 2007; Rahman and Baizer, 2007; Real et al., 2003; Reynhout and Baizer, 1999). In the cerebral cortex, immunoreactivity to these proteins defines different subclasses of GABAergic interneurons (Condé et al., 1994; DeFelipe et al., 1989a,b). The projection neurons of the cortex, the pyramidal cells, do not express those proteins. How general is the finding that calcium-binding proteins mark only inhibitory neurons? In the cerebellar cortex, CB and PV are found in Purkinje cells, which are projection neurons but also inhibitory (Celio, 1990). However, the projection neurons of the inferior olive, which give rise to the excitatory climbing fibers, colocalize CR and CB (Baizer et al., 2011). It is therefore necessary to have additional data to be able to assign a transmitter to the cells in the claustrum that express the calcium-binding proteins.

Monkey

For the macaque monkey, we found that for each of the calcium-binding proteins there were distinctly labeled cells that were uniformly distributed over the entire dorsoventral and anteroposterior extent of the claustrum (Reynhout and Baizer, 1999). These data are summarized in Figures 3.6 and 3.7, which show outline drawings of the claustrum with labeled cells indicated by dots. With PV (Figure 3.6A), stained cells were surrounded by a dense meshwork of stained fibers oriented in all directions. In addition, there were numerous stained puncta. The PV-ir cells were large multipolar cells with large darkly stained somata usually oriented with their long axes parallel to the long axis of the claustrum. For most examples, several processes emerged from the cell bodies.

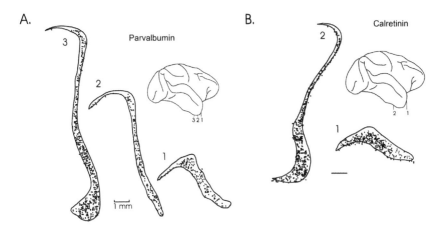

FIGURE 3.6 A: Distributions of parvalbumin-immunoreactive neurons in the claustrum of the monkey shown on drawings made using "MDplot" software. The numbers on the drawing of the lateral view of the brain indicate the levels of the three sections illustrated. B: Distributions of calretinin-immunoreactive neurons in the claustrum of the monkey. The drawings of the lateral view of the brain indicate the levels of the sections 1 and 2. Each *dot* represents one labeled cell. Scale bar = 1 mm. *Redrawn from Reynhout and Baizer, 1999.*

Figure 3.6A shows the density and distribution of PV-ir cells on three coronal sections of the claustrum of the monkey. With CR there were stained neurons against a background of stained fibers and puncta; the CR-ir neurons had elongated cell bodies. Typically two main, beaded dendrites emerged from the poles of the soma with a few subsequent bifurcations. The CR-ir neurons resemble cortical bipolar cells (Peters, 1984) and also share some similarity with cortical double bouquet cells (Condé et al., 1994; Yan et al., 1995). Based on soma size and dendritic morphology we identified three different types of cells with immunoreactivity to CB.

Small, multipolar neurons were the most abundant. They had darkly stained, comparatively small, round or oval somata, and many processes that divided into secondary and tertiary branches arborizing close to the cell body. These cells were evenly distributed throughout the claustrum. The second type of CB-ir cell resembled the PV-ir cells with large somata and several thick, darkly stained processes that were often oriented along the long axis of the claustrum. They were less numerous than the PV-ir cells. Similar cells have been described in Golgi studies of the claustrum of all species studied, including humans (Braak and Braak, 1982; Brand, 1981; LeVay and Sherk, 1981a; Mamos, 1984; Mamos et al., 1986). The third class of CB-ir cells were fusiform cells with elongated cell

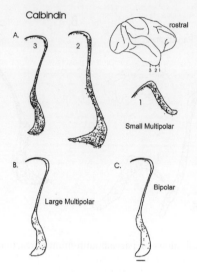

FIGURE 3.7   Distributions of different calbindin-immunoreactive (CB-ir) cells in the claustrum of the monkey. A: distribution of the most common type, the small multipolar cells on three sections through the claustrum. The levels of the sections are shown on the drawing of the lateral view of the brain at the *top right*. B: Distribution of the large multipolar CB-ir cells. C: Distribution of the bipolar CB-ir cells. Scale bar = 1 mm. *Redrawn from Reynhout and Baizer, 1999.*

bodies and with two processes emerging from the poles. The processes continued straight with little branching.

Cat

In IHC studies of the cat brain we used two planes of section, coronal and parasagittal, to look for the possibility of asymmetry in the alignment of the long axes of somata or of dendritic trees (Rahman and Baizer, 2007). Differences in the plane of section can yield very different pictures of cellular arrangements, the most dramatic example being the Purkinje cells of the cerebellar cortex whose dendritic tree is found in a single plane. Immunoreactivity to each of the antibodies tested showed a different subpopulation of neurons. Each subpopulation was distributed throughout the entire claustrum with no apparent grouping or clustering. In general, the results in the cat were similar to the results in the monkey, with a variety of soma sizes and shapes labeled by different calcium-binding proteins. Figure 3.8 shows examples of neurons immunoreactive to each of the calcium-binding proteins in the cat claustrum.

We found PV-ir cells have large somata that were pear shaped, oval, bipolar and multipolar (Rahman and Baizer, 2007). In parasagittal sections, most cell bodies were elongated in the AP direction. These results were in general agreement with Hinova-Palova et al. (2007), who also

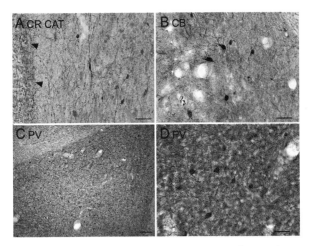

FIGURE 3.8 Appearance and distributions of neurons in the cat claustrum immuno-reactive to calcium-binding proteins. A: calretinin; scale bar = 50 μm. B: calbindin; scale bar = 50 μm. C, parvalbumin; scale C=100 μm. D: Higher magnification image of parvalbumin-ir neurons = 50 μm.

used IHC to look for PV-ir neurons in the cat Cl. They found labeled neurons in a range of sizes throughout the claustrum. Two types of CR-ir neurons were found; one was bipolar with elongated cell bodies and dendrites emerging from either pole (example in Figure 3.8A). The second type was multipolar. There were also many stained puncta throughout the claustrum. With CB there were fewer cells stained than with PV, but all were evenly distributed throughout the claustrum. Most were multipolar, with dendrites emerging in all directions. In the parasagittal plane, somata were still multipolar (examples in Figure 3.8B) but they had a tendency to be elongated parallel to the AP direction.

Rodent

Our data for both cat and monkey were also quite different from what was reported in the rat (Druga et al., 1993). They looked at immunoreactivity to PV and CB in both the dorsal and the ventral claustrum (endopiriform nucleus). Considering only the data from the dorsal claustrum, they found PV-ir somata, in agreement with our observations in cat and monkey, but found only "very weak" CB immunoreactivity, whereas we found many CB-ir neurons. Whether this represents a difference in methods (which antibody was used) or a genuine species difference is unclear. Real et al. (2003) looked at the pattern of expression of the three calcium-binding proteins in the mouse, considering both the dorsal claustrum and the endopiriform nucleus. They found that CB labeled a population of neurons throughout the claustrum, with the density decreasing slightly at more caudal levels. However, they found only scattered cells that were CR-ir, although there was a dense CR-ir neuropil with stained fibers and puncta. (Their CR-free "compartment" was mentioned earlier.) With PV there were also immunoreactive neurons surrounded by stained neuropil and stained axons with varicosities. They also found a darkly stained, embedded patch of neuropil that overlapped the CR-free zone. In colocalization experiments they found that CB and PV were colocalized in only a tiny minority of neurons.

Summary of the Evidence from Expression of Calcium-Binding Proteins

IHC using antibodies for the different calcium-binding proteins in both cat and monkey shows a much greater variety of cell types than was apparent in the Golgi stains (LeVay and Sherk, 1981a). For both cat and monkey, the large cells immunoreactive to each of the calcium-binding proteins resemble the large spiny dendrite cells, the putative projection neurons. The CR-ir bipolar cells we describe are not a good match to any of the Golgi cell types described.

Transmitters in the Claustrum: Markers for Glutamate and GABA

Glutamate and gamma-aminobutyric acid (GABA) are the main excitatory and inhibitory (respectively) transmitters in the brain. In general, projection neurons are glutamatergic and local interneurons are GABAergic, although there are many exceptions for both (e.g. unipolar brush cells of the cerebellum are glutamatergic interneurons, cerebellar Purkinje cells are GABAergic projection neurons (Ghez and Thach, 2000; Mugnaini et al., 2011). There have been studies of markers for glutamatergic neurons in the claustrum, using antibodies to glutamate transporters (VGLUT2, EAAC1) and studies of putative inhibitory interneurons using antibodies for the synthetic enzyme for GABA, glutamic acid decarboxylase (GAD) (Rahman and Baizer, 2007).

In the cat, we found that cells labeled with a marker for glutamatergic neurons, EAAC1, were scattered throughout the claustrum (Rahman and Baizer, 2007). Figure 3.9A shows that in the cerebral cortex, immunoreactivity to this antibody marks the pyramidal cells, the glutamatergic projection neurons (example at *arrow*). Figure 3.9B illustrates the distribution of labeled neurons in the claustrum of the cat. We did not find cells double labeled for EAAC1 and any of the calcium-binding proteins, suggesting that, as in cerebral cortex, neurons immunoreactive to the calcium-binding proteins are different subclasses of inhibitory interneurons and not projection neurons.

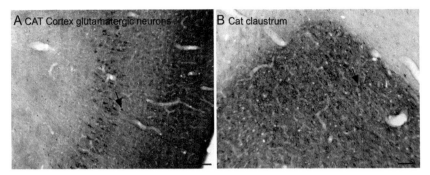

FIGURE 3.9   Glutamatergic neurons in the claustrum. A: Neurons immunoreactive to EAAC1, a glutamate transporter, (example at *arrow*) in the deeper layers of cat cerebral cortex; scale bar = 250 µm. B: EAAC1-ir neurons in cat claustrum. Scale bar = 100 µm.

GABAergic interneurons have been described in the cat, mouse, rat and rabbit (Feldblum et al., 1993; Gomez-Urquijo et al., 2000; Narkiewicz et al., 1988). In our study of the cat, surprisingly, there were very few GAD-ir somata (Rahman and Baizer, 2007).

Subclass of GABAergic Neurons: Markers of Nitric Oxide

The presence of cells that synthesize and release nitric oxide (NO) has also been examined using IHC for the synthetic enzyme for nitric oxide, neuronal nitric oxide synthase (nNOS) or an older histochemical method, NADPH diaphorase histochemistry. In cortex, nNOS-ir neurons are considered to represent a subset of cortical GABAergic interneurons (Gabbott and Bacon, 1995; Valtschanoff et al., 1993) and a reasonable hypothesis is that they are also a subset of GABAergic interneurons in the claustrum.

For the cat, we found that with immunoreactivity visualized by an antibody to nNOS the Cl was darkly stained compared to surrounding white matter (Figure 3.10A). Figure 3.10B illustrates the characteristics and distribution of neurons that were immunoreactive to nNOS. Immunoreactive cells had oval cell bodies about 20–25 µm long with 2–6 dendrites emerging from the cell bodies. Some dendrites had beads or varicosities. There were also many stained puncta and many cells encircled by them.

Since the input to the claustrum from the cerebral cortex is glutamatergic, the presence of nNOS-ir puncta implies complex local connections. In the cat, the cells marked by nNOS might fit the Golgi type II class and are probably interneurons. They are very similar to nNOS-ir cells in the monkey cortex (Smiley et al., 2000). These results are in agreement with two other studies showing nNOS-ir cells in the claustrum of the cat (Hinova-Palova et al., 1997, 2008).

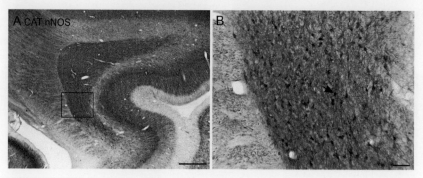

FIGURE 3.10 Neuronal nitric oxide synthase (nNOS) immunoreactivity in the cat claustrum. A: Lower magnification image of the claustrum on a section processed for nNOS-ir. The *rectangle* shows the location of the higher magnification image in B; scale bar = 1 mm. B: nNOS-immunoreactive neurons, example at *arrow*; scale bar = 100 μm.

Nitric Oxide Synthase in the Rodent

The presence of nNOS was also examined in the rodent and a rather different picture emerged. Guirado et al. (2003) studied nitrergic neurons in the mouse claustrum and found two different populations initially distinguished by staining intensity. Darkly stained neurons colocalized GABA and resembled aspiny neurons. They were found throughout the claustrum. The second class, more lightly stained, were more numerous and were not GABAergic. It was proposed that there were some nitrergic neurons that were cortical projection neurons, based on an earlier study that showed that nitrergic neurons could be labeled by retrograde transport from cortex (Kowiański et al., 2001). These results suggest that NO is sometimes found in glutamatergic and sometimes in GABAergic neurons, at least in the rodent. Whether this is also true in the cat or monkey is not known, nor is it clear whether the nNOS-ir cells are a subset of the cells expressing one of the calcium-binding proteins.

Enkephalin-Containing Cells in the Claustrum

An additional neurochemically defined class of cells has been found by Hinova-Palova et al. (2012), who used IHC to describe cells expressing Leu-enkephalin. Cells expressing Leu-enk were found throughout the dorsal claustrum. They were not evenly distributed, however, but appeared as clusters in the dorsal claustrum between stereotaxic levels A12–A16. Such immunoreactive cells were of a variety of sizes and shapes. The relation between these cells and other neurochemically defined classes (i.e. with calcium-binding proteins) was not determined; the class was proposed to include both interneurons and projection neurons (Hinova-Palova et al., 2012).

Neuropeptides in the Rodent Claustrum

A number of other neurochemically defined cell classes have been described in the rodent claustrum, with different cells expressing neuropeptide Y, somatostatin, vasoactive intestinal peptide and cholecystokinin (Kowiański et al., 2001, 2008). All types were distributed throughout the entire claustrum, apparently randomly. Since they were not retrogradely labeled by cortical injections of retrograde tracers, they were considered additional classes of interneurons (Kowiański et al., 2001). Their status as interneurons was supported in a later study showing colocalization of neuropeptides with calcium-binding proteins. Specifically, neuropeptide Y was found colocalized with all three (CB, CR and PV), somatostatin only with CB, and vasoactive intestinal peptide only with CR.

IMPLICATIONS OF MULTIPLE NEUROCHEMICALLY DEFINED CELL TYPES FOR INTRACLAUSTRAL PROCESSING

We saw many CR, nNOS, GAD and PV terminals surrounding neurons throughout the Cl. The source of these terminals is almost certainly from neurons within the claustrum itself since the projection neurons of cortex, the pyramidal cells, do not express any of those markers (Celio, 1990; Smiley et al., 2000; Van Brederode et al., 1990). The idea that neurons in the claustrum have widespread connections within that structure is supported by anatomical data showing that a retrograde tracer injection within the claustrum resulted in many retrogradely labeled neurons within the ipsilateral claustrum.

What is the significance of the different neurochemically defined cell populations? In two species with multiple antibodies we have identified multiple neurochemically distinct cell types that differ in number and appearance. However, each type is distributed throughout the claustrum with no avoidance of the borders of functionally defined regions. This means that the dendrites of a neuron residing at the edge of one functional compartment could extend into, and receive information destined primarily for, a neighboring compartment. This pattern is consistent with the idea of a more global function for information processing in the claustrum than the simple "housekeeping" (tuning of receptive field properties) function proposed years ago by Sherk and LeVay (1983). Indeed Crick and Koch (2005) proposed that the claustrum is critical for perception and consciousness (see commentary in Stevens, 2005). More recently, Smythies et al. (2012) suggested another theory of the function of the claustrum, namely that it acts as a detector of synchrony among multiple inputs and thereby participates in perceptual and cognitive processes across sensory modalities as well as in motor behavior occurring in response to those inputs.

Subcortical Innervation of the Claustrum

Another important determinant of the function of the claustrum may be input from monoaminergic systems. Monoamines are important in regulation of diverse functions including mood, attention, eating, sexual behavior and cognition. While all three monoamines, dopamine, serotonin and norepinephrine (noradrenaline), innervate all areas of monkey cortex, there are differences in the density and laminar distribution of monoaminergic fibers projecting to different cortical regions (Campbell et al., 1987; De Lima et al., 1988; Lewis et al., 1986, 1987; Morrison et al., 1982a,b; Takeuchi and Sano, 1983; Williams and Goldman-Rakic, 1993). These results raise the question of whether there are similar differences in innervation density by any of these systems in different functional zones of the claustrum. Overall, the weight of the evidence suggests that just as the cortex and claustrum differ in the presence/absence of cytoarchitectonically defined functional zones, they also differ in the uniformity of innervation of different zones in the claustrum. Sutoo and colleagues (1994) looked at the cholinergic and catecholaminergic systems in human brain using IHC staining for the synthetic enzyme for acetylcholine (ACh) and the enzyme tyrosine hydroxylase (TH) for dopamine and norepinephrine. Both enzymes are found in axon terminals (Umbriaco et al., 1995) and are therefore good markers for showing dopamine, norepinephrine, and ACh fibers. Immunoreactivity to both enzymes was found in the claustrum, suggesting that it receives both cholinergic and catecholaminergic innervation (Sutoo et al., 1994). This suggestion has been supported in subsequent studies.

Serotonin

Several studies provided evidence for serotonin (5-HT) receptors in the claustrum (Mengod et al., 1996; Pasqualetti et al., 1999; Pompeiano et al., 1994; Wright et al., 1995), suggesting that there might be serotonergic input as well. In the monkey, we found that with 5-HT-ir the claustrum in the monkey was darkly stained compared to the bordering white matter (Baizer, 2001). The staining was equally dense over the entire extent of the claustrum with no obvious regional variations. The stained fibers had varicosities, and the size, shape and spacing of these varied among fibers and even along the length of a single fiber. In the more anterior claustrum, where it appears horizontal in a frontal section, many beaded fibers running roughly parallel to each other could be followed for long distances. In the more caudal and ventral visual claustrum

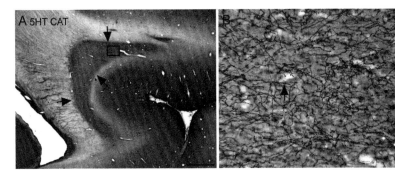

FIGURE 3.11   Serotonergic innervation of the claustrum in the cat. A: The claustrum of the cat (*arrows*) shown on a parasagittal section processed for 5-HT immunohistochemistry. The rectangle shows the location of the higher magnification image shown in B; scale bar = 1 mm. B: The claustrum is covered with a network of fine, beaded fibers running at all directions (example at *arrow*); scale bar = 50 μm.

the pattern was of many short segments of fibers oriented randomly with respect to each other. A similar pattern was seen in the cat; Figure 3.11A shows the appearance of the claustrum on a parasagittal section and Figure 3.11B shows the fine, beaded fibers running at all directions throughout the claustrum.

The distribution of serotonergic fibers in the claustrum is consistent with the idea that single fibers can cross the borders of functional domains since there was no discontinuity of axons suggestive of borders. The pattern of serotonergic innervation of the claustrum is thus different from the cortex in which the various areas differ in the density and laminar distribution of serotonergic input, with an especially heavy input to the primary visual cortex (Berger et al., 1988; De Lima et al., 1988; Kosofsky et al., 1984; Morrison and Foote, 1986; Morrison et al., 1982a; Takeuchi and Sano, 1983; Wilson and Molliver, 1991). Immunoreactivity to 5-HT showed that the cat, like the monkey, receives major serotonergic input. A meshwork of finely beaded fibers and puncta was seen over the whole claustrum. There was no preferential orientation of stained fibers relative to the borders of the claustrum. In parasagittal sections, a few axons could be followed for as far as 200–300 μm.

Adrenergic Innervation of the Claustrum

The transmitter norepinephrine is synthesized by neurons of the locus ceruleus and adrenergic axons widely distributed throughout the cerebral cortex (Campbell et al., 1987; Lewis et al., 1986; Morrison et al., 1979a,b; 1982a,b). Evidence that the claustrum also receives an adrenergic input comes from studies of the distribution of adrenergic receptors of which there are several subtypes including α_{1A}, α_{1B} and α_{1C}. Domyancic and Morilak (1997) looked at the distribution of mRNA for the α_{1A} receptor in the rat brain and found signal in the claustrum, although the part of the claustrum was not specified.

Cholinergic Innervation of the Claustrum

A cholinergic input to the claustrum has been supported by the results of studies in the cat and the primate, including human. Several approaches have been used, ranging from direct assessment of the efficacy of ACh as a transmitter in the claustrum to using anatomical methods to find evidence of its synthesis or the presence of different types of ACh receptors. Cortimiglia et al. (1982) tested the effects of iontophoretically applied ACh on neurons in the cat claustrum. It was excitatory to 40 percent and inhibitory to 47 percent of the 92 neurons tested. Other studies used more indirect approaches. Gill et al. (2007) looked at a primate-specific form of ChAT (82-kDa ChAT) and found it in the claustrum of the human brain. The other form of ChAT (69-kDa), found in known cholinergic areas, was not found in the claustrum.

Cholinergic receptors in the claustrum have been the focus of several investigations. The localization of different nicotinic cholinergic receptors (nAChRs) was assessed by evaluation of the binding of nicotinic agonists, nicotine, cytisine, and epibatidine, and one nicotinic blocker, α-bungarotoxin in the rhesus monkey (Han et al., 2003). They found that there was binding of α-bungarotoxin in claustrum of the macaque monkey, suggesting cholinergic input and the presence of $\alpha7$ nAChRs. Agulhon et al. (1998) used a different approach, looking at the distribution of the mRNA for the $\alpha4$ subunit of nAChR in the human fetal brain. A low to moderate signal was seen in the claustrum of both normal and fragile X brains as well as in the basal ganglia; signal strength was lower than in several other regions including the thalamus and cortex. There was no analysis of the claustrum at different rostrocaudal levels. Winzer-Serhan and Leslie (2005) studied the developmental expression of the $\alpha5$ subunit of the nAChR in the rat. mRNA for the $\alpha5$ subunit in the claustrum was detected at E18 and it continued into adulthood.

Dopaminergic Innervation of the Claustrum

Several studies suggest that there may also be dopaminergic input to the claustrum. As with acetylcholine, several different approaches have been used, with all of the results supporting a dopaminergic input. Cortimiglia et al. (1982) reported that iontophoretically applied dopamine to neurons in the cat claustrum affected cell responses, with 62/85 neurons inhibited, 8/85 excited and 14/85 unresponsive.

There are multiple dopamine receptor subtypes, D1–D5 (review in Missale et al., 1998). The results of studies in several species suggest that the D1, D2 and D3 subtypes are present in the claustrum. Analysis of the distribution of putative G-protein coupled D1 receptors using in situ hybridization in dog or rat brains (Schiffmann et al., 1990) showed mRNA in the claustrum for the D1 receptor subtype. Fuxe et al. (1987) reported faster turnover of the D1 than the D2 receptor in multiple brain sites including the claustrum. mRNA for the D2 dopamine receptor was found in the claustrum in both the rat (Weiner and Brann, 1989) and the primate (Meador-Woodruff et al., 1991). Mijnster and colleagues (1999) used radioligands to map and quantify putative D1 and D2 binding sites in the tree shrew brain. There was signal in the claustrum suggestive of both D1 and D2 receptors, but at a much lower density than in the caudate nucleus. In a study of the human brain, evidence was shown for D3 receptor mRNA at low to moderate levels in the claustrum (Suzuki et al., 1998). For the D5 dopamine receptor, signal was not found in the claustrum of the rat brain (Meador-Woodruff et al., 1992).

AMYGDALA

As a further complication to an already complex anatomical picture, it should be noted that the cortex is the major but not the only source of input to the claustrum. In rodents, there is evidence for reciprocal connections with the claustrum and the amygdala, specifically the magnocellular division of the basal nucleus of the amygdala (Majak et al., 2002). These connections are concentrated in the part of the claustrum that is interconnected with cortical areas associated with emotional processing, especially the ento- and perirhinal cortex and prefrontal cortex. One consequence of the interconnections of the claustrum is that it may help facilitate the spread of seizure activity from the amygdala to other brain areas, especially important in temporal lobe epilepsy (Majak et al., 2002; Zhang et al., 2001). In addition, in the rodent there are projections from the claustrum to the olfactory nucleus, olfactory tubercle, nucleus accumbens, nucleus of the lateral olfactory tract and the amygdala as well as the midline thalamus (Zhang et al., 2001).

SUMMARY: NEUROCHEMISTRY AND THE FUNCTIONS OF THE CLAUSTRUM

There have been almost as many views of the function of the claustrum as there have been scientists writing about it. Our perspective, from the anatomical and neurochemical data discussed above, is that the claustrum is unlikely to serve some basic "housekeeping function" as suggested for visual claustrum (Sherk and LeVay, 1983). The connections, neurochemistry and physiology all suggest a more "global" function.

Note on Methods

The general methods for histology, IHC and injections and analysis of retrograde tracers for the data shown in Figures 3.4–3.7 have been described in detail in the publications from which those figures were adapted (Baizer, 2001; Baizer et al., 1993, 1997; Reynhout and Baizer, 1999). For Figures 3.1–3.3 we photographed archival cresyl-violet and Weil stained celloidin-embedded sections from the cat and the macaque monkey. These slides had been prepared in the laboratory of Dr. Mitchell Glickstein at Brown University. For Figures 3.8–3.11 we created new figures using photomicrographs of sections that had been prepared for our earlier study of the cat claustrum (Rahman and Baizer, 2007).

References

Agulhon, C., Charnay, Y., Vallet, P., Abitbol, M., Kobetz, A., Bertrand, D., et al., 1998. Distribution of mRNA for the alpha4 subunit of the nicotinic acetylcholine receptor in the human fetal brain. Brain Res. Mol. Brain Res. 58, 123–131.

Alexander, G.E., Crutcher, M.D., 1990. Functional architecture of basal ganglia circuits: neural substrates of parallel processing. Trends Neurosci. 13, 266–271.

Arimatsu, Y., Kojima, M., Ishida, M., 1999. Area- and lamina-specific organization of a neuronal subpopulation defined by expression of latexin in the rat cerebral cortex. Neuroscience 88, 93–105.

Arimatsu, Y., Nihonmatsu, I., Hatanaka, Y., 2009. Localization of latexin-immunoreactive neurons in the adult cat cerebral cortex and claustrum/endopiriform formation. Neuroscience 162, 1398–1410.

Baizer, J.S., 2001. Serotonergic innervation of the primate claustrum. Brain Res. Bull. 55, 431–434.

Baizer, J.S., Baker, J.F., 2005. Immunoreactivity for calcium-binding proteins defines subregions of the vestibular nuclear complex of the cat. Exp. Brain Res. 164, 78–91.

Baizer, J.S., Baker, J.F., 2006. Neurochemically defined cell columns in the nucleus prepositus hypoglossi of the cat and monkey. Brain Res. 1094, 127–137.

Baizer, J.S., Robinson, D.L., Dow, B.M., 1977. Visual responses of area 18 neurons in awake, behaving monkey. J. Neurophysiol. 40, 1024–1037.

Baizer, J.S., Desimone, R., Ungerleider, L.G., 1993. Comparison of subcortical connections of inferior temporal and posterior parietal cortex in monkeys. Vis. Neurosci. 10, 59–72.

Baizer, J.S., Lock, T.M., Youakim, M., 1997. Projections from the claustrum to the prelunate gyrus in the monkey. Exp. Brain Res. 113, 564–568.

Baizer, J.S., Sherwood, C.C., Hof, P.R., Witelson, S.F., Sultan, F., 2011. Neurochemical and structural organization of the principal nucleus of the inferior olive in the human. Anat. Rec. (Hoboken) 294, 1198–1216.

Bayer, S.A., Altman, J., 1991. Development of the endopiriform nucleus and the claustrum in the rat brain. Neuroscience 45, 391–412.

Bender, D.B., Davidson, R.M., 1986. Global visual processing in the monkey superior colliculus. Brain Res. 381, 372–375.

Beneyto, M., Prieto, J.J., 2001. Connections of the auditory cortex with the claustrum and the endopiriform nucleus in the cat. Brain Res. Bull. 54, 485–498.

Berger, B., Trottier, S., Verney, C., Gaspar, P., Alvarez, C., 1988. Regional and laminar distribution of the dopamine and serotonin innervation in the macaque cerebral cortex: a radioautographic study. J. Comp. Neurol. 273, 99–119.

Bonin, G.V., Bailey, P., 1947. The Neocortex of Macaca Mulatta. The University of Illinois Press, Urbana, Illinois.

Boussaoud, D., Desimone, R., Ungerleider, L.G., 1992. Subcortical connections of visual areas MST and FST in macaques. Vis. Neurosci. 9, 291–302.

Boyapati, J., Henry, G.H., 1985. The character and influence of the claustral pathway to the striate cortex of the cat. Exp. Brain Res. 61, 141–152.

Braak, H., Braak, E., 1982. Neuronal types in the claustrum of man. Anat. Embryol. (Berl) 163, 447–460.

Brand, S., 1981. A serial section Golgi analysis of the primate claustrum. Anat. Embryol. (Berl) 162, 475–488.

Brodmann, K., Gary, L.J., 2006. Brodmann's Localization in the Cerebral Cortex: The Principles of Comparative Localisation in the Cerebral Cortex based on Cytoarchitectonics. Springer, New York.

Buchanan, K.J., Johnson, J.I., 2011. Diversity of spatial relationships of the claustrum and insula in branches of the mammalian radiation. Ann. N. Y. Acad. Sci. 1225 (Suppl. 1), E30–E63.

Bullier, J., Kennedy, H., Salinger, W., 1984. Bifurcation of subcortical afferents to visual areas 17, 18, and 19 in the cat cortex. J. Comp. Neurol. 228, 309–328.

Campbell, M.J., Lewis, D.A., Foote, S.L., Morrison, J.H., 1987. Distribution of choline acetyltransferase-, serotonin-, dopamine-beta-hydroxylase-, tyrosine hydroxylase-immunoreactive fibers in monkey primary auditory cortex. J. Comp. Neurol. 261, 209–220.

Celio, M.R., 1990. Calbindin D-28k and parvalbumin in the rat nervous system. Neuroscience 35, 375–475.

Chachich, M.E., Powell, D.A., 2004. The role of claustrum in Pavlovian heart rate conditioning in the rabbit (Oryctolagus cuniculus): anatomical, electrophysiological, and lesion studies. Behav. Neurosci. 118, 514–525.

Clascá, F., Avendaño, C., Román-Guindo, A., Llamas, A., Reinoso-Suárez, F., 1992. Innervation from the claustrum of the frontal association and motor areas: axonal transport studies in the cat. J. Comp. Neurol. 326, 402–422.

Condé, F., Lund, J.S., Jacobowitz, D.M., Baimbridge, K.G., Lewis, D.A., 1994. Local circuit neurons immunoreactive for calretinin, calbindin D-28k or parvalbumin in monkey prefrontal cortex: distribution and morphology. J. Comp. Neurol. 341, 95–116.

Cortimiglia, R., Infantellina, F., Salerno, M.T., Zagami, M.T., 1982. Unit study in cat claustrum of the effects of iontophoretic neurotransmitters and correlations with the effects of activation of some afferent pathways. Arch. Int. Physiol. Biochim. 90, 219–230.

Crescimanno, G., Salerno, M.T., Cortimiglia, R., Amato, G., 1989. Claustral influence on ipsi- and contralateral motor cortical areas, in the cat. Brain Res. Bull. 22, 839–843.

Crick, F.C., Koch, C., 2005. What is the function of the claustrum? Philos. Trans. R. Soc. Lond. B. Biol. Sci. 360, 1271–1279.

Cusick, C.G., Scripter, J.L., Darensbourg, J.G., Weber, J.T., 1993. Chemoarchitectonic subdivisions of the visual pulvinar in monkeys and their connectional relations with the middle temporal and rostral dorsolateral visual areas, MT and DLr. J. Comp. Neurol. 336, 1–30.

Davidson, R.M., Bender, D.B., 1991. Selectivity for relative motion in the monkey superior colliculus. J. Neurophysiol. 65, 1115–1133.

Davidson, R.M., Joly, T.J., Bender, D.B., 1992. Effect of corticotectal tract lesions on relative motion selectivity in the monkey superior colliculus. Exp. Brain Res. 92, 246–258.

Dávila, J.C., Real, M.Á., Olmos, L., Legaz, I., Medina, L., Guirado, S., 2005. Embryonic and postnatal development of GABA, calbindin, calretinin, and parvalbumin in the mouse claustral complex. J. Comp. Neurol. 481, 42–57.

De Lima, A.D., Bloom, F.E., Morrison, J.H., 1988. Synaptic organization of serotonin-immunoreactive fibers in primary visual cortex of the macaque monkey. J. Comp. Neurol. 274, 280–294.

DeFelipe, J., Hendry, S.H., Jones, E.G., 1989a. Synapses of double bouquet cells in monkey cerebral cortex visualized by calbindin immunoreactivity. Brain Res. 503, 49–54.

DeFelipe, J., Hendry, S.H., Jones, E.G., 1989b. Visualization of chandelier cell axons by parvalbumin immunoreactivity in monkey cerebral cortex. Proc. Natl. Acad. Sci. USA 86, 2093–2097.

Domyancic, A.V., Morilak, D.A., 1997. Distribution of alpha1A adrenergic receptor mRNA in the rat brain visualized by in situ hybridization. J. Comp. Neurol. 386, 358–378.

Druga, R., 1989. Projections from the claustrum to the occipital cortex in the guinea pig. Folia Morphol. (Praha) 37, 57–63.

Druga, R., Rokyta, R., Benes Jr., V., 1990. Claustro-neocortical projections in the rhesus monkey (projections to area 6). J. Hirnforsch. 31, 487–494.

Druga, R., Chen, S., Bentivoglio, M., 1993. Parvalbumin and calbindin in the rat claustrum–An immunocytochemical study combined with retrograde tracing from frontoparietal cortex. J. Chem. Neuroanat. 6, 399–406.

Duffy, C.J., Wurtz, R.H., 1991a. Sensitivity of MST neurons to optic flow stimuli. I. A continuum of response selectivity to large-field stimuli. J. Neurophysiol. 65, 1329–1345.

Duffy, C.J., Wurtz, R.H., 1991b. Sensitivity of MST neurons to optic flow stimuli. II. Mechanisms of response selectivity revealed by small-field stimuli. J. Neurophysiol. 65, 1346–1359.

Duffy, C.J., Wurtz, R.H., 1995. Response of monkey MST neurons to optic flow stimuli with shifted centers of motion. J. Neurosci. 15, 5192–5208.

Duffy, C.J., Wurtz, R.H., 1997a. Medial superior temporal area neurons respond to speed patterns in optic flow. J. Neurosci. 17, 2839–2851.

Duffy, C.J., Wurtz, R.H., 1997b. Planar directional contributions to optic flow responses in MST neurons. J. Neurophysiol. 77, 782–796.

Edelstein, L.R., Denaro, F.J., 2004. The claustrum: a historical review of its anatomy, physiology, cytochemistry and functional significance. Cell Mol. Biol. (Noisy-le-grand) 50, 675–702.

Eiden, L.E., Mezey, E., Eskay, R.L., Beinfeld, M.C., Palkovits, M., 1990. Neuropeptide content and connectivity of the rat claustrum. Brain Res. 523, 245–250.

Feldblum, S., Erlander, M.G., Tobin, A.J., 1993. Different distributions of GAD65 and GAD67 mRNAs suggest that the two glutamate decarboxylases play distinctive functional roles. J. Neurosci. Res. 34, 689–706.

Felleman, D.J., Van Essen, D.C., 1991. Distributed hierarchical processing in the primate cerebral cortex. Cereb. Cortex 1, 1–47.

Fuxe, K., Agnati, L.F., Merlo Pich, E., Meller, E., Goldstein, M., 1987. Evidence for a fast receptor turnover of D1 dopamine receptors in various forebrain regions of the rat. Neurosci. Lett. 81, 183–187.

Gabbott, P.L., Bacon, S.J., 1995. Co-localisation of NADPH diaphorase activity and GABA immunoreactivity in local circuit neurones in the medial prefrontal cortex (mPFC) of the rat. Brain Res. 699, 321–328.

Ghez, C., Thach, W., 2000. Cerebellum. In: Kandel, E., Schwartz, J., Jessell, T. (Eds.), Principles of Neuroscience, fourth ed. McGraw Hill, New York.

Gill, S.K., Ishak, M., Dobransky, T., Haroutunian, V., Davis, K.L., Rylett, R.J., 2007. 82-kDa choline acetyltransferase is in nuclei of cholinergic neurons in human CNS and altered in aging and Alzheimer disease. Neurobiol. Aging 28, 1028–1040.

Glickstein, M., Cohen, J.L., Dixon, B., Gibson, A., Hollins, M., Labossiere, E., et al., 1980. Corticopontine visual projections in macaque monkeys. J. Comp. Neurol. 190, 209–229.

Gomez-Urquijo, S.M., Gutierrez-Ibarluzea, I., Bueno-Lopez, J.L., Reblet, C., 2000. Percentage incidence of gamma-aminobutyric acid neurons in the claustrum of the rabbit and comparison with the cortex and putamen. Neurosci. Lett. 282, 177–180.

Graybiel, A.M., 1983. Compartmental organization of the mammalian striatum. Prog. Brain Res. 58, 247–256.

Grieve, K.L., Sillito, A.M., 1995. Differential properties of cells in the feline primary visual cortex providing the corticofugal feedback to the lateral geniculate nucleus and visual claustrum. J. Neurosci. 15, 4868–4874.

Gross, C.G., Bender, D.B., Rocha-Miranda, C.E., 1969. Visual receptive fields of neurons in inferotemporal cortex of the monkey. Science 166, 1303–1306.

Gross, C.G., Rocha-Miranda, C.E., Bender, D.B., 1972. Visual properties of neurons in inferotemporal cortex of the Macaque. J. Neurophysiol. 35, 96–111.

Guirado, S., Real, M.A., Olmos, J.L., Davila, J.C., 2003. Distinct types of nitric oxide-producing neurons in the developing and adult mouse claustrum. J. Comp. Neurol. 465, 431–444.

Guldin, W.O., Markowitsch, H.J., Lampe, R., Irle, E., 1986. Cortical projections originating from the cat's insular area and remarks on claustrocortical connections. J. Comp. Neurol. 243, 468–487.

Han, Z.Y., Zoli, M., Cardona, A., Bourgeois, J.P., Changeux, J.P., Le Novere, N., 2003. Localization of [3H]nicotine, [3H]cytisine, [3H]epibatidine, and [125I]alpha-bungarotoxin binding sites in the brain of Macaca mulatta. J. Comp. Neurol. 461, 49–60.

Hashikawa, T., Rausell, E., Molinari, M., Jones, E.G., 1991. Parvalbumin- and calbindin-containing neurons in the monkey medial geniculate complex: differential distribution and cortical layer specific projections. Brain Res. 544, 335–341.

Hinova-Palova, D., Edelstein, L., Paloff, A., Hristov, S., Papantchev, V., Ovtscharoff, W., 2008. Neuronal nitric oxide synthase immunopositive neurons in cat claustrum – a light and electron microscopic study. J. Mol. Histol. 39, 447–457.

Hinova-Palova, D., Edelstein, L., Papantchev, V., Landzhov, B., Malinova, L., Todorova-Papantcheva, D., et al., 2012. Light and electron-microscopic study of leucine enkephalin immunoreactivity in the cat claustrum. J. Mol. Histol..

Hinova-Palova, D.V., Paloff, A.M., Usunoff, K.G., Dimova, R.N., Yossifov, T.Y., Ivanov, D.P., 1988. Reciprocal connections between the claustrum and the auditory cortical fields in the cat. An experimental study using light- and electron microscopic anterograde degeneration methods, and the horseradish peroxidase retrograde axonal transport. J. Hirnforsch. 29, 255–278.

Hinova-Palova, D.V., Paloff, A., Christova, T., Ovtscharoff, W., 1997. Topographical distribution of NADPH-diaphorase-positive neurons in the cat's claustrum. Eur. J. Morphol. 35, 105–116.

Hinova-Palova, D.V., Edelstein, L.R., Paloff, A.M., Hristov, S., Papantchev, V.G., Ovtscharoff, W.A., 2007. Parvalbumin in the cat claustrum: ultrastructure, distribution and functional implications. Acta Histochem. 109, 61–77.

Hoover, J.E., Strick, P.L., 1999. The organization of cerebellar and basal ganglia outputs to primary motor cortex as revealed by retrograde transneuronal transport of herpes simplex virus type 1. J. Neurosci. 19, 1446–1463.

Hubel, D.H., Wiesel, T.N., 1968. Receptive fields and functional architecture of monkey striate cortex. J. Physiol. 195, 215–243.

Illing, R.B., Graybiel, A.M., 1986. Complementary and non-matching afferent compartments in the cat's superior colliculus: innervation of the acetylcholinesterase-poor domain of the intermediate gray layer. Neuroscience 18, 373–394.

Kaas, J.H., Hackett, T.A., Tramo, M.J., 1999. Auditory processing in primate cerebral cortex. Curr. Opin. Neurobiol. 9, 164–170.

Kemp, J.M., Powell, T.P., 1970. The cortico-striate projection in the monkey. Brain 93, 525–546.

Kennedy, H., Bullier, J., 1985. A double-labeling investigation of the afferent connectivity to cortical areas V1 and V2 of the macaque monkey. J. Neurosci. 5, 2815–2830.

Kondo, H., Tanaka, K., Hashikawa, T., Jones, E.G., 1999. Neurochemical gradients along monkey sensory cortical pathways: calbindin-immunoreactive pyramidal neurons in layers II and III. Eur. J. Neurosci. 11, 4197–4203.

Kosofsky, B.E., Molliver, M.E., Morrison, J.H., Foote, S.L., 1984. The serotonin and norepinephrine innervation of primary visual cortex in the cynomolgus monkey (Macaca fascicularis). J. Comp. Neurol. 230, 168–178.

Kowiański, P., Timmermans, J.P., Moryś, J., 2001. Differentiation in the immunocytochemical features of intrinsic and cortically projecting neurons in the rat claustrum – combined immunocytochemical and axonal transport study. Brain Res. 905, 63–71.

Kowiański, P., Moryś, J.M., Dziewiątkowski, J., Wójcik, S., Sidor-Kaczmarek, J., Moryś, J., 2008. NPY-, SOM- and VIP-containing interneurons in postnatal development of the rat claustrum. Brain Res. Bull. 76, 565–571.

LeVay, S., 1986. Synaptic organization of claustral and geniculate afferents to the visual cortex of the cat. J. Neurosci. 6, 3564–3575.

LeVay, S., Sherk, H., 1981a. The visual claustrum of the cat. I. Structure and connections. J. Neurosci. 1, 956–980.

LeVay, S., Sherk, H., 1981b. The visual claustrum of the cat. II. The visual field map. J. Neurosci. 1, 981–992.

Lewis, D.A., Campbell, M.J., Foote, S.L., Morrison, J.H., 1986. The monoaminergic innervation of primate neocortex. Hum. Neurobiol. 5, 181–188.

Lewis, D.A., Campbell, M.J., Foote, S.L., Goldstein, M., Morrison, J.H., 1987. The distribution of tyrosine hydroxylase-immunoreactive fibers in primate neocortex is widespread but regionally specific. J. Neurosci. 7, 279–290.

Lock, T.M., Baizer, J.S., Bender, D.B., 2003. Distribution of corticotectal cells in macaque. Exp. Brain Res. 151, 455–470.

Macchi, G., Bentivoglio, M., Minciacchi, D., Molinari, M., 1981. The organization of the claustroneocortical projections in the cat studied by means of the HRP retrograde axonal transport. J. Comp. Neurol. 195, 681–695.

Majak, K., Pikkarainen, M., Kemppainen, S., Jolkkonen, E., Pitkanen, A., 2002. Projections from the amygdaloid complex to the claustrum and the endopiriform nucleus: a Phaseolus vulgaris leucoagglutinin study in the rat. J. Comp. Neurol. 451, 236–249.

Mamos, L., 1984. Morphology of claustral neurons in the rat. Folia Morphol. (Warsz) 43, 73–78.

Mamos, L., Narkiewicz, O., Moryś, J., 1986. Neurons of the claustrum in the cat; a Golgi study. Acta Neurobiol. Exp. (Wars) 46, 171–178.

Mathur, B.N., Caprioli, R.M., Deutch, A.Y., 2009. Proteomic analysis illuminates a novel structural definition of the claustrum and insula. Cereb. Cortex 19, 2372–2379.

McCourt, M.E., Boyapati, J., Henry, G.H., 1986. Layering in lamina 6 of cat striate cortex. Brain Res. 364, 181–185.

Meador-Woodruff, J.H., Mansour, A., Civelli, O., Watson, S.J., 1991. Distribution of D2 dopamine receptor mRNA in the primate brain. Prog. Neuropsychopharmacol. Biol. Psychiatry 15, 885–893.

Meador-Woodruff, J.H., Mansour, A., Grandy, D.K., Damask, S.P., Civelli, O., Watson Jr., S.J., 1992. Distribution of D5 dopamine receptor mRNA in rat brain. Neurosci. Lett. 145, 209–212.

Medina, L., Legaz, I., Gonzalez, G., De Castro, F., Rubenstein, J.L., Puelles, L., 2004a. Expression of Dbx1, Neurogenin 2, Semaphorin 5A, Cadherin 8, and Emx1 distinguish ventral and lateral pallial histogenetic divisions in the developing mouse claustroamygdaloid complex. J. Comp. Neurol. 474, 504–523.

Medina, L., Legaz, I., Gonzalez, G., De Castro, F., Rubenstein, J.L., Puelles, L., 2004b. Expression of Dbx1, Neurogenin 2, Semaphorin 5A, Cadherin 8, and Emx1 distinguish ventral and lateral pallial histogenetic divisions in the developing mouse claustroamygdaloid complex. J. Comp. Neurol. 474, 504–523.

Mengod, G., Vilaro, M.T., Raurich, A., Lopez-Gimenez, J.F., Cortes, R., Palacios, J.M., 1996. 5-HT receptors in mammalian brain: receptor autoradiography and in situ hybridization studies of new ligands and newly identified receptors. Histochem. J. 28, 747–758.

Mijnster, M.J., Isovich, E., Flugge, G., Fuchs, E., 1999. Localization of dopamine receptors in the tree shrew brain using [3H]-SCH23390 and [125I]-epidepride. Brain Res. 841, 101–113.

Minciacchi, D., Granato, A., Barbaresi, P., 1991. Organization of claustro-cortical projections to the primary somatosensory area of primates. Brain Res. 553, 309–312.

Minciacchi, D., Granato, A., Antonini, A., Tassinari, G., Santarelli, M., Zanolli, L., et al., 1995. Mapping subcortical extrarelay afferents onto primary somatosensory and visual areas in cats. J. Comp. Neurol. 362, 46–70.

Missale, C., Nash, S.R., Robinson, S.W., Jaber, M., Caron, M.G., 1998. Dopamine receptors: from structure to function. Physiol. Rev. 78, 189–225.

Miyasaka, N., Hatanaka, Y., Jin, M., Arimatsu, Y., 1999. Genomic organization and regulatory elements of the rat latexin gene, which is expressed in a cell type-specific manner in both central and peripheral nervous systems. Brain Res. Mol. Brain Res. 69, 62–72.

Miyashita, T., Nishimura-Akiyoshi, S., Itohara, S., Rockland, K.S., 2005. Strong expression of NETRIN-G2 in the monkey claustrum. Neuroscience 136, 487–496.

Morecraft, R.J., Geula, C., Mesulam, M.M., 1992. Cytoarchitecture and neural afferents of orbitofrontal cortex in the brain of the monkey. J. Comp. Neurol. 323, 341–358.

Morrison, J.H., Foote, S.L., 1986. Noradrenergic and serotoninergic innervation of cortical, thalamic, and tectal visual structures in Old and New World monkeys. J. Comp. Neurol. 243, 117–138.

Morrison, J.H., Molliver, M.E., Grzanna, R., 1979a. Noradrenergic innervation of cerebral cortex: widespread effects of local cortical lesions. Science 205, 313–316.

Morrison, J.H., Molliver, M.E., Grzanna, R., Coyle, J.T., 1979b. Noradrenergic innervation patterns in three regions of medial cortex: an immunofluorescence characterization. Brain Res. Bull. 4, 849–857.

Morrison, J.H., Foote, S.L., Molliver, M.E., Bloom, F.E., Lidov, H.G., 1982a. Noradrenergic and serotonergic fibers innervate complementary layers in monkey primary visual cortex: an immunohistochemical study. Proc. Natl. Acad. Sci. USA 79, 2401–2405.

Morrison, J.H., Foote, S.L., O'Connor, D., Bloom, F.E., 1982b. Laminar, tangential and regional organization of the noradrenergic innervation of monkey cortex: dopamine-beta-hydroxylase immunohistochemistry. Brain Res. Bull. 9, 309–319.

Mower, G., Gibson, A., Robinson, F., Stein, J., Glickstein, M., 1980. Visual pontocerebellar projections in the cat. J. Neurophysiol. 43, 355–366.

Mugnaini, E., Sekerkova, G., Martina, M., 2011. The unipolar brush cell: a remarkable neuron finally receiving deserved attention. Brain Res. Rev. 66, 220–245.

Nakamura, H., Gattass, R., Desimone, R., Ungerleider, L.G., 1993. The modular organization of projections from areas V1 and V2 to areas V4 and TEO in macaques. J. Neurosci. 13, 3681–3691.

Narkiewicz, O., Nitecka, L., Mamos, L., Moryś, J., 1988. The pattern of the GABA-like immunoreactivity in the claustrum. Folia Morphol. (Warsz) 47, 21–30.

Obst-Pernberg, K., Medina, L., Redies, C., 2001. Expression of R-cadherin and N-cadherin by cell groups and fiber tracts in the developing mouse forebrain: relation to the formation of functional circuits. Neuroscience 106, 505–533.

Olson, C.R., Graybiel, A.M., 1980. Sensory maps in the claustrum of the cat. Nature 288, 479–481.

Pasqualetti, M., Ori, M., Castagna, M., Marazziti, D., Cassano, G.B., Nardi, I., 1999. Distribution and cellular localization of the serotonin type 2C receptor messenger RNA in human brain. Neuroscience 92, 601–611.

Pérez-Cerdá, F., Martinez-Millán, L., Matute, C., 1996. Anatomical evidence for glutamate and/or aspartate as neurotransmitters in the geniculo-, claustro-, and cortico-cortical pathways to the cat striate cortex. J. Comp. Neurol. 373, 422–432.

Perkel, D.J., Bullier, J., Kennedy, H., 1986. Topography of the afferent connectivity of area 17 in the macaque monkey: a double-labelling study. J. Comp. Neurol. 253, 374–402.

Peters, A., 1984. Bipolar cells. In: Peters, A. Jones, E.G. (Eds.), Cerebral Cortex. Cellular Components of the Cerebral Cortex, vol. 1. Plenum Press, New York.

Picard, N., Strick, P.L., 2001. Imaging the premotor areas. Curr. Opin. Neurobiol. 11, 663–672.

Pirone, A., Cozzi, B., Edelstein, L., Peruffo, A., Lenzi, C., Quilici, F., et al., 2012. Topography of Gng2- and NetrinG2-expression suggests an insular origin of the human claustrum. PLoS One 7, e44745.

Pompeiano, M., Palacios, J.M., Mengod, G., 1994. Distribution of the serotonin 5-HT2 receptor family mRNAs: comparison between 5-HT2A and 5-HT2C receptors. Brain Res. Mol. Brain Res. 23, 163–178.

Prensa, L., Richard, S., Parent, A., 2003. Chemical anatomy of the human ventral striatum and adjacent basal forebrain structures. J. Comp. Neurol. 460, 345–367.

Preuss, T.M., 2011. The human brain: rewired and running hot. Ann. N. Y. Acad. Sci. 1225 (Suppl. 1), E182–191.

Rahman, F.E., Baizer, J.S., 2007. Neurochemically defined cell types in the claustrum of the cat. Brain Res. 1159, 94–111.

Real, M.A., Davila, J.C., Guirado, S., 2003. Expression of calcium-binding proteins in the mouse claustrum. J. Chem. Neuroanat. 25, 151–160.

Real, M.A., Davila, J.C., Guirado, S., 2006. Immunohistochemical localization of the vesicular glutamate transporter VGLUT2 in the developing and adult mouse claustrum. J. Chem. Neuroanat. 31, 169–177.

Reblet, C., Alejo, A., Blanco-Santiago, R.I., Mendizabal-Zubiaga, J., Fuentes, M., Bueno-Lopez, J.L., 2002. Neuroepithelial origin of the insular and endopiriform parts of the claustrum. Brain Res. Bull. 57, 495–497.

Reynhout, K., Baizer, J.S., 1999. Immunoreactivity for calcium-binding proteins in the claustrum of the monkey. Anat. Embryol. (Berl) 199, 75–83.

Saint-Cyr, J.A., Ungerleider, L.G., Desimone, R., 1990. Organization of visual cortical inputs to the striatum and subsequent outputs to the pallido-nigral complex in the monkey. J. Comp. Neurol. 298, 129–156.

Salin, P.A., Bullier, J., Kennedy, H., 1989. Convergence and divergence in the afferent projections to cat area 17. J. Comp. Neurol. 283, 486–512.

Sandell, J.H., Graybiel, A.M., Chesselet, M.F., 1986. A new enzyme marker for striatal compartmentalization: NADPH diaphorase activity in the caudate nucleus and putamen of the cat. J. Comp. Neurol. 243, 326–334.

Schiffmann, S.N., Libert, F., Vassart, G., Dumont, J.E., Vanderhaeghen, J.J., 1990. A cloned G protein-coupled protein with a distribution restricted to striatal medium-sized neurons. Possible relationship with D1 dopamine receptor. Brain Res. 519, 333–337.

Sherk, H., 1986. The claustrum and the cerebral cortex. In: Jones, E.G. Peters, A. (Eds.), Cerebral Cortex, vol. 5. Plenum, New York. pp. 467–499.

Sherk, H., LeVay, S., 1981. The visual claustrum of the cat. III. Receptive field properties. J. Neurosci. 1, 993–1002.

Sherk, H., LeVay, S., 1983. Contribution of the cortico-claustral loop to receptive field properties in area 17 of the cat. J. Neurosci. 3, 2121–2127.

Shima, K., Hoshi, E., Tanji, J., 1996. Neuronal activity in the claustrum of the monkey during performance of multiple movements. J. Neurophysiol. 76, 2115–2119.

Sloniewski, P., Usunoff, K.G., Pilgrim, C., 1986. Retrograde transport of fluorescent tracers reveals extensive ipsi- and contralateral claustrocortical connections in the rat. J. Comp. Neurol. 246, 467–477.

Smiley, J.F., McGinnis, J.P., Javitt, D.C., 2000. Nitric oxide synthase interneurons in the monkey cerebral cortex are subsets of the somatostatin, neuropeptide Y, and calbindin cells. Brain Res. 863, 205–212.

Smith, J.B., Alloway, K.D., 2010. Functional specificity of claustrum connections in the rat: interhemispheric communication between specific parts of motor cortex. J. Neurosci. 30, 16832–16844.

Smythies, J., Edelstein, L., Ramachandran, V., 2012. Hypotheses relating to the function of the claustrum. Front. Integr. Neurosci. 6, 53.

Snider, R.S., Niemer, W.T., 1961. A Stereotaxic Atlas of the Cat Brain. University of Chicago Press, Chicago.

Stevens, C.F., 2005. Consciousness: crick and the claustrum. Nature 435, 1040–1041.

Strick, P.L., 1986. The organization of thalamic inputs to the "premotor" areas. Prog. Brain Res. 64, 99–109.

Striedter, G.F., 2005. Principles of Brain Evolution. Sinauer Associates, Sunderland, Mass..

Sutoo, D., Akiyama, K., Yabe, K., Kohno, K., 1994. Quantitative analysis of immunohistochemical distributions of cholinergic and catecholaminergic systems in the human brain. Neuroscience 58, 227–234.

Suzuki, M., Hurd, Y.L., Sokoloff, P., Schwartz, J.C., Sedvall, G., 1998. D3 dopamine receptor mRNA is widely expressed in the human brain. Brain Res. 779, 58–74.

Takeuchi, Y., Sano, Y., 1983. Immunohistochemical demonstration of serotonin nerve fibers in the neocortex of the monkey (Macaca fuscata). Anat. Embryol. (Berl) 166, 155–168.

Tokuno, H., Tanji, J., 1993. Input organization of distal and proximal forelimb areas in the monkey primary motor cortex: a retrograde double labeling study. J. Comp. Neurol. 333, 199–209.

Umbriaco, D., Garcia, S., Beaulieu, C., Descarries, L., 1995. Relational features of acetylcholine, noradrenaline, serotonin and GABA axon terminals in the stratum radiatum of adult rat hippocampus (CA1). Hippocampus 5, 605–620.

Ungerleider, L.G., Desimone, R., Galkin, T.W., Mishkin, M., 1984. Subcortical projections of area MT in the macaque. J. Comp. Neurol. 223, 368–386.

Updyke, B.V., 1993. Organization of visual corticostriatal projections in the cat, with observations on visual projections to claustrum and amygdala. J. Comp. Neurol. 327, 159–193.

Valtschanoff, J.G., Weinberg, R.J., Kharazia, V.N., Schmidt, H.H., Nakane, M., Rustioni, A., 1993. Neurons in rat cerebral cortex that synthesize nitric oxide: NADPH diaphorase histochemistry, NOS immunocytochemistry, and colocalization with GABA. Neurosci. Lett. 157, 157–161.

Van Brederode, J.F., Mulligan, K.A., Hendrickson, A.E., 1990. Calcium-binding proteins as markers for subpopulations of GABAergic neurons in monkey striate cortex. J. Comp. Neurol. 298, 1–22.

Walker, A.E., 1938. The Primate Thalamus. University of Chicago Press, Chicago.

Webster, M.J., Bachevalier, J., Ungerleider, L.G., 1993. Subcortical connections of inferior temporal areas TE and TEO in macaque monkeys. J. Comp. Neurol. 335, 73–91.

Weiner, D.M., Brann, M.R., 1989. The distribution of a dopamine D2 receptor mRNA in rat brain. FEBS Lett. 253, 207–213.

Wiesendanger, M., Wise, S.P., 1992. Current issues concerning the functional organization of motor cortical areas in nonhuman primates. Adv. Neurol. 57, 117–134.

Williams, S.M., Goldman-Rakic, P.S., 1993. Characterization of the dopaminergic innervation of the primate frontal cortex using a dopamine-specific antibody. Cereb. Cortex 3, 199–222.

Wilson, M.A., Molliver, M.E., 1991. The organization of serotonergic projections to cerebral cortex in primates: regional distribution of axon terminals. Neuroscience 44, 537–553.

Winzer-Serhan, U.H., Leslie, F.M., 2005. Expression of alpha5 nicotinic acetylcholine receptor subunit mRNA during hippocampal and cortical development. J. Comp. Neurol. 481, 19–30.

Wright, D.E., Seroogy, K.B., Lundgren, K.H., Davis, B.M., Jennes, L., 1995. Comparative localization of serotonin1A, 1C, and 2 receptor subtype mRNAs in rat brain. J. Comp. Neurol. 351, 357–373.

Wurtz, R.H., 1996. Vision for the control of movement. The Friedenwald Lecture. Invest. Ophthalmol. Vis. Sci. 37, 2130–2145.

Wurtz, R.H., Duffy, C.J., 1992. Neuronal correlates of optic flow stimulation. Ann. N. Y. Acad. Sci. 656, 205–219.

Yan, Y.H., van Brederode, J.F., Hendrickson, A.E., 1995. Developmental changes in calretinin expression in GABAergic and nonGABAergic neurons in monkey striate cortex. J. Comp. Neurol. 363, 78–92.

Zeki, S.M., 1974. Functional organization of a visual area in the posterior bank of the superior temporal sulcus of the rhesus monkey. J. Physiol. 236, 549–573.

Zhang, X., Hannesson, D.K., Saucier, D.M., Wallace, A.E., Howland, J., Corcoran, M.E., 2001. Susceptibility to kindling and neuronal connections of the anterior claustrum. J. Neurosci. 21, 3674–3687.

4

Development and Evolution of the Claustrum

Luis Puelles

Department of Human Anatomy and Psychobiology,
School of Medicine, University of Murcia, Spain

INTRODUCTION

The claustrum, sitting underneath its dominant neighbor, the cortex, remains a little known part of the mammalian telencephalon. Due to various circumstances, which we now are starting to understand, it has been difficult to visualize its development. Where exactly, and how, is the claustrum formed? How can it be related to the cortex, but so dissimilar? Moreover, notions about a plausibly corresponding histogenetic process in non-mammalian vertebrates have also been quite limited, largely because it was not clear whether a claustrum homolog exists in non-mammals. The insular cortical environment characteristic of the mammalian claustrum has not been recognized so far in non-mammals. It has even been maintained that monotremes (prototheria) do not have a claustrum (see below), suggesting implicitly that this formation perhaps first emerged in the now extinct precursors of marsupials (metatheria) and modern mammals (eutheria), since both diverging lineages do have a claustrum. This imprecise and scarcely investigated scenario has changed and diversified with the advent of gene markers in recent years, but numerous obscure points persist. There is also what I regard as a last-minute surprise, owing to very recent data suggesting that the development of the claustrum is indeed related to that of the insular cortex, an initial concept that had been largely abandoned in recent times (Puelles et al., submitted).

In the following sections, I will examine first the older views about claustral development, which include some outdated ideas but also some sound data that need to be contemplated in any modern interpretation. I will also deal with the hypothesis that monotremes lack a claustrum,

J. Smythies, L. Edelstein and V. Ramachandran (Eds):
The Claustrum.

DOI: http://dx.doi.org/10.1016/B978-0-12-404566-8.00004-0

before covering the novel developmental and comparative scenario created by the availability of molecular markers. This leads to a section where I present some of the new data we have collected by combining a molecular marker of the claustrum with analysis of local radial glial pattern in mouse embryos. In the final section, the comparative scenario of the claustrum across the sauropsidian–mammalian evolutionary transition is reexamined summarily under the new light thrown by these recent data.

RESEARCH ANTECEDENTS

Various ideas about the developing mammalian claustrum were in circulation before the advent of molecular labeling approaches. In this chapter, comparative aspects will be included in the cases of authors who extrapolated their own conceptions regarding mammals to other vertebrate species.

Preconceived ideas about claustral development were implicit in early controversies about the ascription of the mammalian claustrum either to the pallium or to the subpallium. This uncertainty extended at least into the mid-1970s. I remember, during my formative years, attending a talk about the "basal ganglia" by the noted American expert Ann Graybiel; she included the claustrum in an overall schema of basal ganglia, but did not mention it in her talk. In discussion later, I asked whether the claustrum belonged to the subpallium or to the pallium, and she replied that she was not sure about the answer.

The pallium and the subpallium represent the two primary subdivisions of the telencephalic territory (Figure 4.1), which had been clearly identified by the early part of the last century (Ariens Kappers, 1921; Edinger, 1908; Herrick, 1910; His, 1904). Nowadays these two domains are known to have differential gene expression codes and to produce mainly excitatory versus inhibitory neurons, respectively (Swanson and Petrovich, 1998; Puelles et al., 2000). These initially distinct populations subsequently partially intermix via tangential migrations in both senses in all vertebrates; we now know that the mammalian claustrum, like the cortex and the pallial amygdala, is populated largely by excitatory neurons (the pallial feature), but it also contains up to 20 percent of migrated subpallial cells, representing various types of inhibitory interneurons (e.g., Dávila et al., 2005; Legaz et al., 2005). Both pallium and subpallium are subdivided developmentally into distinct sectors that produce characteristic derivatives in all vertebrates, and these different parts have, historically, raised diverse possibilities about claustral origins.

Medial (hippocampal), dorsal (neocortical), and lateral and ventral (olfactory) sectors of the cortical pallium (MPall, DPall, LPall, VPall; Figure 4.2A) are distinguished nowadays in both amniotes and anamniotes,

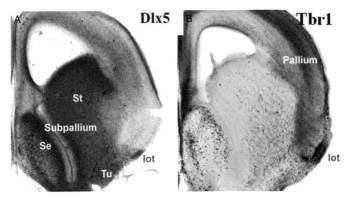

FIGURE 4.1   Modern delimitation of subpallial (A) and pallial (B) telencephalic territories, illustrated by in situ reaction (*blue color*) against *Dlx5* (A) or *Tbr1* (B) mRNA in an E15.5 mouse embryo; markers are expressed mainly in postmitotic cells of the mantle zone. Note extension of both pallium and subpallium into the septum. The sinusoidal pallio-subpallial boundary starts medial to the lateral ventricular angle (indistinct, due to artefactual compression of tissue) and ends medial to the lateral olfactory tract (lot) at the brain surface. Both sections are counterstained with anti-calretinin immunoreaction, which shows a few interneurons plus labeled fibers, particularly of the lot. Other stained fibers are seen coursing via the internal capsule into the subcortical white matter. Se, septum; St, striatum; Tu, olfactory tuberculum.

according to the comparative analysis undertaken by Puelles et al. (1999, 2000) with gene markers, and abundant correlative results from other authors accrued afterwards. This schema includes pallial parts of the septum and the amygdala (Figures 4.2B, 4.3–4.4). As we will see below, this standard tetrapartite pallium model seems to need revising, with LPall now being associated specifically with claustro-insular formations rather than to claustro-olfactory ones (Figure 4.5E). Sometimes the claustrum is mentioned in association with the amygdala, as part of the concept of a *claustro-amygdaloid complex*, which would extend within the telencephalon from rostral (claustrum, endopiriform nucleus) to caudal (pallial amygdala) (see Figures 4.2B and 4.3). This complex includes all *pallial nuclear masses* found deep to olfactory or insular/perirhinal and amygdaloid cortical formations. As we will see below, there are reasons to dissociate the claustro-amygdaloid complex into separate pallial domains, without negating the pallial nuclear nature of their respective constituents. It is important to understand that "pallium" is not synonymous with "cortex" (a frequent misconception), insofar as both cortical and nuclear structures can be pallial. Similarly, "subpallium" is not synonymous with "nuclei," since some corticoid structures are subpallial (the olfactory tuberculum). In any case, the imminent revision of the pallium model mentioned above questions the continuity of the claustro-amygdaloid nuclear complex and suggests that some of its parts are molecularly distinct and developmentally independent formations.

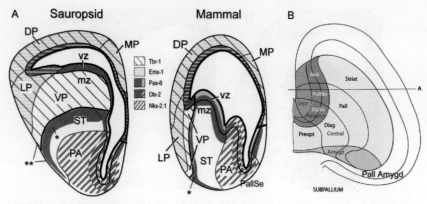

FIGURE 4.2 Schematic drawings showing in (A) pallial and subpallial molecular subdivisions postulated in the tetrapartite pallial model of Puelles et al. (2000), comparing the conserved topologic pattern in reptiles and birds (sauropsids) and mammals, and divergent relative growth and morphogenesis (e.g. thickness, absolute size and presence of ventricular bulges) of the different pallial or subpallial sectors. The *thick black line* indicates the pallio-subpallial boundary in both cases. (The gene markers used for the color code are listed in the center.) Note the ventricular zone (vz) and mantle zone (mz) may show particular labeling features. DP, dorsal pallium; LP, lateral pallium; MP, medial pallium; PA, pallidum; PallSe, paraseptal pallidal subdomain; ST, striatum; VP, ventral pallium. *Asterisks* mark migrating populations (*Pax6+ cells; **Emx1+ tangentially migrating olfactory cortex neurons). B: whole right mouse telencephalon is depicted from above (as if transparent) to show detailed distribution and organization of the four parallel (or concentric) subpallial domains (Striat, striatum; Pall, pallidum; Diag, diagonal area; Preopt, preoptic area). Color code indicates their common pattern of subdivisions along the so-called septo-amygdaloid axis (Sept, septal; Parasept, paraseptal; Central (sic); Amygd, amygdaloid). Note approximate position of pallial amygdala (Pall. Amygd, within colorless pallium), caudal to the subpallial amygdala (*green*). Transverse line *A* indicates approximate level where the schemata in (A)–of the left brain half–were taken (to understand why some subpallial components do not appear there; note septal region has no name tag in A, but paraseptal PallSe domain lying under the lateral ventricle is marked for cross-reference). Sections in Figure 4.1 pass more rostrally, through the accumbens striatal paraseptal component, Acb).

The subpallium in its turn subdivides into striatal, pallidal, diagonal and preoptic primary sectors, all of them extending from the ventricle to the corresponding pia (St, Pal, Dg, POA in Figure 4.2B; see alternative terminology in Flames et al., 2007). These four domains are stretched along the septoamygdaloid axis. Part of the Dg–its reticular intermediate stratum–used to be identified as the area innominata, or more recently as the anterior entopeduncular area (AEP; e.g., Puelles et al., 2000); these two traditions both disregarded the clear-cut radial relationships of the innominate-entopeduncular area with the subpial diagonal band nuclei and part of the periventricular bed nucleus of the stria terminalis; the resulting complete radial unit is alluded to now by the term "diagonal area," which, accordingly, fully separates at all radial levels the pallidum

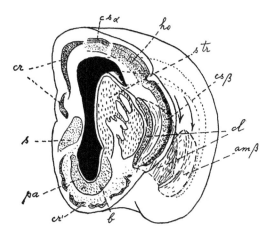

FIGURE 4.3   Tri-dimensional drawing from Kuhlenbeck (1924) illustrating pallial and subpallial components in a generalized "lower" mammal, and his idea of a joint origin of the striatum and the claustrum within his D1 pallial region, as well as the caudal continuity of the claustrum with the pallial amygdala (amβ, nucleus amigdalae beta). Note there is an *arrow* marking where claustral neurons come from in the ventricular zone, which is the small domain medial to the lateral ventricular angle which we now conceive to be pallial (compare ventral pallium of Figure 4.2), whereas the underlying striatal area is clearly subpallial (compare Figure 4.1). Interestingly, the claustrum is depicted as lying deep to olfactory cortex (csβ, cortex striatus beta), that is, refers rather to an endopiriform formation; the insula is not identified, but would lie ventrally within the neocortex (ho, cortex homogeneticus of Brodmann). This shows he clearly supported an origin independent from the insula.

from the preoptic area (e.g., Allen Developing Mouse Brain Atlas). The subpallium also includes septal and amygdaloid portions; the subpallial continuity of some amygdaloid components with more rostral elements is exploited in the concept of "extended amygdala" (Figure 4.2B; Swanson and Petrovich, 1998; Puelles et al. 2000; Medina and Abellán, 2012).

Due to the consistent topographic relationship of the adult mammalian claustrum with the deep aspect of the insula, a hypothetic pallial origin of the claustrum was attributed originally to the *insular cortex*, of which it seemed to represent a deep layer or sublayer (Meynert, 1868, 1872; Figure 4.5A). The insula may be classified either as a distinct ventral part of the DPall (i.e., its ventralmost or transitional part, next to prepiriform olfactory allocortex; Puelles et al., 2000; Figure 4.5A–D), or as a cortical part of the LPall, transitional in structure between the respectively iso- and allocortical components of the DPall and VPall (this is a novel position, which I will defend below; see Figure 4.5E). Alternatively, a subpallial origin of the claustrum was postulated within the subpallial *striatum*; this occurred largely because all telencephalic nuclei were wrongly assumed to derive from the ganglionic eminences, where the so-called "basal" subpallial ganglia arise

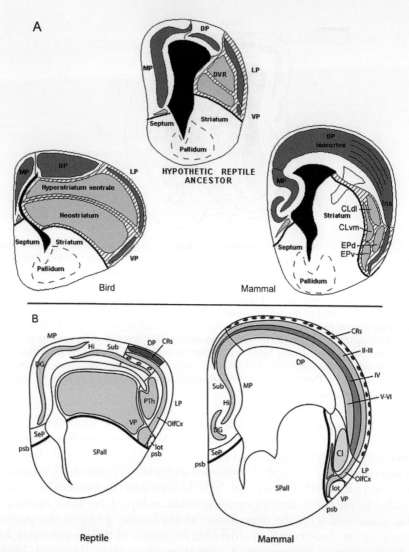

FIGURE 4.4 Two views of pallial subdivisions compared between sauropsids and mammals at an interval of 10 years. A: shows a schema from Puelles (2001), illustrating the bipartition discovered in Puelles et al. (2000) of the classic "lateral pallium" (*obliquely striped* area in the pallium) into the new lateral pallium (LP) and ventral pallium (VP). Both domains have a hypopallial structure (display a superficial cortical structure, thought to be olfactory, and deep pallial nuclei underneath), jointly forming in reptiles and birds the "dorsal ventricular ridge" (DVR) of Johnston (1916), held to be homologous with the mammalian claustrum by Holmgren (1925). The modest ventricular protrusion of the DVR in reptiles contrasts with the much larger one seen in birds. Extrapolation of this structure to mammals requires that the corresponding mammalian domain (in which the derived cell populations translocate nearer to the brain surface) be also divided into two parallel LP and

(Figure 4.5B). These early contradictory opinions about claustral origin were eventually substituted by belief in an *intermediate pallial source*, deemed independent from both the basal ganglia and the neighboring insular cortex. This extra intermediate pallial territory roughly corre- lates with the subsequently postulated VPall and LPall sectors (Figure 4.5C,D); note that in this conception the insula was still thought to belong to DPall, and the claustrum was held to be related superficially to the olfactory cortex. This third position implied that the apparent intimate relationship of the claustrum with the insula was misleading, as was supposedly corroborated by a claustro-olfactory organization of radial glial processes in this area (Misson et al., 1988, 1991; Valverde and Santacana, 1994). Since the concept of a separate VPall (Figures 4.2A and 4.5D) only arose around the year 2000 (Smith-Fernandez et al., 1998; Puelles et al., 1999, 2000), earlier studies did not attempt to dis- tinguish VPall derivatives from LPall ones, and systematically lumped together, or confused one with another, different cell masses that now can (and should) be separated. The confusion in the field also caused some authors to conceive (wrongly) mixed claustral origins in both the pallium and subpallium (e.g. Bisconte and Marty, 1975), or to postulate a contribution of pallial claustral cells to the striatum (e.g. Kuhlenbeck, 1973; Hirata et al., 2002). Autoradiographic studies of the birthdates and migration pattern of developing claustral neurons in the context of the neighboring olfactory and insular cortex versus the subpallial stria- tum also tended to support a more or less independent pallial origin. I will now consider these various notions in more detail.

◄ FIGURE 4.4 VP domains (compare corroborating data in Figure 4.2). The claustral com- plex appears divided into LP and VP dorsal (claustrum proper parts) and ventral (endop- iriform nuclei) components. Note position assigned to the insula within the DP (dorsal pallium). B: reproduces a recent schema from the Puelles (2011) review, comparing reptil- ian and mammalian pallial domains; interest was centered on evolution of the neocortex (partly in relationship with the emergence of Cajal-Retzius tangential migrations into the molecular layer, and other pallio-pallial migrations), but an updated view of the LP and VP domains in tetrapods is also represented. In reptiles, the DVR bulge is attributed mainly to VP, whereas LP hypothetically partly extends past the lateral ventricular angle, under the major part of the olfactory cortex, containing as a deep nucleus the classic so-called "pal- lial thickening" (PTh), a possible homolog of the claustrum in mammals; the former VP counterpart of the claustrum lying deep to the lot is left unnamed (no longer thought to be claustral). This schema is now fully revised in the new concept of lateral pallium and ventral pallium presented in this chapter (Figures 4.5E, 4.16) and elsewhere (submitted). CLdl, dorsolateral claustrum; CLvm, ventromedial claustrum; DG, dentate gyrus; EPd, dor- sal endopiriform nucleus; EPv, ventral endopiriform nucleus; hi, hippocampus; lot, lateral olfactory tract; psb, palliosubpallial boundary; SeP, pallial septum; sub, subiculum.

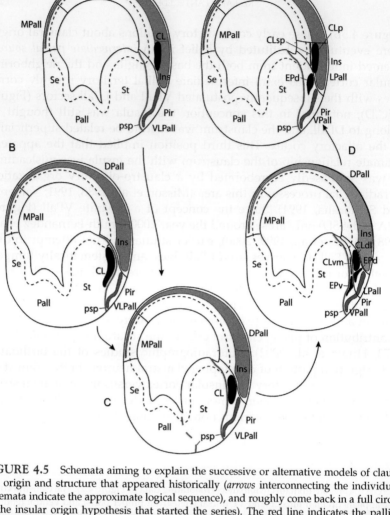

FIGURE 4.5 Schemata aiming to explain the successive or alternative models of claustral origin and structure that appeared historically (*arrows* interconnecting the individual schemata indicate the approximate logical sequence), and roughly come back in a full circle to the insular origin hypothesis that started the series). The red line indicates the pallio-subpallial boundary, and claustral elements are filled in black. A: the claustrum is held to delaminate from the deep strata of the insula; B: the claustrum originates from the dorsal part of the lateral ganglionic eminence (striatum), whose boundary against the pallium is wrongly assumed to lie at the "corticostriatal" angle of the ventricle; C: the claustrum is produced at the pallial matrix present immediately medial to that angle, jointly with the olfactory cortex neurons (VLPall, ventrolateral pallium) and independently from the insula; D: this model is essentially the same as in (C), but the former VLPall is divided into independent VPall and LPall pallial domains (compare Figure 4.2), causing both the claustrum and endopiriform nuclei lying deep to olfactory cortex to be divided into ventropallial and lateropallial moieties (CLdl, CLvm; EPd, EPv); E: new model defended here, after corroborating the insular origin of the claustrum (note its subdivision into subplate and principal parts; CLsp, CLp); the LPall is redefined, departing from (A), so both the claustrum and insula are ascribed to it (insula is no longer a DPall component); moreover, the whole olfactory cortex falls now in the VPall; a lateropallial claustral cell population drawn in *black*, like the claustrum, is postulated to invade tangentially the adjacent VPall domain, acquiring an endopiriform status (EPd). For simplicity, other VPall structures are not represented.

The Claustrum as an Insular Cortical Stratum

Meynert (1868, 1872), in his pioneering studies on the telencephalic cortex, thought that the claustrum was a deep layer of the insular cortex (forming part of its fifth layer, nowadays known as isocortical layers 5 and 6). According to Spiegel (1919), this influential opinion was shared at the time by Krause, Schwalbe, Gegenbaur, Wernicke, and Honnegger. Implicitly, it was assumed that the claustrum develops jointly with the insula (see Figure 4.5A). This idea acquired full force in the work of Brodmann (1909), who regarded the claustrum as a special sublayer split off from the insular multiform layer 6, generating what he identified as layer 7 (as we will see below, this is different from present-day layer 6b, also known as subplate or layer 7). Ariens Kappers (1921), Spatz (1921), and Rose (1927, 1928) supported the Meynert/Brodmann idea. Ramon y Cajal (1902), while describing the claustrum jointly with the insular cortex, nevertheless voiced the reservation that "... the fact that in the claustrum there are no vertically situated fusiform cells, whose radial processes cross the more external [insular] layers and terminate in the plexiform [molecular] layer, leads to the assumption that this gray formation is independent of the insular cortex" (my translation). He also disliked the idea of a striatal nature, and thus probably thought that the claustrum was a pallial nucleus rather than a cortical layer (Figure 4.5C). Golgi studies of claustrum neurons undertaken in various mammals usually describe the majority elements as spiny spindle-shaped or pyramidal cells, without implying a radial direction of their "apical dendrites" (which are not apical and are devoid of a terminal bouquet) towards the insular molecular layer, or any other pial direction. Contrarily, these dendrites tend to be oriented parallel to the external and extreme capsules, that is, orthogonal to the topologic radial dimension of the insula (e.g., Brand, 1981; Braak and Braak, 1982; Mamos et al., 1986). A further early argument adduced against an insular ascription of the claustrum was that the extreme capsule that separates the claustrum from the insula in large-brained animals was thought to be a cortico-cortical association tract, and it seemed odd that it should course between two deep layers of the insula, as implied by the insular origin hypothesis, rather than through the white matter underneath (Landau, 1919).

The Claustrum as Part of the Subpallium (Basal Ganglia)

Authors dealing with development of basal ganglia in human embryos, such as Landau (1919), Kuhlenbeck (1924, 1927), Kodama (1927) and Macchi (1951), expressed the opinion that the claustrum derives from the embryonic lateral ganglionic eminence, that is, the striatal primordium (see Figure 4.5B). However, both Landau and Kuhlenbeck later changed their views (see below). This interpretation was clearly facilitated by the widespread erroneous belief at that time that the lateral angle of the lateral ventricle marks a cortico-striatal boundary. In several cases (e.g., Kuhlenbeck, 1924, 1927; see Figure 4.3) it was noticed that, according to its position deep to the insula, the claustrum should originate selectively from the matrix territory immediately underneath the ventricular angle, which was then thought to be ganglionic (i.e. striatal), though we now know it is pallial in molecular signature (Puelles et al., 2000; see Figures 4.1 and 4.2A). The claustral primordium indeed appears roughly under this angular locus, clearly separated from both the striatum and the prospective insula, when first distinguished in 3-mm-long human embryos (Kodama, 1927; his Figure 2; see also Bayer and Altman, 2006; Figures 4.5C and 4.6). The rationale of this group of authors thus was based on a wrong assumption, the simplistic one that the telencephalon is *only* divided into cortex and basal nuclei (forming the so-called subcortical domain), so that all non-cortical formations had to be interpreted as subcortical and belonging to the basal ganglia. This rationale dismissed the existence of pallial nuclei, a point that became clearer subsequently, particularly due to advanced study of the amygdala, whose pallial nuclear portion is massive (see Figure 4.3). This minority view on a striatal origin of the claustrum was clearly disproved subsequently by chemoarchitectonic and hodologic criteria (e.g., by lack of an acetylcholinesterase (AChE)- and dopamine-rich neuropil in the claustrum, a typical characteristic of the striatum; for review, see MacLean, 1990, Striedter, 1997), as well as by recent genoarchitectonic studies (lack of subpallial gene markers; Puelles et al., 1999, 2000; see Figure 4.1B).

The Claustrum as a Non-Insular Pallial Derivative

In line with Ramon y Cajal's (1902) sceptic position about the insular origin hypothesis (cited above), De Vries (1910), Spiegel (1919), Landau

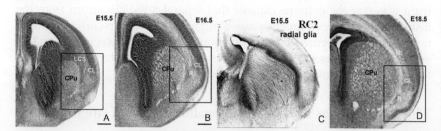

FIGURE 4.6   Visualization in Nissl material (A,B,D) of the mouse claustrum (CL) as it becomes detectable. It is compared to the transient structures named by Bayer and Altman (1991b): the lateral cortical stream (LCS) and "reservoir" (R), which supposedly hold and transfer postmitotic neurons into the claustrum. A: E15.5 embryo; B: E16.5 embryo; C: E15.5 embryo immunoreacted with the RC2 antibody against radial glia, showing the packed glial bundle that accompanies the LCS along the pallio-subpallial boundary; D: E18.5 embryo. The landmarks represented by the lateral olfactory tract, anterior commissure and external capsule can be distinguished. CPU, caudoputamen (striatum); 6b, cortical layer 6b (subplate).

(1923a,b, 1936), Holmgren (1925), Von Economo and Koskinas (1925), Kuhlenbeck (1973, 1977) and Filimonoff (1966) eventually held that the claustrum is a *pallial nucleus*, whose development is independent from the insular cortex, and takes place within the matrix and mantle pallial space that appears intercalated between the insula and the striatum, and is covered by the olfactory cortex superficially (Figure 4.5C).

De Vries (1910) observed in human embryos that the predominant orientation of claustral cell bodies that appeared to emanate from the proliferative ventricular zone (note this assumption turned out to be wrong; at the stages he was studying, claustral neurons were far beyond their postmitotic period) was parallel to both the pallio-subpallial boundary (with striatal neurons clearly on the other side of the boundary) and the insular cortex. This observation was held to exclude in principle both striatal and insular origins. De Vries (1910) also performed a parallel comparative analysis of the adult claustrum among major mammalian orders (primates, ungulates, carnivores, lagomorphs, rodents, insectivores, marsupials) and concluded that the claustrum emerges evolutionarily as an independent transitional pallial field deep to the rhinal sulcus, in a distinct pallial area intercalated between the insular and the prepiriform cortical primordia. Deep cells from this area would become increasingly internalized under the insula by differential growth, as well as by subpial compression (pinching) of this area between the expanding insular and prepiriform domains, thus building at some point the internalized claustrum. According to this interpretation, the claustrum would practically constitute a pallial sector by itself, intercalated between olfactory and insular cortex sectors; this sector would secondarily suffer internalization of the majority of its population (the claustrum), losing contact with the brain surface (in retrospect, it is difficult to imagine how this can occur). However, the subsequent studies of Landau (1923a,b, 1936), Holmgren (1925), Filimonoff (1966) and Kuhlenbeck (1973, 1977) discarded this particular idea of an extra intercalated claustral pallial domain and, instead, conceived the idea of the claustrum as developing in the pallial sector that contains superficially the prepiriform cortex (the piriform cortex proper is found more caudally, alongside the amygdala). Its origin was thus independent from both the striatum and the insula (see Figure 4.5C).

Holmgren (1922, 1925) studied descriptively the development of the pallium and subpallium in anamniotes and amniotes. He concluded that outside of the isocortical primordium there is a shared pallial developmental unit, identifiable in mammals as the claustro-amygdalo-piriform complex which lies ventral to the neocortical plate, next to the pallio-subpallial boundary. According to him, this unit would be homologous to the homotopic telencephalic domain variously known in other vertebrate forms as epistriatum (Edinger, 1908; Johnston, 1908), hypopallium (Elliot Smith, 1919), dorsal ventricular ridge (Johnston, 1916, 1923) or epibasal nucleus

(Kuhlenbeck, 1924, 1927, 1973). This domain is represented schematically as VLPall in our Figure 4.5C. The pallial versus subpallial (or mixed) nature of this territory was initially controversial, but it has been definitively established as being pallial by modern molecular and comparative evidence (Figures 4.1 and 4.2A; Striedter, 1997; Smith-Fernandez et al., 1998; Puelles et al., 1999, 2000, 2007; Reiner et al., 2004). Other authors subsequently aligned themselves under Holmgren's position, including Landau (1923b, 1936), Källen (1951), Filimonoff (1966) and, more recently, MacLean (1990) and Puelles (Puelles et al., 1999, 2000, 2004, 2007; Puelles, 2001, 2011); at this point the VLPall was divided into VPall and LPall components, for reasons explained below; see Figures 4.2, 4.4A,B, and 4.5D.

Kuhlenbeck (1924, 1927, 1973, 1977) likewise performed over the years a comparative analysis of telencephalic development in tetrapods. He concluded that the pallium of amphibia and sauropsida was divided into three sectors D1–D3 (from lateral to dorsomedial). These parts roughly correspond to the subdivisions VLPall (with lumped VPall and LPall), DPall and MPall, respectively (see Figure 4.5C). In reptiles and birds, the derivatives of D1 were identified as *epibasal nucleus* formations (referring to their position covering the "basal" subpallial formations). This author held that in mammals a thin dorsal part of the D1 pallial sector was the source of the claustrum and the pallial amygdala (see Figure 4.3), and also contributed cells to the olfactory and insular cortex, whereas the larger ventral part of D1 participated in forming the striatum (a mixture of the models in Figure 4.5B,C–see the *dashed red boundary* in Figure 4.5C, wrongly allowing the striatum to be pallial; clearly his mammalian ventral D1 corresponded to the lateral ganglionic eminence, whereas his dorsal D1 was the VPall, whose matrix lies under the lateral ventricular sulcus; see Figures 4.2 and 4.5D). This also implied the wrong assumption, so common at the time, that the striatum ends at the lateral ventricular sulcus. (In all his studies Kuhlenbeck systematically used the ventricular sulci as structural boundaries in the neural wall, a paradigm disproved by much subsequent molecular evidence, and practically discontinued nowadays). In any case, Kuhlenbeck (1973, 1977) repeatedly expressed his opinion that the mammalian claustrum plus the prepiriform cortex (both extending caudally into the amygdala; see Figure 4.3) are field-homologous to the sauropsidian epibasal complex (the dorsal ventricular ridge; DVR), which is entirely pallial, and thus was essentially in agreement with Holmgren (1925) and the models in Figures 4.4 and 4.5C.

Autoradiographic Studies of Claustral Neurogenesis and Migration

Hinds and Angevine (1965) studied claustral neurogenesis in the mouse using H³-thymidine-autoradiography and found that claustral neurons are very precocious and chronologically continuous with deep cells of the insular cortex (cited by Bisconte and Marty, 1975). Similar studies undertaken by Smart and Smart (1977, 1982) found mouse claustrum labeled as of E11, with few if any cells born at E13. Bisconte and Marty (1975) likewise observed that rat claustral populations are produced early, between E11.5 and E14.5, with a peak at E12.5 (see their Figures 2a,b, 9, and related text). Curiously, these latter authors wrongly conjectured on the basis of chronoarchitectonic considerations that the claustrum has mixed pallial and subpallial origins. Moreover, Bisconte and Marty (1975) observed that the claustrum can be strictly distinguished from the insular cortex, due to a clearcut radial dissociation of sets of deep claustral and superficial insular cells, both born at E13.5 (see their Figure 9); this observation has not yet received a satisfactory explanation (I suggest that a tangentially migrated cell contingent may underlie these data). Valverde and Santacana (1994) also found that rat claustral neurons are born as early as E11.5, being produced jointly with pyramidal neurons of the prepiriform cortex and the even more precocious layer 6b neocortical neurons (see also Valverde et al., 1989, 1995). These diverse authors however did not identify or even discuss the ventricular site(s) where claustral neurons are produced, nor its dependent versus independent status with regard to the insula.

In previous periods, the claustrum was often divided into dorsal and ventral parts. Loo (1931) eventually named the ventral portion *endopiriform nucleus* (EP; referred to by the abbreviation END in Chapter 2). He differentiated it from the dorsal claustrum (now *claustrum* proper; CL) by its topography deep to the piriform cortex, as opposed to the juxtainsular topography of the claustrum. Bayer and Altman (1991b) studied the development of the rat EP and CL formations on the basis of H^3-thymidine-autoradiographic data. Surprisingly, their cell birthday data for the EP and CL are not the same as those of the earlier authors who also worked on rat neurogenesis (cited above).

In the first place, Bayer et al. (1991) and Bayer and Altman (1991a) had reported in the rat a cortical migratory stream that they described as being deflected into a tangential (circumferential) course; in other words, this corticopetal stream was deemed not to be oriented radially, which is the typical route for immature postmitotic cortical neurons reaching the cortical plate. This so-called *lateral cortical stream* (LCS) was held to contribute migrating neurons to lateral and ventrolateral parts of the cortex primordium. This conclusion about a non-radial course of migration was reached despite the fact that the stream in question coincides with a dense bundle of radial glial fibers (e.g., Gadisseux et al., 1989; Misson et al., 1988, 1991; Valverde and Santacana, 1994; these works were cited by Bayer and Altman, 1991a, themselves; see Figure 4.6C). According to the interpretation of these authors, this peculiar phenomenon occurs because the pallial proliferative ventricular zone retracts dorsomedially from the area occupied by the cortical plate at the ventrolateral brain surface, as they observed in tridimensional reconstructions over various stages of development. However, the fact that one may artificially separate the ventricular zone from the cortical plate in a reconstruction does not prevent them from remaining in intimate structural unity, due to the well-known neuroepithelial structure of the brain wall. Not realizing that this morphogenetic process would result in the affected radial glial cells being deformed, without losing their topological radial disposition, Bayer and Altman (1991a,b) jumped to the conclusion that postmitotic neurons destined for ventrolateral cortical areas (and, implicitly, the claustrum) would no longer reach them radially. Therefore, they postulated that these neurons must use a non-radial detour, the LCS, which lies parallel to the pallio-subpallial boundary (LCS; Figure 4.7). These authors did not realize either that the pallio-subpallial boundary is topologically radial itself, since it separates distinct pallial and subpallial sectors of the telencephalic neuroepithelium (Figures 4.1, 4.2, and 4.6C), so the LCS must be likewise radial, at least where it is parallel to that limit. The radial glial fibers coinciding in their trajectories with the LCS have their cell bodies in the matrix neighborhood that lies ventral to the lateral ventricular angle (the palliostriatal angle of Filimonoff, 1966). This area was interpreted by Bayer and Altman as containing the ventricular zone of the "dorsal cortex," that is, DPall. However, this matrix area is now widely ascribed to VPall (Figures 4.2A, and 4.5D; Bielle et al., 2005; Medina et al., 2004; Puelles et al., 2000; Teissier et al., 2010). Following these glial processes, which appear tightly bundled together within the thin VPall domain, adjacent to the pallio-subpallial boundary (see Figure 4.6C), the LCS was found to skirt around the striatum, ending at a transient cell repository, named the *reservoir* by Bayer and Altman (R; see Figure 4.7). The reservoir is visible at appropriate intermediate stages roughly at the locus where the anterior commissure meets the external capsule, medial to the primordium of the claustrum, and deep to both the insula and the prepiriform cortex (R; Figure 4.6A,B,D). The latter formations were reported to receive some neurons from the reservoir, and/or directly from the LCS, though this point was not well documented (see Figure 4.7). The claustrum was not mentioned in these publications (Bayer et al., 1991; Bayer and Altman, 1991a).

However, in a separate work (Bayer and Altman, 1991b), it was conjectured that the LCS and the reservoir contribute migrating neuroblasts to the developing EP and CL as well as to the insular and prepiriform cortices (see Figure 4.7). As regards the cell birthdays reported by these authors, the EP cells are most precocious; they were said to be born between E12 and E17 (see Figure 4.7); actually this is a surprisingly long period for a precocious population, which contrasts with previous neurogenetic data reported in the mouse and rat, as summarized above. Peak production for

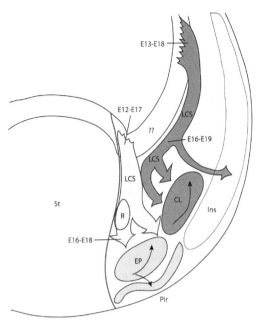

FIGURE 4.7 Sketch of the fundamental concepts about endopiriform (*light gray*) and claustrum (*dark gray*) development in the rat brain that were reported by Bayer and Altman (1991b). The more precocious EP formation was conjectured to be produced between E12 and E17 at the "palliostriatal angle" (our VPall matrix). It migrates into position deep to the olfactory (piriform) cortex (Pir) between E16 and E18, via the "lateral cortical stream" (LCS), possibly waiting partially at the "reservoir" (R), and then moving forward to their target. This interpretation invoking a VPall origin is valid for the EPv nucleus (compare Figure 4.10C), but not for the larger EPd (see text of Figure 4.11A,B). In contrast, the claustrum population was conjectured to arise far away in the neocortical ventricular zone between E13 and E18. These cells were postulated to migrate circumferentially along the LCS, and later incorporate into the claustrum between E16 and E19, contributing also some collateral cells to the insula. The *thin arrows* inside EP and CL imply reported neurogenetic gradients. The *double question marks* at the ventricular zone indicate an area whose progeny was not determined (maybe the bulk of the insula was thought to originate there). The claustral (plus EPd) pattern proposed is inconsistent with our present observations (compare Figure 4.16).

ventromedial EP apparently occurred at E14, as compared to peak production for the dorsolateral EP at E15 (a ventrodorsal neurogenetic gradient was suggested). The EP did not show any longitudinal neurogenetic gradient (implying a localized source), and its production slightly precedes that of prepiriform cortex neurons, which are nearly as precocious (compare Bayer, 1986; Valverde and Santacana, 1994). In contrast, the rat CL population was reported to have later birthdates, between E13 and E18 (a similarly long period, compared to earlier data), with peak production at E15-E16. In this case, a caudorostral gradient comparable to that found in the insular cortex was noticed.

The claustral data on CL/EP birthdates of Bayer and Altman (1991b) thus contrast rather markedly with those obtained also in the rat by Bisconte and Marty (1975) and Valverde and Santacana (1994). Those other authors found peak CL production at E12.5, as opposed to E15/16. This discrepancy may be partly diminished, but not totally explained, by considering staging or methodological differences. Notably, Bayer and Altman (1991b) labeled the day on which the dams were found sperm-positive as E1, whereas commonly the day a vaginal plug is found is labeled E0 or E0.5 (this would allow correcting all their data, if so desired, by at least half a day, or perhaps one full day). Bisconte and Marty (1975) dated gestations from coitus time minus 8 hours (following criteria of Witschi, 1956), so that their neuronal birthday data may be labeled some 4 hours ahead of standard methods. In addition, both Bisconte and Marty (1975) and Valverde and Santacana (1994) used so-called flash-labeling (single dose of H^3-thymidine administered), whereas Bayer and Altman (1991b) used partial accumulative labeling, by giving a second H^3-thymidine dose after 24 hours (note it is by no means clear that a single dose lasts for a full 24 hours, so that with this method, essentially, two peaks of label are obtained at an interval of 24 hours, separated by a period with decreasing signal in the blood, as the dam's liver metabolizes the marker). The conclusions on birthdates adopted by Bayer and Altman in all their studies performed with this method nevertheless are based on determining the proportion of neurons born at each day, deduced from quantitative and statistic treatment of the changing labeling indexes obtained across the series of experiments. Such conclusions are difficult to compare with studies reporting simply labeling patterns obtained after giving a single dose of marker.

My impression is that these methodological differences do not explain the recorded difference between CL neurogenetic peaks at E15/16 (Bayer and Altman, 1991b) versus E12.5 (Bisconte and Marty, 1975; Valverde and Santacana, 1994), so that other circumstances need to be considered to cover this angle. For instance, the criteria used to identify the claustrum in the sections may have had a systematic error that meant deep insular neurons were included in the claustrum, or perhaps they did not discriminate glial cells well enough, resulting in birthdates that seemed to be later than they actually were. I am inclined to give more credence to the sets of rat data of Bisconte and Marty (1975) and Valverde and Santacana (1994). They corroborate each other, and also seem to be strictly consistent with the excellent mouse data of Smart and Smart (1977, 1982). Moreover, these authors apparently have a clearer idea of the radial organization of the relevant territory.

Bayer and Altman (1991b) did not investigate in their sequential-survival analysis of EP and CL migration, and histogenesis, any CL cells labeled earlier than E15, or any EP neurons born earlier than E14, an analysis that might have shown different results. According to the other studies, Bayer and Altman (1991b) did not, as they thought, study the peak production wave but only the latest cells entering the claustral complex, if at all. A further consideration is that these authors only identified the claustral primordium at the deep place where it is found after E15.5 in the mouse (see Figure 4.6A), ignoring any earlier states of the primordium; this could be done if the birthdates were not much separated from the moment when one sees the primordium, but that was not the situation in this case. This inconsistency with the precocious birthdates reported for these neurons was also committed by all earlier students of the developing claustrum, motivated no doubt by the dominating image of the adult claustrum as a deep structure. However, being among the earliest neurons produced, the claustral neurons must first aggregate at the brain surface, as I will show below. All the sequential-survival analysis of Bayer and Altman (1991b) aimed to find how cells born around E15 arrived at the deep locus where the claustrum becomes visible later (see Figure 4.7). In my opinion, the dubious data and false assumptions apparent in this line of analysis discredit the conclusions.

None of these students of claustral neurogenesis clarified the position of their "labeled claustrum," or of the "lateral cortical stream" relative to distinct histogenetic pallial sectors (i.e. DPall, LPall, or VPall). As regards neuroepithelial origins, the sequential-survival analysis of Bayer and Altman (1991b) led these authors to postulate that EP neurons are produced at the matrix area of the palliostriatal angle (interpretable as our present VPall; see Figure 4.5D; compare Bielle et al., 2005; Medina et al., 2004; Puelles et al. 1999, 2000; Teissier et al., 2010), whereas they conjectured that CL neurons originate more dorsally from the ventricular zone of the dorsal isocortex (their schematic Figure 16 suggests they were referring to the DPall) and secondarily incorporate into the lateral part of the LCS (see Figure 4.7). These cortical cells reportedly would follow the LCS to reach the reservoir, from where they would transfer into the CL or the insula (or both). Whether such final distribution occurred radially or non-radially (i.e., tangentially) was not discussed by Bayer and Altman (1991b). However, if we accept that the LCS is an essentially ventropallial transient radial entity, according to its inner radial glial structure (see Figure 4.6C), then non-radial invasion mechanisms need to be considered for final arrival of neurons from the reservoir at the claustrum and the insula (see Figure 4.7).

A Possible Double Set of Claustral Derivatives

It was subsequently suggested that the claustral nuclear complex (CL, EP) is actually subdivided into VPall and LPall moieties, each having CL and EP portions more or less distant radially from the respective prepiriform cortex portions at the brain surface (see Figure 4.5D). The CL would be subdivided into dorsolateral and ventromedial parts–CLdl, CLvm–and the EP appears subdivided into dorsal and ventral parts–EPd, EPv. The CLdl and EPd were postulated to belong to LPall, whereas the CLvm and EPv were ascribed to VPall (see Figures 4.4 and 4.5A; Gorski et al., 2001; Medina et al., 2004; Puelles et al., 1999, 2000). Subsequently, most authors have systematically disregarded the postulated, sparsely populated VPall part of the claustrum complex (best distinguished with molecular markers, interstitially to the external capsule). The literature generally applies the name "claustrum" only to the more visible and compact CLdl (see Figure 4.5D). Given this strong trend, and the novel developments considered below, which raise doubts about the claustral nature of the population that was called CLvm, I think it is convenient for simplicity and clarity to retain the term claustrum (CL) exclusively for the main dorsolateral claustral part, and rename the "CLvm" nucleus found within VPall as *bed nucleus of the external capsule* (BEC). This term emphasizes the distinct topography and interstitial distribution of its cells relative to the external capsule, as well as the convenience of differentiating semantically BEC from CL. This does not mean that the CL is no longer bipartite, as I explain in the next paragraph. Basically, the CLdl itself has two parts.

Classical evidence supporting a double developmental constitution of the claustral nuclear complex was first published by Hochstetter (1919) in his study of the developing human brain, as illustrated in an embryo of 87 mm CRL (e.g. his Figures 116, 117); this result was not mentioned in the text. This dual feature also figured prominently in the analysis of the claustrum of diverse insectivores undertaken by Haller (1908, 1909), Narkiewicz and Mamos (1990) and Dinopoulos et al. (1992), but is less obvious in other small mammals (see Urban and Richard, 1972, or Manocha et al., 1967, their Figures 10–12). The clarifying study of Narkiewicz and Mamos distinguished in all cases a lateral and thicker "pars principalis claustri" from a thin, denser "lamina profunda claustri." The latter was thought to be continuous with cortical layer 6b, or subplate (CLsp). This in principle associates the more superficial principal CL (CLp) part to layer 6a. A discrepancy emerged here between our developmental ascription of the bipartite claustrum to VPall and LPall and the conclusion of Narkiewicz and Mamos (1990) about the relationship between the lamina profunda claustri and cortical layer 6b. Clearly, we interpreted this subplate component as the CLvm part of VPall. I am now convinced that the Polish authors were right, though this does not negate the existence in this area of extra VPall derivatives, which were not sufficiently understood before (for instance, the BEC; this means that the previous two parts have now grown into three parts, two of them properly claustral, and belonging to LPall, and a separate, non-claustral VPall derivative; see Figure 4.5E). I presently postulate that the bipartite CL (CLp, CLsp) belongs to the insular pallial sector, which can be held to represent the new LPall; once the CLvm is converted to BEC, the remaining bipartite claustrum proper seems devoid of radial relationships with olfactory cortex, now fully encompassed by the new VPall (see Figures 4.5E and 4.16).

The line of evidence that led us to support a bipartite claustral complex divided into LPall and VPall emerged from the postulate of claustrum field homology with the sauropsidian DVR (Holmgren, 1925; Källén, 1951; Kuhlenbeck, 1973, 1977; Striedter, 1997). The reptilian and avian DVR represents an evolutionarily enlarged ventrolateral pallial territory rather than being evolutionarily reduced, as is typical of the supposed mammalian counterpart (Puelles, 2001; Puelles et al., 2000). The DVR clearly shows a complete bipartition into distinct VPall and LPall components (*Emx1*-negative versus *Emx1*-positive parts; Puelles et al., 2000; see Figures 4.2A and 4.4); these two structural components were initially misidentified as "neostriatum" and "ventral hyperstriatum" (see Figure 4.4A; review in Striedter, 1997). Recent collective reformulation of avian pallial nomenclature changed these names to "nidopallium" and "mesopallium" (Reiner et al., 2004), or VPall and LPall (Puelles et al., 2007), respectively (see Figure 4.4B). In principle, the mentioned field homology would not be tenable if the corresponding territory in mammals were not subdivided likewise into two parts (Puelles, 2001, 2011; Puelles et al., 2000, 2007). Insofar as the claustrum was held to be field homologous with the bipartite sauropsidian DVR, and corresponding *Emx1*-positive and *Emx1*-negative subdomains were also present in the mouse, Puelles et al. (1999, 2000) proposed the division of the claustrum into dorsolateral and ventromedial parts, which supposedly derived from LPall and VPall, respectively (CLdl, CLvm; see Figure 4.5D). The presence of the derivatives of the ventropallial migratory stream seemed supportive of this solution. As mentioned, our present interpretation of the CL bipartition is in agreement with Narkiewicz and Mamos, 1990, and in this new schema both visible claustral parts plus the migrated EPd belong to LPall (see Figures 4.5E and 4.16). The *Emx1*-negative VPall derivatives (such as BEC and EPv) are not thought to be "claustral" according to this new thinking; they might be regarded as "pallial extended amygdala," since they extend caudalwards into that nuclear complex (Figures 4.8–4.10; Medina et al., 2004), whereas the lateropallial claustrum does not. This obviously changes the homology conclusions as well, since in the new scenario only the claustrum can be homologous to the *Emx1*-positive part of the DVR (treated in the last section).

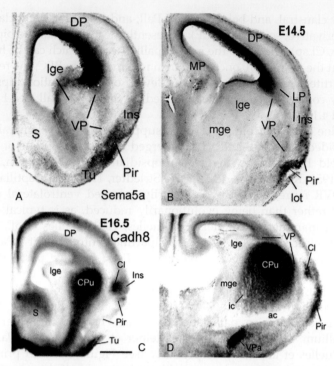

FIGURE 4.8 Examples of molecular characterization of differential ventral pallium (VPall) versus lateral pallium (LPall) derivatives, extracted from the work of Medina et al. (2004). A,B: illustrate *Sema5a* expression at rostral and middle section levels at E14.5 in the VPall migratory stream (VP) and its derivatives aggregating at this stage superficially at the olfactory (piriform) cortex (Pir); the claustrum is fully negative (compare C,D); superficial labeled cells seen at the negative insula (Ins) and beyond in the subpial neocortex are probably the ventropallial Cajal-Retzius cells that migrate subpially into the cortex (see Bielle et al., 2005). Note also expression of the marker in a large part of the pallial ventricular zone, including the VPall area extending medial to the ventricular lateral angle, but not in the ganglionic eminences (lge, mge). C,D: show *Cadh8* expression in the lateral pallium at rostral and middle section levels at E16.5. Signal is distributed through the dorsal pallium (DP), becoming weaker at the insula (Ins), but the lateropallial claustral primordium (Cl) expresses it strongly, in a selective pattern. The postulated lateropallial migratory stream (marked LP in B) is not labeled. Interestingly, the cortical subplate is not marked, nor any potential endopiriform nucleus neurons (compare *Nr4a2* pattern in Figures 4.12 and 4.13). The striatal mantle is also independently strongly positive, and we see a clearcut negative gap separating it from the claustrum, that must contain the LCS or VPall migratory stream (VP; compare A,B). D: there appears also a subpial positive population, estimated to correspond to Hirata cells migrating towards the olfactory (piriform) cortex in VPall (Pir; compare Figure 4.16). ac, anterior commissure; ic internal capsule; Tu, olfactory tuberculum (subpial striatum).

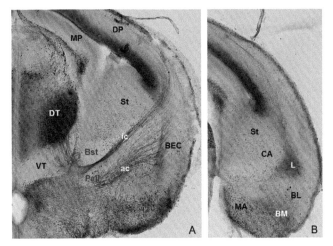

FIGURE 4.9   Middle and caudal coronal sections through an embryonic mouse brain at E18.5, hybridized for *Sema5a*, a marker of ventral pallium (VPall) derivatives (though the marker is present in the VPall, LPall and DPall ventricular zone, only the VPall mantle zone elements continue expressing this gene). Preparations counterstained with anti-calretinin immunoreaction (*brown colored* cells and fibers). A: shows the weakly labeled (*blue*) VPall migratory stream extending into the bed nucleus of the external capsule (BEC), interstitial to the external capsule, part of whose fibers come from the anterior commissure (ac). Calretinin+ thalamo-cortical fibers cross the reticular nucleus (unnamed within VT) and then pass through the pallidal (Pall, Bst) and striatal (St) subpallium, penetrating via the internal capsule into the cortical white matter. The olfactory cortex is weakly labeled. B: is a more caudal section through the subpallial (CA, MA) and pallial (L, BL, BM) amygdala. A very similar and wider VPall migratory stream extends from the ventricular zone into the lateral amygdalar nucleus (L), and some label appears also at the basomedial amygdalar nucleus (BM). This illustrates that there is a rostrocaudal continuity of the VPall ventricular and mantle zones between anterior levels, where the claustrum is present (unlabeled here), and the amygdala at the back.

THE APPARENT LACK OF A CLAUSTRUM IN MONOTREMES

The development and evolution of the claustrum also needs to be discussed with regard to a point that has usually been presented as an isolated issue; namely the statement by some authors that there is no claustrum in monotremes (platypus and echidna; e.g. Abbie, 1940; Butler et al., 2002; Divac et al., 1987). This conclusion poses theoretic difficulties that have rarely been fully addressed (but see Ashwell et al., 2004; Butler et al., 2002). Because a claustrum as such was not classically described in non-mammals (though hypotheses pointing to the sauropsidian DVR as a homolog were advanced by Holmgren, 1925, Källén, 1951, and Kuhlenbeck, 1927, 1973, 1977), the suggested absence of a claustrum in monotremes did not appear to be an evolutionary exception to the authors

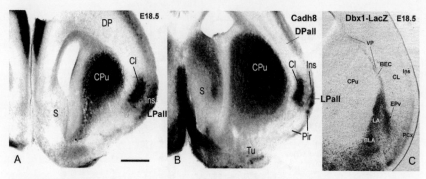

FIGURE 4.10   Comparison of lateral and ventral pallium (LPall and VPall) molecular labeling in E18.5 mouse embryonic brain. A,B: show rostral and middle section levels reacted for *Cadh8* (extracted from Medina et al., 2004). The pattern is similar to that in Figure 4.8 at E16.5, but upper insular layers show increased signal (Ins). The claustrum (Cl) continues specifically labeled, as is the striatal mantle (CPu). Note absence of a labeled LPall migratory stream. C: Example of *Dbx1-LacZ* mouse brain at E18.5, illustrating by blue beta-galactosidase reaction product the presence of neuronal progeny of the VPall ventricular progenitors that selectively expressed *Dbx1* (the postmitotic neurons do not express it, but carry the transgenic marker). The VPall migratory stream is still visible (tagged VP), though very thin at this stage, coming out from the VPall matrix area at the ventricular angle. Cell density increases at the reservoir, marked here as BEC, since its cells are held to be already differentiating locally as the "bed nucleus of the external capsule." Farther along the stream, the ventral endopiriform nucleus population (EPv) is more abundant than that at BEC. The EPv lies deep to the ventral half of the similarly labeled olfactory cortex. The number of labeled olfactory cells decreases towards the negative insula (Ins). There is no label at all at the claustrum (compare with B). VPall progeny is also abundant in the pallial amygdala (LA, lateral amygdalar nucleus; BLA, basolateral amygdalar nucleus). DP, dorsal pallium; Pir, olfactory (piriform) cortex; Tu, olfactory tuberculum (subpial striatum).

supporting this position. They simply stated the fact that they could not see a claustrum where it was expected (inside white matter deep to the insula) in adult echidna or platypus. Possibly this was interpreted as a sign of divergent evolution of extant early mammals (prototheria) from the line that evolved into modern mammals. Notably, the claustrum is uniformly present in metatheria (marsupials), and eutheria (modern mammals).

There are three evolutionary possibilities to be considered in this respect. First, although it is true that with conventional staining methods a CL is not readily observable at the expected position in adult monotremes, perhaps it is hidden by particular histogenetic circumstances (e.g. inclusion of separately originated claustral cells within the insular cortex, or lack of separation from it, if they primarily arise together). In that case the cryptic CL may still be detectable in adults if appropriate molecular and experimental labeling methods are used, as achieved in the afrotherian hedgehog tenrec (Künzle and Radtke-Schuller, 2000, 2001). In that case the evolutionary problem would disappear. The second possibility is that the primordial reptilian claustrum homolog was selectively lost in the monotreme lineage, but was maintained during parallel evolution of the ancestors of marsupial mammals (and subsequently of modern mammals). This case would still validate the field homology proposed by Holmgren (1925) and others, holding that a claustral domain of the pallium is present in both anamniotes and amniotes. However, this case would require explanation of the mechanism that caused selective loss of the CL in monotremes. Finally, the third possibility is that the absence of a claustrum in monotremes is a consequence of the lack of a CL homolog in non-mammals (implying crucial defects in the existing homology hypotheses). This case would require an explanation of how a claustrum first evolved, apparently in ancestral marsupials. The three arguments depend strongly on a solid demonstration that the claustrum is really lacking in monotremes.

In her detailed analysis of the *platypus* brain, Hines (1929) marked twice with a question mark what she thought might be the claustrum in horizontal sections (in Figure 34 on her plate 55, and Figure 62 on plate 68), although no mention of this structure was made in the text. The tentative claustrum appears largely immersed within the deep insular cortex, implying lack or very scarce development of the extreme capsule that should separate these structures. Interestingly, Ziehen (1908) had previously illustrated similar horizontal brain sections of adult echidna, and something that looks very much like Hines' (1929) claustral structure appears under the insular cortex at levels just external to the bifurcation of the anterior commissure; this potential claustrum is partly detached from the insula by a subtle extreme capsule (unlabeled in Ziehen, 1908; his Figures 23, 24). Kuhlenbeck (1977) identified the claustrum in echidna at a similar position, with "x"s placed upon a section copied from Addens and Kurotsu (1936) and on a further section taken from Hines (1929) (Kuhlenbeck's Figures 226B and C, respectively). I tend to agree with this fragmentary classic line of evidence that there is indeed a claustrum in the monotremes, though it apparently is hardly distinguishable from the insular cortex with cytoarchitectonic or myeloarchitectonic methods (see some Golgi data in Schuster, 1910).

Ashwell et al. (2004) studied the putative claustrum and endopiriform nuclei in echidna and platypus adult specimens, considering in particular the respective populations of (inter)neurons expressing calcium-binding proteins (calbindin, calretinin, or parvalbumin). They identified sets of potential claustral and endopiriform neurons lying interstitially in the white matter underneath the insular and piriform cortexes, respectively. In agreement with results in rodents from Druga et al. (1993), Real et al. (2003) and Kowiański et al. (2004), they found parvalbumin cells selectively in what they identified as claustrum, and calretinin cells in the endopiriform nucleus. An oddity nevertheless present in the Ashwell et al. (2004) report is that the postulated endopiriform nucleus inconsistently separates the claustrum from the insular cortex (e.g., see their Figures 1 and 4). These Ashwell et al. (2004) data, jointly with the classic ones cited above, probably can be also interpreted in the sense that all the neurons found inside the white matter may represent a layer 6b component (identified as "claustrum" by Ashwell et al., 2004) and the endopiriform nucleus, whereas the principal claustrum lies hidden inside the insular cortex and was not identified by these authors. The semi-hidden claustrum they also illustrated for platypus (their Figure 10a) agrees with that interpretation. A further note of caution when searching for the claustrum in these primitive mammals is that we also have to distinguish any candidates from the derivatives of the VPall, not so far considered in the scarce specific literature (though the presence of olfactory bulb and associated piriform cortex proves it exists). Evolutionary closeness to reptiles, seen in various morphogenetic features, would suggest that the VPall can be large in monotremes. Any hypotheses need to be tested, preferably using selective molecular markers for the *intrinsic lateropallial cell populations* of layer 6b, CL and EP, as well as of VPall derivatives, rather than attending solely to extrinsic interneuronal phenotypes.

A partially cryptic situation of the CL, largely hidden within the insular cortex, was reported in bats (Humphrey, 1936–some of her figures were also copied by Kuhlenbeck, 1977, see his Figure 227A–D; see also the bat brain atlas of Schneider, 1966, and his Figures 6–10 and 47–49). So-called "basal" insectivores display a cytoarchitectonically distinct claustrum, but this lacks separation from the insula by an extreme capsule (Narkiewicz and Mamos, 1990). The claustrum of the tree shrew appears partly fused with the insular cortex (Tigges and Shantha, 1969). A more or less hidden claustrum was identified using chemoarchitectonic and hodologic criteria in the afrotherian tenrec (*Echinops telfairi*), whose pallial structures display a very primitive form (minimal neocortex area). This investigation revealed its cryptic position within the rhinal/insular cortex (Künzle and Radtke-Schuller, 2000, 2001).

Similarly to Narkiewicz and Mamos (1990), who had included a tenrec in their list of basal insectivores, Künzle and Radtke-Schuller, (2000, 2001) pointed out that the principal claustrum in the tenrec is separated from the external capsule by the cortical layer 6b (this means our CLp + CLsp parts; see Figure 4.5E; compare also Figure 4.16). I argue that this subplate element, the "lamina profunda claustri" of Narkiewicz and Mamos (1990), is probably constant in all mammals, irrespective of the fact that large increments in cortical and white matter may disguise it (see Figure 4.5E; see also the clearcut layer 6b deep claustral component in the hedgehog; Sanides, 1969; Dinopoulos et al., 1992).

All these comparative considerations suggest that the mammalian claustrum emerges evolutionarily and developmentally from the insular cortex, with which it stays partly connected (through lack of an extreme capsule) in various primitive forms, including basal insectivores, bats, afrotherians, marsupials and monotremes. We are thus, I believe, in the first evolutionary scenario sketched above, so that the existence of a claustrum homolog in non-mammalian forms can be expected, though the homology rationale needs to be adapted to the strong evidence that the pallial origin of the claustral population is insular (see model in Figure 4.5E, rather than that in Figure 4.5C,D).

THE ADVENT OF MOLECULAR MARKERS

Specific study of the developing claustrum with molecular markers was first reported by Puelles et al. (1999, 2000), from a perspective of the molecular characterization of the telencephalon in mice and chicken embryos. These studies aimed to distinguish shared patterns of pallial and subpallial molecular regionalization in avian and mammalian models. Various gene markers were mapped by in-situ hybridization on serial sections, and the morphological interpretation of the observed patterns was thought to be consistent with the pioneering views of Holmgren (1925). Initially, the classic tripartite sectorial division of the pallium was considered–medial or hippocampal cortex, dorsal or general cortex, and lateral claustro-olfactory cortex (see Figure 4.5C; reviewed by Striedter, 1997). It was found independently by Smith-Fernandez et al. (1998), in a similar comparative molecular mapping study, that there exists a particular pallial domain that lacks expression of the transcription factor *Emx1* (the latter is expressed in the rest of the pallium; moreover, all the pallium is also marked by *Pax6* signal in the ventricular zone and *Tbr1* expression in the mantle zone; Puelles et al., 2000). Smith-Fernandez et al. (1998) thought that the *Emx1*-negative sector found next to the striatum was transient and contributed via migrations to the formation of the amygdala; no specific name was proposed for it. Puelles et al. (1999, 2000) corroborated these observations, but concluded that the new subdivision–named ventral pallium (VPall)–was permanent, although it is subsequently compressed between its neighbors (which reduces its visibility). The VPall extended throughout the length of the olfactory pallium, and seemed to be a useful concept (jointly with that of the adjacent pallial sector, the lateral pallium (LPall)), for understanding the comparative development of the claustroamygdaloid complex and the olfactory cortex (see Figures 4.2A, 4.4, and 4.5D). The insula was thought to belong to the neighboring dorsal pallium (DPall), following the line of thought of earlier authors that rejected the idea of a claustral connection with the insula. These studies also highlighted the sharp molecular transition between VPall and striatum, called the pallio-subpallial boundary, which passes lateral to the striatum and medial to the lateral olfactory tract at the marginal stratum of VPall (Figure 4.1A,B). This boundary was traced to septal and amygdalo-preoptic ending regions (see Figure 4.2B).

Whereas the area where the claustral primordium becomes visible cytoarchitectonically is quite narrow in the mouse (see Figure 4.6), being compressed between the insular cortex, laterally, and the striatum, medially, the corresponding chicken territory is voluminous, and shows a distinct subdivision into two parallel radial domains, as already mentioned above. Analysis of pallial selective markers, such as the transcription factors *Pax6* and *Tbr1*, confirmed that these two avian territories are indeed pallial, as had been previously conjectured on the basis of comparative developmental considerations (Holmgren, 1925; Kuhlenbeck, 1973; review in Striedter, 1997). Lack of *Emx1* expression was restricted in the avian and reptilian DVR to the classic neostriatum (renamed nidopallium by Reiner et al., 2004, or VPall by Puelles et al., 2000), whereas the classic ventral hyperstriatum (renamed mesopallium, or LPall, by the same sets of authors) did express *Emx1*. This highlighted a fundamental genoarchitectonic heterogeneity that was previously unsuspected within a pallial region that had been classically lumped under the conventional idea of "lateral pallium" (classic tripartite pallial model). Due to the assumption that the claustral complex is related radially to the olfactory cortex (see Figure 4.5C), both domains were described as containing, superficially, portions of the olfactory cortex; curiously, there are *Emx1*-positive subpopulations in both regions, which is also the case in mammals (Puelles, 2001; Gorski et al., 2002). I explain this in more detail below. This analysis of the apparent partition of the classic sauropsidian "lateral pallium" (see Figure 4.5C) into two molecularly distinct sectors called LPall and VPall (Figures 4.4 and 4.5D) was consistent with comparable mouse molecular mapping data (Puelles et al., 1999, 2000), and was soon generally corroborated in several tetrapods and anamniotes. As a result, the "claustral" nuclear primordia were distributed among these two adjacent pallial sectors. The thicker *dorsolateral claustrum* and the *dorsal endopiriform nucleus*, which express *Emx1* (Puelles et al., 2000; Gorski et al., 2002; Medina et al., 2004), were held to derive from the LPall (this was the dorsal or insular claustrum of earlier literature, including both its pars principalis and lamina profunda; see Narkiewicz and Mamos, 1990). A thinner *ventromedial claustrum* and the *ventral endopiriform nucleus* were *Emx1*-negative and were held to derive from the VPall (Puelles et al., 1999, 2000); this latter part is now redefined as non-claustral BEC and associated EPv (see Figure 4.16).

This analysis was subsequently expanded by Medina et al. (2004), who applied additional differential markers of VPall and LPall to the developing mouse claustroamygdaloid complex as a whole (*Emx1* and *Cadh8* for the LPall, according to Korematsu and Redies, 1997 and Obst-Pernberg et al., 2001, and *Dbx1*, *Ngn1* and *Sema5a* for the VPall). Interestingly, *Dbx1* selectively labels the ventricular zone of the VPall, but not its derivatives in the mantle, thus providing strong evidence of a primary molecular difference already at the matrix territory (Bielle et al., 2005; Hirata et al., 2002; Yun et al., 2001). Various other differential markers of this proliferative pallial zone were reported subsequently (Assimacopoulos et al., 2003; Kim et al., 2001; Puelles, 2011). VPall derivatives were negative for *Emx1* and *Cad8*, and expressed *Ngn1* and *Sema5a* (see Figures 4.8A,B, 4.9, and 4.10). This included the ventromedial claustrum (or BEC) and associated EPv, plus lateral, basomedial, amygdalo-hippocampal and anterior/posteromedial cortical parts of the amygdala and correlative parts of the olfactory cortex (see Figures 4.8A,B and 4.9). In contrast, LPall derivatives were limited to the dorsolateral (principal) claustrum, a posterior part of the endopiriform nucleus, and the basolateral and posterolateral cortical parts of the amygdala, with dorsal parts of the olfactory cortex.

Subsequently, the VPall derivatives have been further illuminated by analysis of *Dbx1-LacZ* or *Dbx1-Cre* transgenic mouse lines (Bielle et al., 2005; Hirata et al., 2009 Teissier et al., 2010; Puelles et al., unpublished observations). Ventropallial progeny did not invade the claustrum proper, or the insula, and was found to be restricted rostrally–in the claustrum area–to olfactory cortex, EPv and the population interstitial to the external capsule that I now call BEC (this seems to relate to the early "reservoir"; see Figure 4.16). Caudally, the pallial amygdala was largely composed of VPall progeny (see Figure 4.10; Hirata et al., 2009). Bielle et al. (2005) discovered that, early on, the VPall sector also produces an early population of Cajal-Retzius neurons that migrate tangentially and subpially into the neocortex (invading layer 1; see review in Puelles, 2011; *green arrow* in Figure 4.16). More recently, Teissier et al. (2010) identified a substantial additional stream of mouse VPall cells that migrate tangentially at E14 via a subventricular route into the neocortex. These cells, which I call *Pierani cells*, referring to Alessandra Pierani, the principal investigator behind their discovery (see Figure 4.16), uniformly invade all cortical regions and layers, forming a dispersed population of glutamatergic neurons that amounts to 4 percent of the total cortical population. Curiously, most of them die postnatally, but they nevertheless exert at early stages (during their migration) an important mitogenic effect on the cortical proliferative compartments (ventricular and subventricular zones; see Teissier et al., 2012).

The molecular analysis corroborating the distinctness as a pallial sector of the VPall has been influential in recent years, leading to widespread usage of the corresponding tetrapartite pallial model (e.g. Lindsay et al., 2005; Remedios et al., 2004; Stenman et al., 2003; Teissier et al., 2010, 2012; Tole et al., 2005), and also to its extrapolation to gnathostomes in general (Nieuwenhuys, 2009). The presence of a tetrapartite pallium in agnatha is still a controversial issue, but not because of a lack of VPall or LPall; it has been questioned whether these animals have a DPall (Pombal et al., 2009; Puelles, 2001).

During the last 10 years, numerous other gene markers have been mapped in the pallium, and the CLp, CLsp, EPd, EPv and BEC nuclei are comprised variously in the patterns observed, as are the insular and olfactory cortexes. I cannot insert a review of all the data here, although they do merit attention. The model postulated for this region by Puelles et al. (1999, 2000) has not been subjected to specific public critical analysis (other than corroborating the hypothesis that the molecularly defined VPall exists in other vertebrates, including man; Lindsay et al., 2005), though a number of authors so far eschew adopting it as a paradigm. In any case, no alternative viable model has been proposed. However, irrespective of its surprisingly wide success as a useful paradigm in the field, this model was based only on a few molecular markers, and rested on some assumptions that now appear in a different light. The model clearly rejects the possibility that the lateropallial part of the claustrum may be developmentally related to the insula, a conclusion already strongly suggested by previous literature in basal mammals, cited above, owing to the cryptic claustral structures found inside the insular cortex in various primitive species. In this regard, it was simply assumed by both proponents and followers of the Puelles et al. (1999, 2000) pallial model that the insula does not participate in claustral development. It was further assumed that the olfactory cortex, conceived in a subpial position relative to the bipartite claustral complex, was likewise divided into ventral and dorsal parts that relate specifically to VPall and LPall (Figures 4.2A, 4.4, and 4.5D).

It is known that there are subtle differences between the dorsoventral parts of the olfactory cortex (the most obvious one being that only the ventral part carries the olfactory tract), irrespective of the fact that they may both turn out to originate within the VPall, at least in large part, as *Dbx1*-labeled progeny studies and other markers suggest (see Figures 4.8, 4.9, and 4.10; see also Hirata et al., 2002, who used a different marker). Apparently, *Dbx1*-lineage tracing does not label all the neurons in the olfactory cortex (it probably labels predominantly deeper cells; Hirata et al., 2002). Another apparent origin of olfactory cortex neurons is the overlying cortical pallium (possibly largely the DPall; Tomioka et al., 2000), where at early stages a cell population marked selectively with the lot1 antibody is produced that subsequently migrates tangentially ventralwards along the marginal layer, finally aggregating superficially within the VPall-derived olfactory cortex. In the conclusive schema of Figure 4.16, I entered these lot1 cells as *Hirata cells* (migration represented by a *red arrow*). These superficial lot1+ cells were shown to act as signposts for the oriented growth of the lateral olfactory tract (Hirata et al., 2002; Sato et al., 1998, Tomioka et al., 2000). It is because they aggregate particularly at the ventral part of the olfactory cortex that the lateral olfactory tract grows selectively through that part. This result possibly explains why some olfactory cortex neurons express *Emx1*, held to be a marker for LPall, DPall and MPall, but not of VPall (Gorski et al., 2002; Puelles et al., 2000).

There persists a need to arrive at a global theory of claustral development that is consistent with the diversity of anatomic, neurogenetic, migratory and molecular data and conceptual viewpoints that have accrued so far. Most notable is the bewildering discovery of a handful of tangential pallio-pallial migratory mechanisms—for example, several types of Cajal-Retzius cells (Puelles, 2011), plus Hirata cells and Pierani cells—that has come after the earlier surprise on the subpallio-pallial migrations of inhibitory interneurons (Anderson et al., 1997; Legaz et al., 2005; Marín and Rubenstein, 2003; Xu et al., 2004). The pallio-pallial migrations transfer cell populations from a specific progenitor area to one or more different functional environments, where they permanently or transiently produce certain effects. We thus have reexamined claustral development with a specific interest in investigating a possible relationship with the insula, and consequent independence from the olfactory cortex, questioning as well whether any pallio-pallial migrations arise from the claustral progenitor domain (see suggestion in Puelles, 2011; Puelles et al., submitted).

RECENT GENOARCHITECTONIC ANALYSIS OF CLAUSTRUM DEVELOPMENT AND ITS RADIAL TOPOLOGY SUPPORTS A RELATIONSHIP WITH THE INSULA

Medina et al. (2004) explicitly postulated in the mouse a radial *latero-pallial migratory stream* coursing into CLp that should be parallel and lateral to the *ventropallial migratory stream*; both of them would roughly compose the lateral cortical stream of Bayer and Altman (1991b; see Figure 4.7), irrespective of the differential definition of the cell sources and spatial topology (compare Figure 4.16). However, it has been an unnerving fact that such an LPall stream has not been observed in either standard Nissl-stained material or any gene expression pattern reported so far. I include here a brief description of mouse telencephalic cytoarchitecture at middle telencephalic level across three stages (see Figure 4.6A,B,D). A thick ventropallial migratory stream clearly courses lateral to the pallio-subpallial boundary at E15.5 (unlabeled in Figure 4.6A; compare Figure 4.8A,B); its thickness appears reduced at E16.5, when the "reservoir" becomes evident at the confluence of the anterior commissure and the external capsule (see Figure 4.6B). Figure 4.6C illustrates the dense packet of radial glial processes that accompanies the cell stream. Finally, at E18.5, the ventropallial stream is much reduced and mainly shows a remnant of the reservoir (see Figure 4.6D). Across these stages, the prepiriform cortex seen at this middle telencephalic level becomes densely populated, and the isocortical plate becomes increasingly layered. A similar E18.5 section from a *Dbx1-LacZ* mice specimen, in which VPall derivatives are labeled with a blue dye, clearly shows the much reduced ventropallial stream of blue cells limiting the striatum, the aggregation of cells at the reservoir and neighboring lateral amygdala, and the presence of blue cells in the underlying prepiriform cortex–but none in the claustrum (see Figure 4.10C; Puelles et al., in preparation; material kindly provided by A. Pierani; see also Bielle et al., 2005; Hirata et al., 2009). The earlier stages of the VPall stream also stand out in E14.5 sections reacted in situ for *Sema5a* (see Figures 4.8 and 4.9; Medina et al., 2004), in agreement with the data of Hirata et al. (2002).

Returning to the Nissl material, the primordium of the claustrum (LPall) starts to be distinguishable from the insular cortical plate at E15.5, and does not seem to be connected to the VPall stream (see Figure 4.6A). At E16.5, the CL appears better separated from both the VPall stream and the insula by the incipient external and extreme capsules; interestingly, it appears continuous with the primordium of the isocortical layer 6b (see Figure 4.6B). Finally, at E18.5, both external and extreme capsules are clearly visible, and the CL appears distinctly continuous with the isocortical layer 6b (see Figure 4.6D). At earlier stages than E15.5 the claustral primordium cannot be identified cytoarchitectonically (i.e. cannot be separated from the insular cortical primordium, though the population must be hidden inside it, due to its precociousness). The observed pattern lacks the expected radially migrating stream of LPall cells, which hypothetically should extend from the ventricular zone strictly above the *Dbx1*-positive VPall area into the CL primordium detectable at E15.5 and the corresponding part of the olfactory cortex (Medina et al., 2004). Note that our rationale for postulating the LPall migratory stream in that investigation was essentially similar to that suggested by Bayer and Altman (1991b; see Figure 4.7).

Recently we have examined in detail the developing mouse claustrum between E12.5 and postnatal stages (we knew that its neurons are born very early; see autoradiographic data reviewed above; I also advanced above a discussion of the alternative viewpoint of Bayer and Altman, 1991b). We used a rather specific (but not exclusive) molecular marker, *Nr4a2* (previously known as *Nurr1*). This orphan nuclear receptor has been widely studied due to its important role in the formation of the mesodiencephalic dopaminergic cells of the substantia nigra, but is known to be strongly expressed as well in other less investigated brain sites, including the claustrum and endopiriform nucleus, as well as the isocortical layer 6b and some other neuronal subpopulations in specific cortical areas (Arimatsu et al., 2003; Perlmann and Wallén-Mackenzie, 2004). Our first effort aimed to determine whether *Nr4a2* is a satisfactory developmental marker of the CL/EP, as suggested by its very sharp adult expression pattern. We thus studied *Nr4a2* expression at stages E12.5, E13.5, E14.5, E16.5, E18.5, P4 and P37 in sagittal, horizontal and coronal section planes.

At E12.5, there was no pallial *Nr4a2* signal in two out of five specimens studied. The three more advanced embryos showed, superficially, a sparse labeled population in what we believe is the LPall (just overlying the well-mapped primordium of the olfactory cortex; Sato et al., 1998; Tomioka et al., 2000); this incipient aggregate was best developed at caudal telencephalic levels, but was also present more rostrally (Figure 4.11A–C).

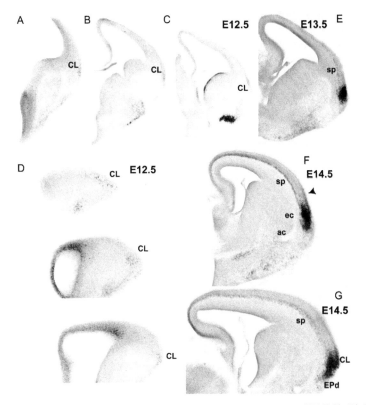

FIGURE 4.11   Pallial expression of *Nr4a2* in mouse embryos at stages E12.5 (A–D), E13.5 (E) and E14.5 (F,G). The preparations are 100 μm-thick vibratome slices processed floating. A–C: A rostrocaudal series of coronal sections shows earliest expression at the claustral primordium (CL), ventrally to the suprainsular cortical plate, and dorsally to the olfactory cortex. The positive cells are slightly more numerous caudally. There is separate expression in the subpallium. D: Three lateromedial sagittal sections illustrate the distribution of early labeled cells in a longitudinal band (CL). E: One day later the superficial CL patch has increased significantly, and labeled cells connected to its deep aspect have invaded the DPall at subplate (sp) level; these cells have advanced significantly towards the cortical convexity, where the incipient cortical plate is much thinner. F: shows a coronal section at E14.5; some *Nr4a2*-negative cells start to arrive superficially, covering the ventral part of the labeled CL; the subplate layer is fully developed, and there appears a marginal migration out of the upper aspect of the CL that invades the overlying neocortex (*arrowhead*). G: shows a sagittal section at the same stage, passing through the rostral part of the CL primordium (lateral and slightly caudal to the anterior olfactory area). At this level a sizeable group of labeled cells sorts ventrally out of the CL and advances with a leading edge into the underlying VPall domain, deep to olfactory cortex, starting to form the dorsal endopiriform nucleus (EPd). ac, anterior commissure; ec, external capsule.

In sagittal sections the labeled cells formed a longitudinal band that did not reach the caudal pole of the telencephalon (Figure 4.11D). The *Nr4a2*-positive cells were clearly limited to the ventral rim of the incipient cortical plate, just above the olfactory cortex primordium, which separates them from the pallio-subpallial boundary. We believe this is the first recognizable primordium of the claustrum (and possibly some related subplate cells), fully three days before the nucleus is visible cytoarchitectonically (see Figure 4.6A). It was a bit of a surprise to see the claustrum so early at the brain surface, where one expects the cortical plate, but on reflection, this should follow from what we know about its precocious birthdates (discussed above). Since claustral neurons jointly with subplate neurons are among the first born in this area (Valverde et al., 1989, 1995), they must first aggregate subpially, and should subsequently acquire a deeper position as younger neurons are added superficially to the cortical plate. Remarkably, the *Nr4a2* marker is not expressed at all within the underlying ventricular and subventricular zones; it seems to be activated only after the cells become postmitotic, and occupy the immature mantle layer.

One day later, at E13.5, a thicker *Nr4a2*-positive patch was observed at the same superficial locus, occupying quite distinctly the whole thickness of the ventral rim of the cortical plate (Figure 4.11E); the cortical plate cells placed dorsal to the claustral primordium (DPall on top of LPall in our model; see Figure 4.5E) generally do not express *Nr4a2* at all (see exceptions below). Moreover, the claustral band lengthened in a rostral direction. Rostrally, the claustral primordium was observed to stop roughly over the lateral part of the anterior olfactory nucleus. Simultaneously, additional *Nr4a2*-positive cells extended at E13.5 dorsally from the deep aspect of the claustral primordium, penetrating strictly under the incipient cortical plate, superficial to the subventricular and intermediate zones. These cells practically reached the hemispheric convexity in the most advanced embryos (see Figure 4.11E); here they lay almost marginally, due to the scarcity of the local cortical plate cell population. The less advanced E13.5 embryos showed fewer dorsally displaced labeled subplate cells, which gives the impression that they migrate out of the claustral primordium. This possible DPall-invading formation clearly corresponds in position to the subplate, or future layer 6b, as was corroborated at later stages. We have not determined whether all layer 6b neurons are *Nr4a2*-positive, but this is the impression obtained as development continues.

At the next stage examined, E14.5, the labeled claustral patch of superficial cells was found again, and was larger, though now some unlabeled cells were found superficial to the positive neurons, particularly most ventrally, at the primordium of the agranular insula (see Figure 4.11F,G). These unlabeled superficial cells probably represent newly immigrated non-claustral cells starting to form the insular part of the cortical plate (their radial migration presumably passes through the CL). Some labeled cells of the claustral patch start to disperse in a ventral direction into the subjacent VPall domain, deep to the prepiriform cortex. This process, which will give rise to the dorsal EP nucleus, is more advanced at the rostral part of the claustrum, just above and behind the anterior olfactory area (EPd; see Figure 4.11G). The subplate population was again distinctly labeled, contrasting with no labeling at all in the subventricular and intermediate zones (see Figure 4.11F,G). The subplate extends medially at this stage into the interhemispheric side, caudally into the entorhinal area, and deep to the claustrum itself, from which it appears separated by a thin clear space (see Figure 4.11); this ventralmost part of the subplate contrasts with the unlabeled reservoir. Finally, E14.5 embryos also showed a marginal sheet of *Nr4a2*-labeled cells that spread dorsally from the claustro/insular primordium into the overlying (parietal) part of the cortical plate (*arrowhead*; see Figure 4.11F). Caudally in the hemisphere, a separate pattern of *Nr4a2* expression starts at the hippocampal-parahippocampal region (Arimatsu et al., 2003). This is held to be unrelated to the claustrum, and will not be covered further here.

At E16.5 the scenario only showed minor changes. The *Nr4a2*-positive claustral patch was larger than two days before, both in height and thickness, but this growth seems proportional to that of the whole pallium and may simply be due to maturation, as suggested as well by the increased cyto- and genoarchitectonic separation from the insula (see Figures 4.6B, 4.8C,D, and 4.12A). The incipient *Nr4a2*-negative insular cortex found superficial to the claustrum is much thicker rostrally and showed a sprinkling of *Nr4a2*-positive cells (not shown). The earlier separation between the deep side of the claustrum and the underlying subplate is made indistinct by a partial confluence of both sets of cells. The labeled cells dispersed ventrally out of the claustrum into the VPall appeared well developed rostrally and started to be seen at section levels through the anterior commissure. In general, these invaded cells were restricted to the prospective dorsal endopiriform area, eschewing the reservoir and the olfactory cortex (see Figure 4.12A). The marginal tangential migration, which progressed in a dorsal direction into the immature parietal cortex, was thicker and about twice as long as at E14.5 (*arrowhead*; see Figure 4.12A).

At this stage we checked the course of radial glial fibers passing in coronal sections across the *Nr4a2*-labeled claustrum (immunoreacted with the RC2 antibody from the Developmental Hybridoma Bank, Iowa, on top of the *Nr4a2* ISH reaction). These glial processes were formerly assumed to cross the CLp dorsoventrally and end in the dorsal part of the olfactory cortex (Medina et al., 2004; Puelles et al., 1999, 2000), although this had never been observed specifically. We then tested this possibility against a potential mediolateral intraclaustral course with a

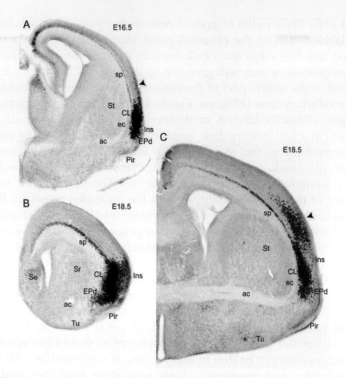

FIGURE 4.12   Pallial expression of *Nr4a2* in coronal sections of mouse embryos at stages E16.5 (A) and E18.5 (B,C). A: At middle section levels (ac, anterior commissure), both the negative insula and the positive dorsal endopiriform nucleus (EPd) are larger than before. The *arrowhead* indicates that the marginal migratory stream has increased in length and population. The subplate is unchanged. B: At a rostral level the E18.5 CL seems massive and many labeled cells disperse superficially within the covering insular cortex, particularly dorsally. The EPd migration is completed, since the medialmost part of the olfactory (piriform) cortex (Pir) is never invaded (occupied by the negative EPv). C: At the level of the anterior commissure the EPd formation is now better developed than at E16.5 (compare A), and the marginal migration has stopped within the suprajacent parietal cortex and is being covered by unlabeld younger cortical plate cells that have reached the surface. ec, external capsule; Se, septum; St, striatal subpallium;Tu, olfactory tuberculum (subpial striatum).

pial end at the insular surface. After various trials, which gave slightly positive but inconclusive results (owing to the apparent obliquity of coronal sections relative to the relevant fibers), we sectioned some specimens obliquely, at an angle of about 30–45 degrees relative to the brain midline. These preparations beautifully showed that radial glial fibers coursing dorsolateral to the ventropallial dense glial packet (Figure 4.13A,B) traverse the *Nr4a2*-labeled claustrum mediolaterally into the insular cortical primordium (see Figure 4.13C,D). None of them proceeded into olfactory cortex, which was served by the fasciculated radial glial fibers passing through the VPall migratory stream.

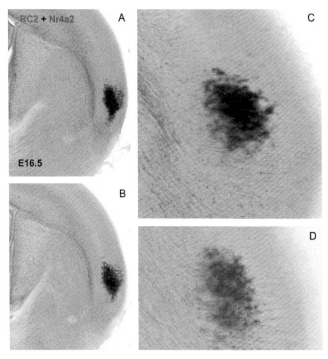

FIGURE 4.13   Combination of selective molecular identification of the claustral primordium using *Nr4a2* in situ hybridization on floating sections at E16.5 in the mouse with secondary RC2 immunoreaction of the radial glial cells in the same sections. This aimed to test the hypothesis that the claustrum relates to the insula against the hypothesis that it relates to the olfactory cortex. In the first case, radial glial fibers are expected to pass from left to right across the claustrum (from the depth into the insular pia); in the second case they should pass from the top–the ventral pallium (VPall) matrix at the ventricular angle–towards the olfactory cortex at the bottom (compare Figure 4.10C). These sections were oriented in a plane oblique by some 45 degrees relative to the brain midline, found empirically to label the glial fibers in the insular-claustrum area most efficiently. A,B: are two low-magnification images, to show the general layout; the dense glial bundle of the VPall territory (see also Figure 4.6C) lies at some distance from the claustrum (blue patch); note the ventral spike of migrating EPd blue cells starting to invade the neighboring VPall territory deep to the olfactory cortex (the completely unlabeled superficial area is the lateral olfactory tract). C,D: higher-magnification detail images of the claustrum and overlying insula to the right (not the same sections; these are a bit more caudal, where the orientation was optimal). It can be clearly seen that radial glial fibers approaching the claustrum from the depth (to the left and above) course through it into the insula. No fiber in any of the sections in several brains was seen to cross the claustrum ventrally towards the olfactory cortex. Similar data were obtained in E14.5 and E18.5 specimens as well.

At E18.5, the claustrum appeared already fully separated from the brain surface by a population of unlabeled insular cells that was nearly as thick (see Figure 4.12B,C). From Figure 4.12B, it can be seen that the rostral part of the insula displays a heavy sprinkling of *Nr4a2*-positive cells, whereas such cells are restricted to deep insular levels more caudally (see Figure 4.12C). The labeled subplate cell population was more massive, and was still continuous ventrally with the deep aspect of the claustrum (see Figure 4.12B,C). The labeled cells that dispersed from the claustrum into the underlying EPd area were now also more numerous caudally, practically filling the area next to the dorsal half of the prepiriform cortex (which remained unlabeled itself, like the subjacent ventral half, the EPv area, and the reservoir; see Figure 4.12C). This invasive pattern was more marked rostrally, above and behind the anterior olfactory area (see Figure 4.12B). Finally, the still immature parietal cortical primordium displayed a labeled cell stratum at intermediate radial level, which was covered by a more superficial stratum of unlabeled neurons (the latter probably are the prospective layers 2–4). These *Nr4a2*-labeled elements inside the parietal cortical plate were slightly spread-out into the neighboring strata, as well as dorsally (*arrowhead*; see Figure 4.12C). They represent the derivatives of the earlier marginal stream of migration described as advancing in a dorsal direction out of the claustrum at E14.5 and E16.5 (arrowheads in Figures 4.11F and 4.12A). The fact that these elements adopt a deeper topography at E18.5 suggests that the tangential migration stopped shortly after E16.5, allowing younger supragranular cells to incorporate radially into the parietal superficial cortical plate, covering the tangentially migrated ones.

At P4 the upper layers of the insular cortex were in place. The *Nr4a2*-positive claustrum attained its definitive deep position under the insula, remaining just superficial to the subplate layer of labeled cells. We detected a tendency of the mature claustral cells to separate into distinct dorsal and ventral clumps of cells, although these remained interconnected by a cloud of sparser elements (not shown). The infragranular insular layers were sprinkled with abundant *Nr4a2*-labeled cells; a few such cells also appeared in the supragranular insular layers, particularly dorsally, at the primordium of the granular insular sector. The labeled EPd population was very distinct, though it still remained in contact with the claustrum proper; these cells occupied the dorsal endopiriform region specifically, and only invaded sparsely the overlying layers 3 and 4 of the prepiriform cortex. They continued eschewing the EPv area under the reservoir completely, as well as the reservoir itself (less distinct at this stage, transformed into the BEC; compare with Figure 4.10C). A *Nr4a2*-positive infragranular population was still present in the parietal cortex, with only sparse cells found at supragranular levels (not shown).

At P37 the scenario changed, mainly by showing a more mature-looking insular cortex (layers 2/3 becoming quite distinct) and a dorso-ventrally bipartite (dorsal and ventral core parts) CLp formation, partly separated from the CLsp and the migrated EPd population (CLsp, CLpv, CLpd, EPd; Figures 4.14A–C and 4.15B). The marginally migrated sub-granular *Nr4a2*-positive population within the parietal cortex appeared

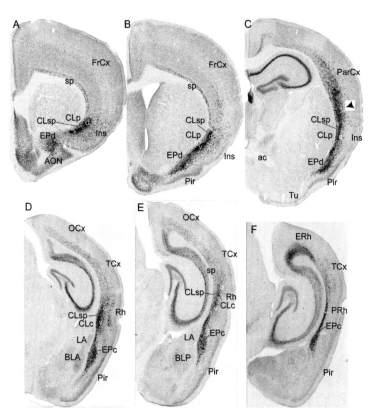

FIGURE 4.14 Coronal sections showing expression of pallial *Nr4a2* in the adult P37 mouse (A–F). A–C: correspond to sections through the principal claustrum (CLp), which now appears partly separated into dorsal and ventral condensations or core portions (*white v/d* tags), surrounded by other more dispersed cells. The claustral subplate component is distinctly continuous with layer 6b (sp; CLsp). The dorsal endopiriform nucleus (EPd) is fully developed and ends rostrally above the lateral anterior olfactory nucleus (AON in A). More caudally the marginally migrated population (*arrowhead*) can be seen, which now occupies an infragranular position in layers 6a and 5; the cell number decreases dorsally. D–F: correspond to sections through the smaller caudal claustrum (CLc), found next to the rhinal cortex (Rh; note a layer 5 labeled intrarhinal population in D, probably of claustral origin as well); the CLc appears connected ventrally with the massive caudal endopiriform nucleus (EPc), which lies deep to the olfactory (piriform) cortex (Pir) all the way to its caudal end. No labeled cells enter the pallial amygdala. The anterior and posterior basolateral amygdalar nuclei (BLA, BLP) show a slightly stronger background signal than the rest of the pallial and subpallial amygdala. There also seems to exist a sparse invasion of neurons into the infragranular layers of the temporal cortex (TCx). The CLc ends as the entorhinal cortex level is approached (ERh) and the EPc adopts a deeper position and also diminishes; there is a sharp boundary at the amygdalar capsule (D–F). FCx, frontal cortex; Ins, insula; OCx, occipital cortex; ParCx, parietal cortex; PRh, perirhinal cortex.

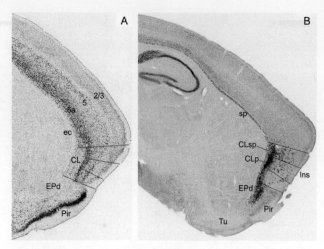

FIGURE 4.15   Sagittal sections showing either expression of *Rprm* at P14 (A; image downloaded from the public Allen Developing Mouse Brain Atlas; annotations mine) or expression of *Nr4a2* in the adult P37 mouse (B). A: the *Rprm* marker (reprimo, TP53-dependent G2 arrest mediator candidate) clearly labels the neocortical layers 6a, 5 and 2/3 predominantly. As these extend into the insula, they can be followed individually. CL is negative, except for some background signal; layer 6a band becomes thinner and denser superficial to the CL (similar for layers 5 and 2/3); layer 6b, or subplate, is not labeled at all, either in the neocortex or under the claustrum. The dorsal endopiriform nucleus (EPd) shows even less signal than the CL (no background either), whereas the prepiriform layer 2 is strongly positive. B: the principal claustrum (CLp) is accompanied by the labeled EPd (not fully separated from CLp) and the claustral subplate population, continuous with layer 6b (sp, CLsp). The overlying three sectors of the insula show a dorsoventrally decreasing content of dispersed *Nr4a2* cells at infra- and supragranular layers. The ventral agranular insula is nearly devoid of such cells (compare with A). ec, external capsule; Ins, insula; sp, subplate; Tu, olfactory tuberculum (subpial striatum).

substantial in the adult, and can be compared with similar cells mapped by Arimatsu and collaborators with *Nr4a2*, as well as with latexin (*arrowhead*; see Figure 4.14C; Arimatsu and Ishida, 2002; Arimatsu et al., 1999a,c 1994, 2003, 2009). Layer 6b, the subplate, remained *Nr4a2*-positive (see Figures 4.14 and 4.15B). We checked that the insular cortex covering the claustrum contained separate representation of layers 6a and 5 (which are thus not represented by the claustrum itself). Data extracted from the Allen Developing Mouse Brain Atlas (www.developingmouse.brain-map. org) about the expression of *Rprm*, a selective gene marker for isocortical layers 6a, 5 and 2/3, which leaves the claustrum essentially unlabeled, revealed that the insula indeed has an extension of layers 6a and 5 covering the CLp (see Figure 4.15A). The insular layer 6a actually has denser *Rprm* labeled cells than other regions of the isocortex, probably because the population is somewhat compressed by the mass of the claustrum; an insular continuation of layer 5 is likewise distinct. Our material was not favorable for detecting the expected extreme capsule layer, which, supposedly, is barely present in rodents.

At the caudal transition of the insular cortex into rhinal cortex the *Nr4a2*-positive claustrum becomes progressively reduced in size, but still can be followed nearly to the border of the entorhinal cortex. This portion can possibly be distinguished as the *caudal claustrum* (CLc; see Figure 4.14D,E). At anterior rhinal levels there appears a distinct layer of labeled cells in the subgranular part of the rhinal cortex, apparently within layer 5 (see Figure 4.14D). This disappears more caudally. Accompanying the CLc, there is a much larger ventral *Nr4a2*-positive population that occupies an endopiriform position, though it partly remains fused with the CLc via a positive cell bridge. This I propose to call the *caudal endopiriform nucleus* (EPc; Figure 4.14D–F). The EPc can be followed in a caudal direction under the whole piriform cortex, and alongside the lateral nucleus of the amygdala. It reaches the amygdalo-piriform transition area with fewer cells (see Figure 4.14D), and it continues to diminish in size, adopting a periventricular position as it approaches the entorhinal area caudally (see Figure 4.14E,F). The EPc is always distinctly separated from all nuclear or corticoid amygdaloid formations. This caudal claustro-endopiriform formation apparently represents the derivative of the caudal tail of the initial *Nr4a2*-positive band identified at early stages. The EPc, similarly to the EPd rostrally, contains labeled cells that have migrated *tangentially* from the overlying claustrum (LPall) into the VPall territory (Figure 4.16). This secondary translocation is strongly supported by the absence of these EP cells at early stages. They become visible, first rostrally and later caudally, at E14.5 (very rostrally, they are barely visible), and during the period E16.5–E18.5–P4. Note that Bayer and Altman (1991b; see Figure 4.7) interpreted the EP formations as a radially migrated VPall phenomenon, wholly distinct from the claustrum, and they did not observe any longitudinal gradient in their birthdates. The generative caudorostral gradient they described for the claustrum is not consistent with our observations on a rostrocaudal differentiation pattern with the *Nr4a2* marker. There was the same contrast when looking at the insula.

DISCUSSION OF THESE RESULTS

These new developmental data strongly support the conclusion that the development of the claustrum complex (at least of its majoritary *Nr4a2*-expressing elements) is topographically and causally associated to the insular pallium and its caudal extension into rhinal cortex. Interestingly, the possibility of an insular origin of the claustrum was recently suggested again on the basis of immunocytochemical mapping of Gng2 and Ntr2G2 markers in adult man by Pirone et al. (2012).

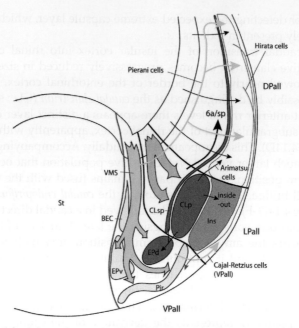

FIGURE 4.16 Schematic presentation of present conclusions about claustral-insular development and composition, including the redefined ventral, lateral, and dorsal pallial domains (VPall, LPall, DPall). VPall is defined by the olfactory cortex and the VPall migration stream (*green*), which contributes locally to the bed nucleus of the external capsule (BEC), ventral endopiriform nucleus (EPv) and olfactory (piriform; Pir) populations (*light gray shade*), and possibly also to some other cells dispersed among the large migrated dorsal endopiriform nucleus (EPd) nucleus. According to Bielle et al. (2005) there are ventropallial Cajal-Retzius cells that depart from the VPall domain subpially (*small green arrow*). Data from Teissier et al. (2010) revealed that some deep VPall cells move tangentially via a subventricular route into the neocortex, invading the cortical plate in a dispersed way (*deep green arrows* marked as "Pierani cells", referring to the principal investigator of this important investigation). On the other hand, the new LPall is defined by the histogenetic conjunction of the claustrum (CLsp, CLp), forming first a superficial stratum, but finally a deep nucleus, and the subsequently inside-out developed, six-layered structure of the insula (*orange* postmitotic elements coming out of the LPall progenitors; derivatives in darker gray shades). Various cell cohorts labeled similarly as the CL by the *Nr4a2* marker exit the original primordium and invade radially or tangentially other pallial domains (*black arrows*). The subplate cells (layer 6b) seem to expand dorsally from the CLsp shortly after E12.5. The EPd separates ventrally from the CLp as of E14.5 and invades the subjacent subpiriform VPall in a rostrocaudal gradient, using the space not occupied by the EPv (caudally, the EPc does the same, separating partially from the CLc; not shown in the schema; see Figure 4.14). Other cells disperse superficially (radially) into the insular cortex, mainly in its granular and disgranular portions. Finally, a further *Nr4a2*-positive population migrates marginally into the neighboring parietal cortex (between E14.5 and E16.5), and ends up in the infragranular layers, where I suggest they be called Arimatsu cells (see citations in text). Cell populations in the VPall are further complemented via pallio-pallial marginal tangential migration by the DPall-originated Hirata signpost cells, first recognized by their immunoreaction with the lot1 antibody (Sato et al., 1998; Tomioka et al., 2000). Obviously, once the insula-claustrum complex is ascribed to LPall, the DPall definition is modified by loss of the insula, which formerly was thought to belong to it.

Previously in the literature, only the careful work undertaken on the tenrec claustrum by Künzle and Radtke-Schuller (2000, 2001) had mentioned that the claustrum extends beyond the insula proper into a caudal area overlain by rhinal cortex.

Exactly as conceived initially by Meynert (1868, 1872), we have now shown that the claustrum detaches from the lower stratum of the insular pallial domain whenever the extreme capsule develops; otherwise it remains partially embedded within the insula. These considerations suggest that the extreme capsule is an optional intracortical fiber-rich layer, which may become traversed as well by longitudinal and transverse cortico-cortical axons, as is the general case of layers 6a and 6b, otherwise known as layers VI and VII (Vandevelde et al., 1996; Thomson et al., 2010). This conclusion therefore agrees beautifully with the comparative data commented on above regarding the indistinct external limit of the claustrum in monotremes, some insectivores, and bats. A capsular delimitation is no longer necessary for confident identification of the claustrum, since the claustral population turns out to be a molecularly distinct unit of the claustro-insular LPall domain, with or without capsular separation, and this is true from E12.5 onwards in the mouse.

In a way, such capsular detachment is never more than apparent, insofar as layer 6b neurons always remain underneath the claustrum. This is clearly visible in insectivores (Dinopoulos et al., 1992; Narkiewikz and Mamos, 1990; Sanides, 1969) as well as in the afrotherian tenrec (Künzle and Radtke-Schuller, 2001, 2002; Künzle et al. 2002). In basic agreement with these authors' conceptions, I more precisely interpret their cell group H in tenrec as the claustrum proper (CLp), and the underlying cell group K as the claustral layer 6b component (CLsp). Note also that in the Künzle et al. (2002) report, retrogradely labeled cells projecting to the parabrachial nucleus mapped within layer 5 of the somatosensory cortex, provide a reminder of the distribution of the marginally migrated *Nr4a2*-positive cells in subgranular mouse parietal cortex. This population–identified in Figure 4.16 as *Arimatsu cells*–was extensively described by Arimatsu and colleagues after studying latexin and Nurr1-positive cells, as was cited above. As far as I know, these authors apparently never suspected that this peculiar population might arise in the neighboring claustrum.

New Concept of LPall and VPall

The formation of the claustrum and related *Nr4a2*-positive pallio-pallial migrated populations is a notable feature that is lacking in the neocortex in general. This suggests that the insula and underlying claustrum jointly constitute a special part of the pallium. This point is underlined in the presently postulated conception of this unit as representing the true LPall sector (Figure 4.16; Puelles et al., submitted), which is now considered to be more different from the VPall than it was before. The previous idea of the VPall/LPall domains (Puelles, 2001, 2011; Puelles et al., 1991, 2000) has clearly been invalidated by our crucial observation about the radial glial processes that traverse the claustrum and the insula jointly, independently from the olfactory cortex and the VPall in general (if we except the EP cells invading the latter secondarily). This point about the radial glial framework associated with the claustrum and the insula was not known before.

Notably, the insula builds a six-layered cortical structure on top of the claustrum, with subtly different granular, disgranular, and agranular sectors disposed from dorsal to ventral, respectively (see Figure 4.15A,B). Layer 4 cells are reportedly mainly present in the granular and disgranular portions, where dispersed *Nr4a2*-positive cells are also most abundant (see Figure 4.15B). The claustrum, though apparently immersed within layer 6a, is *not* a part of it, or at least is not identical with it, as judged, for instance, by its lack of the *Rprm* molecular marker characterizing among other elements the layer 6a neurons (see Figure 4.15A), or its selective expression of *Nr4a2* (Arimatsu et al., 2003; present data), *Gng2* (Mathur et al., 2009; Pirone et al., 2012) and *Cbln2* (Miura et al., 2006). To this may be added the argument of Ramon y Cajal (1902) that claustral neurons do not participate in insular columnar circuitry or radial dendritic texture, a point which has been corroborated by later observations. The fact that the insular progenitor area first develops a deep claustral nuclear entity and then adds the six-layered cortex, whereas the rest of the neocortex apparently only does the second step, can be attributed to its possible differential molecular identity as a distinct pallial sector, namely the LPall. In a different pallial sector the histogenetic rules may be differentially regulated. It would be more difficult to understand this drastic insular histogenetic peculiarity–having a deep nucleus between 6a and 6b–as an areal difference within the DPall. A further alternative possibility can be conceived; namely that the claustrum with its radial glial framework primarily forms a cortex-less nuclear domain, and the components of the six insular layers later migrate tangentially from the DPall into the LPall, thus invading secondarily the purely claustral sector. Such a process might explain the gradual change in cellularity across granular, disgranular and agranular insular sectors. However, the first *Nr4a2*-negative insular cells seen at E14.5 were observed at the ventral end of the prospective insula, which seems to weigh against that hypothesis. Moreover, the reported ventrodorsal gradient in insular neurogenesis relative to the overlying neocortex (Bayer and Altman, 1991a) would be consistent instead with a LPall origin of its cells. Nevertheless, it seems worthwhile to examine this important point experimentally.

As announced above, the present data motivated us to revise the definitions of VPall and LPall presented by Puelles et al. (2000). It is clear now that the claustrum proper develops jointly with the insula within a pallial sector that is both molecularly distinct and heterochronic in terms of histogenesis. It can no longer be said that the claustrum develops *out* of the insula, irrespective of what the Nissl- or myelin-stained preparations suggest, as we now understand that its formation strictly precedes that of the insula. The insula develops on top of the claustrum. In a comparative viewpoint, we can envisage a vertebrate that only has a claustrum homolog at its LPall, and no insula on top of it, or one that has insula-homologous cells *deep* to it (because of differences in stratification; outside-in, instead of inside-out).

Moreover, at least four independent pallio-pallial migrations selectively originate from the claustral primordium (the subplate, EPd and EPc ones, and the Arimatsu cells entering the parietal cortex–Figure 4.16). The labeled layer 5 rhinal cortex neurons might represent a fifth case, and we ignore lesser phenomena here and there, such as the cells sprinkled within the insula proper. None of these phenomena are known for areas of DPall, apart from the very different Hirata cells (see Figure 4.16).

The CL primordium does not seem to participate at all in the formation of the dorsal part of the olfactory cortex, as had been surmised previously (Hirata et al., 2002). It now appears that the entire olfactory cortex develops within the VPall and receives its own early-born cells from the reservoir and the ventropallial migratory stream, as was suggested by Medina et al. (2004) and demonstrated experimentally by the *Dbx1-LacZ* progeny data (see Figure 4.10C; Bielle et al., 2005; Teissier et al., 2010). The small ventropallial cell population that was formerly included in the "ventromedial claustrum" (Puelles et al., 1999, 2000) corresponds most probably to the final form of the reservoir (plus EPv), as indicated by its persistent intimate interstitial relationship with both the external capsule and the transition into it of many anterior commissure fibers. I suggest that these dispersed cell groups should no longer be assimilated to the claustrum concept, since their molecular profile and mode of development is typically ventropallial (Figure 4.10C). As mentioned on p.136n, they should be named henceforth the bed nucleus of the external capsule (BEC), and the ventral endopiriform nucleus (EPv), for the benefit of anatomic and semantic clarity.

There is, in contrast, a more substantial endopiriform cell population that originates jointly with the claustrum within the LPall sector, and later invades tangentially the neighboring VPall area deep to the olfactory cortex, forming the dorsal endopiriform nucleus (EPd) (Figure 4.16). Similar cells extend ventrally from the caudal claustrum lying underneath the rhinal cortex, generating the caudal endopiriform nucleus (EPc) associated with piriform cortex, and neighboring the amygdala, which itself remains independent from *Nr4a2*-positive formations. The LPall-derived EPd and EPc occupy accordingly a postmigratory VPall-related position in the light of present analysis. Ventromedial to the dorsal EP there appears the ventral EP described in the literature, which I believe is a VPall derivative intercalated between the reservoir and the olfactory cortex (compare Medina et al., 2004, and see Figure 4.10C).

A more detailed analysis is needed to explore the caudal frontier of the LPall as defined by the present set of molecular markers (for instance, does the caudal end reach the postrhinal area?).

It is remarkable that there is no expression of $Nr4a2$ in the amygdaloid region. Notably, the basolateral and posterolateral cortical amygdaloid nuclei were previously held to belong to the LPall, due to the expression of $Emx1$ and lack of VPall markers (Medina et al., 2004), but it does not show a significant $Nr4a2$ signal (though there is higher background reaction in it than elsewhere in the amygdala; see Figure 4.14D,E). This raises the need to reexamine the molecular nature and progenitor ascription of the pallial sector that contains the pallial amygdala, or at least of the part formerly held to belong to LPall.

In the present revised pallial model, I ascribe the whole olfactory cortex and no part of the claustrum proper to the VPall (irrespective that, the EPd and EPc populations invade the VPall from the LPall). The LPall contains the insula together with the principal and subplate-related claustrum, as well as the perirhinal cortex (and possibly postrhinal cortex; see Burwell et al., 1995) together with the minor CLc component accompanying it (partly fused with EPc). Analogous claustral formations are not identifiable under the so-called medial limbic cortex that appears beyond the postrhinal area (e.g. retrosplenial, cingulate, and infralimbic areas). This indicates that the LPall only exists as such (claustro-cortical complex) at the lateral aspect of the pallium (a further advantage of the proposed name). It may well be the case that the new LPall does not participate in the pallial amygdala, though this conclusion needs more detailed analysis.

The pallial sector developing the insular transitional six-layered cortex (LPall) is clearly distinct from the DPall one (where the standard isocortical structure develops), since only the former develops the singular *nuclear* claustrum structure (as first noticed by Ramon y Cajal, 1902). Note that a cortex formation with underlying nuclear structure was defined as *hypopallium* by Elliot Smith (1919). Both VPall and LPall are hypopallial in this sense, whereas the DPall and MPall are not, at least primarily (we can conceive the claustrum-derived subplate as a tangentially migrated deep nucleus, rather than as a layer). It is probably parsimonius (in the sense that simpler hypotheses are preferable to complicated ones) to explain all these now somewhat clearer patterns as evidence of distinct MPall, DPall, LPall and VPall sectors of the pallium, with assorted pallio-pallial tangential migrations across their mutual boundaries thrown in for complexity.

About the Mode of Development of the Claustrum

In retrospect, the whole specific embryologic literature had wrongly discerned the earliest appearance of the claustral primordium, only identifying it when it became apparent in its final deep position, under the incipient extreme capsule. Clearly, that stage must now be seen as a late event in claustral development, and it is not at all related to the migration of postmitotic young claustral neurons out of the ventricular zone into the primordium (as believed by Bayer and Altman, 1991b; Filimonoff, 1966; Holmgren, 1925; Källén, 1951; Kuhlenbeck, 1924, 1927, 1973; Landau, 1923a,b, 1936; de Vries, 1910; and other authors including myself; e.g. Figure 4.7). There is no question either of a "migration" of an early superficial claustrum into a deeper position. The known cell birthdates corroborate that this outside-in apparent translocation of the claustrum must be passive, due to younger insular neurons migrating across the claustrum into their definitive superficial position, exactly as happens with younger neurons relative to the deep cortical neurons elsewhere in the neocortex. We found that there are radial glial processes that can guide prospective insular neurons migrating radially toward the superficial insula primordium across the preexistent claustrum.

There is a short period in early corticogenesis in which the claustrum lies subpially, occupying the future position of the insula and the rhinal cortex. This novel conclusion, which was implicit in the published birthdating results, but apparently not realized before by anyone, has now been demonstrated thanks to the selectiveness of the *Nr4a2* labeling pattern. This analysis explains that no claustral stream of migrating postmitotic neurons was ever observed to lead into the deep claustrum primordium, once this was histologically visible (the cells probably cease to stream into the claustrum 2 days earlier in the mouse, according to autoradiographic birthdate data). The part of the lateral cortical stream of Bayer and Altman (1991b) that they thought contributed to the claustrum systematically bypasses the claustral primordium when it is *lying at the brain surface*, or *hidden*, undistinctive in standard material, within the incipient insula (see Figure 4.7).

Prospective claustral cells express the marker *Nr4a2* immediately upon exiting the early ventricular/subventricular zone and enter the incipient mantle layer at E12.5. Strictly speaking, at this moment, the insular cortex does not yet exist, as is also true of the future DPall neocortex, if we leave out the prospective subplate and Cajal-Retzius neurons, equally precocious as the claustrum, but surprisingly both produced apparently outside of the neocortical ventricular zone (Puelles, 2011; Valverde et al., 1989, 1995; present results). By stage E14.5, when the *Nr4a2*-negative insular and rhinal cortical cells start to accumulate superficially to the claustrum, the total claustral cell population is probably already present in the thick superficial patch expressing *Nr4a2* that was formed up to that stage. The observed precocity of claustral cells is fully consistent with the autoradiographic birthdating data of Smart and Smart (1977, 1982), Bisconte and Marty (1975), Valverde and Santacana (1994). As commented on above, the studies of Bayer and Altman (1991a,b) apparently missed this point.

The relatively later claustral origins described by these authors are not consistent with the present molecular and temporal data (see comments above about the EP and CL gradients). It is difficult to understand why these authors failed to follow earlier labeled cell cohorts sequentially, a circumstance that led them to miss the superficial initial position of the CL primordium.

A Possible New Concept of the Subplate (Layer 6b or VII)

The similarity in birthdate profile of the claustral and cortical layer 6b cells, which was noticed in the earlier neurogenetic studies, remarkably coincides with the notion that Nr4a2-positive subplate cells seem to arise jointly with claustral cells within the LPall. Subsequently, they apparently migrate tangentially from there into the DPall, selectively entering at a tangent into the subplate stratum; the latter apparently is invaded completely from lateral to medial between E13.5 and E14.5. However, our descriptive observations strictly do not allow us, to discard the possibility that the observed pattern might result instead from temporally and spatially graded upregulation of Nr4a2 in preexistent subplate cells intrinsic to the DPall cortical primordium. The conjectured subplate migration suggested by the available data will require experimental corroboration.

The subplate migration hypothesis is interesting in that it subtly provides simpler assumptions (thereby adding parsimony) to the well-established conception of mammalian cortical histogenesis of Marin-Padilla (1971, 1978, 2011), according to which an early-born cortical cell population first forms a "primordial plexiform layer," which is subsequently cleaved into layer 1 and layer 6b (subplate) by radial migratory invasion of the younger cells of the cortical plate. It was already demonstrated in the meantime that precocious layer 1 neurons–Cajal-Retzius cells–are all produced at the periphery of the telencephalic pallium and migrate tangentially into the cortical marginal stratum (review in Puelles, 2011). Present data now suggest that the deep precocious cells of the subplate may arise at the lateral border of the cortex, namely at the claustro-insular LPall domain, from where they would migrate tangentially to occupy the neocortical subplate in DPall. Obviously, being pallial in origin, the subplate neurons are uniformly excitatory. Our Nr4a2 data suggest that the subplate starts to be formed under the cortical plate at E13.5, that is, shortly *after* the cortical plate starts to be formed by locally generated elements. This can be explained within our hypothesis by the time needed for their tangential migration from LPall into DPall; these data are not so easily explainable if the subplate population is intrinsic to the cortical primordium. Marin-Padilla's primordial marginal and subplate cells are thus indeed the oldest neurons of the mature cortex, but both types are probably not autochthonous and may come from elsewhere, thus arriving with some delay at their final sites. The logic of autoradiographic studies of cortical layer development, having been long centered in radial movements, never contemplated these tangential movement possibilities. In principle, the presently revised model of cortical histogenesis (see Figure 4.16), mixing autochthonous with non-autochthonous cell cohorts, simplifies the genetic regulation of neurogenesis occurring at the DPall ventricular zone, since we postulate that only *cortical plate cells* need to be produced and patterned locally, whereas the subplate cells are generated in the LPall, and the Cajal-Retzius layer 1 cells come from various other sources. The emergence of the subplate is placed in a different light, and the new perspective may also have an interesting impact on ideas about how the mammalian cortex evolved. This argument expands the previous intuited concept of pallio-pallial tangential migrations, introduced explicitly quite recently (Puelles, 2011), though specific instances of these phenomena were already in front of us before (e.g. Hirata cells, see p. 146 and legend of Figure 4.16).

The Claustral Subplate as the Second Claustral Component

Another related point confirmed by our investigation is that the subplate also extends primarily underneath the claustrum, as observed in E13.5 and E14.5 embryos. This explains, I believe, the bipartite claustrum described by Narkiewickz and Mamos (1990) in insectivores, since their *pars principalis claustri* corresponds to the claustrum proper, whereas the *lamina profunda claustri* represents the claustral subplate, as was duly pointed out by these authors. The same pattern can be discerned in the tenrec claustral structure described by Künzle and Radtke-Schuller (2000, 2001) and Künzle et al. (2002), as was commented on above (see pp. 142 and 159), and in the hedgehog bipartite claustrum with differential connections described by Dinopoulos et al. (1992).

Several gene markers are known that selectively label both layer 6b and some cells deep and extending slightly ventral to the principal dorsal claustrum, while leaving the latter largely unlabeled (e.g. *Cplx3*, *Ctfg*, *Rail14*; see Allen Developing Mouse Brain Atlas, GENSAT, EUROEXPRESS, or similar databases). Other markers, such as *Gng2*, label only cells found inside the claustrum proper, as reported by Mathur et al. (2009). In addition, these authors identified a lamina intercalated between the CL and the external capsule that showed strong AChE activity and weak parvalbumin immunoreaction, in contrast to the CL proper (this must be the subplate component). They also remarked that the claustrum is surrounded by "layer 6" neurons, which is consistent with the 6b plus 6a components identified here around the claustrum proper. A similar comparison can be made with the "core" and "shell" claustral domains underlined by Dávila et al. (2005), Guirado et al. (2005), Legaz et al. (2005), and other chemoarchitectonic studies cited above; the claustral shell clearly corresponds to the subplate component. Guirado et al. (2005) found Projections upon the claustral shell, as opposed to none over the claustral core (CLp); thalamic input is a general feature of the subplate (Thomson et al., 2010).

Interestingly, other markers such as *Ntng2* and *Nnat* selectively label a subset of claustrum cells plus the migrated subgranular parietal neurons (Arimatsu cells , see Figure 4.16), while leaving the subplate unlabeled. These elements presumably represent a third claustral subpopulation that produces collaterally the marginally migrated *Nr4a2*-positive Arimatsu cells. These cells are probably born later than the subplate and deeper claustral cells since they are first seen migrating dorsally in a subpial position at E14.5. This component apparently correlates with described latexin-positive neurons identified in the claustrum, endopiriform nucleus and parietal deep cortical layers, many of which co-express *Nr4a2*, formerly known as *Nurr1* (Arimatsu et al, 1999c, 1994, 2003, 2009). The experimental neurogenetic study of Arimatsu et al. (1994) discovered that all latexin-positive neurons in the subgranular parietal cortex were born at the same day (E15 in the rat). I am postulating here that these cells originate from the claustral primordium within the LPall (at a later stage than the subplate cells), and subsequently invade the overlying parietal cortex via a marginal tangential migration coursing in a dorsal direction. The subpial migration is visible between stages E14.5 and E16.5 (this dating is consistent with these cells being born at E13.5 in the mouse and E15 in the rat). Other latexin-positive neurons present in the subplate, claustrum and endopiriform nucleus have somewhat earlier birthdates (Arimatsu et al., 2003). This indicates that molecularly diverse subsets of LPall claustral neurons may be produced sequentially. Perhaps the subplate elements are the earliest ones, as suggested by data from Valverde et al. (1989, 1995), followed by the *Gng2*-positive principal claustrum elements, and ending with the claustro-parietal latexin-positive cells. The *Nr4a2*-positive EPd population that apparently migrates out from the early claustral patch remains unlabeled with all the selective markers cited above, except latexin (no data available for EPc). Note this also implies a short-range pallio-pallial tangential migration; in this case, LPall elements invade the VPall. These EP cells may be born earlier than the principal claustrum cells, as suggested by the birthdate analysis in the rat of Bayer and Altman (1991b), but for some reason they migrate relatively late, after the parietal-invading population of Arimatsu cells is ahead. The incipient EPd migration was observed at E16.5 (with sparse labeled cells present at E14.5 far rostrally) and probably finished at E17.5, given the intermediate position of the cells at E18.5.

Such differential molecular and temporal profiles reveal that various cell populations contained initially within the claustral primordium are primarily molecularly heterogeneous, or become so as time passes by.

The strict localization of the marginal migration of Arimatsu cells at parietal cortex level (plus other phenomena, such as the layer 5 cells in rhinal cortex, or the stronger invasion of the insula rostrally) suggests that there is also differential anteroposterior regulation of the production of specific claustral neuronal subtypes. A primary molecular heterogeneity at the level of LPall progenitors as compared to DPall progenitors, possibly correlative with subtle heterochronic birthdate dynamics (Arimatsu et al., 1999b; Takiguchi-Hayashi, 2001), would easily explain the subsequent differential migratory and maturation behavior of these different claustral subpopulations and may also underpin any differential connectivity properties (Arimatsu and Ishida, 2002; Arimatsu et al., 1999a; Bai et al., 2004).

VPall Derivatives are not Claustral

The VPall migratory stream and associated glial bundle (Medina et al., 2004) is strictly eschewed by all claustral and migrated dorsal endopiriform *Nr4a2*-positive cells developing initially within the LPall. We already know that the VPall stream does not practically contribute neurons to either the claustrum or the insula, as demonstrated by *Dbx1*-derived progeny (Figure 4.10C; Bielle et al., 2005; Hirata et al., 2002, 2009; Teissier et al., 2010). There are nevertheless ventropallial *Dbx1*-derived Cajal-Retzius neurons that pass subpially across LPall into the DPall (Figure 4.16; Bielle et al., 2005), some of which probably remain in the layer 1 of the insula. There is probably also a pallio-pallial contribution to the claustro-insular complex by the *Dbx1*-derived Pierani cell subpopulation that migrates tangentially via the subventricular stratum and invades all layers of the neocortex (Teissier et al., 2010; *green arrows* in Figure 4.16; note the latter contingent only sums up to about 4 percent of the whole cortical population, and is transient, since it largely disappears by cell-death postnatally. However, neither Bielle et al. (2005) or Teissier et al. (2010) described a significant invasion of *Dbx1*-derived progeny into the claustrum and insula. In contrast, VPall derivatives that migrate radially via the lateral cortical stream (ventropallial migratory stream) aggregate at the reservoir (BEC), the underlying EPv area, and the prepiriform cortex, eschewing significant participation in the claustro-insular LPall domain (see Figure 4.10C). Similar data were collected before with other markers, such as *Ngn1* and *Sema5a* (Medina et al. 2004; see also Hirata et al., 2002). These results are consistent with the presently reexamined trajectories of ventropallial and lateropallial radial glial fibers (Figures 4.13 and 4 16).

The Amygdala Problem

The present *Nr4a2* data raise a new problem, which affects the conceptual unity of the claustroamygdaloid complex, examined recently by Medina et al. (2004) and Martínez-García et al. (2012), among other sources. Indeed, whereas the *Dbx1*-labeled VPall extends backwards into an amygdaloid caudal field (see Figure 4.9; as already summarily contemplated by Puelles et al., 1999, 2000), the *Nr4a2*-labeled claustro-insular LPall clearly does not extend backwards into the amygdala (see Figure 4.14). In the previous non-insular definition of the LPall, this sector was associated to the dorsal part of the olfactory cortex, and was accordingly thought to follow that structure back into the amygdaloid region. Importantly, the basolateral and the posterolateral cortical amygdalar nuclei were held to be lateropallial in origin, on the basis of their expression of *Emx1* (which is absent in VPall derivatives), and lack of typical VPall markers (Medina et al., 2004). However, we now see that these amygdalar structural elements are not associated to the caudal end of the *Nr4a2*-positive claustro-endopiriform domain, represented exclusively by the caudal EP nucleus (the latter lies outside the amygdala proper, deep to the whole piriform cortex). The insular cortex with the underlying principal claustrum is continuous caudally with the rhinal cortex (eventually, also postrhinal cortex) and the associated caudal claustrum, but not with the amygdaloid formation. The issue of the partial *Emx1* expression pattern in the pallial amygdala will need to be reexamined.

A GLANCE AT THE EMERGING EVOLUTIONARY SCENARIO

Some results presented here suggest novel evolutionary ideas about the reptilian-mammalian transition in claustral position and structure. I assume that the main pallial domains are shared as regards topological position and fundamental molecular profile among all gnathostome vertebrates, even if they have developed very differently histologically. Extrapolation to sauropsidian forms (see Figure 4.4) of the redefined boundaries of the mammalian VPall and LPall domains immediately indicates that the likely field homolog of the lateropallial claustrum/insula complex (leaving aside tangential migrations that may or may not obtain in other vertebrate lineages) might be found strictly in the mesopallium of birds and the corresponding upper part of the reptilian anterior dorsal ventricular ridge. Interestingly, Reiner et al. (2011) recently found that the central nuclear core of the chicken dorsal mesopallium clearly expresses *Cbln2* (cerebellin 2). They compared this pattern with mammalian expression in the neocortex without finding a meaningful correlation, but did not notice that *Cbln2* is also sharply expressed in the mammalian claustrum, particularly at postnatal stages (Miura et al., 2006; see also Allen Developing Mouse Brain Atlas; www.developingmouse. brain-map.org). Such lack of attention to the claustrum has contributed to the lack of progress in reconciling the different comparative scenarios (see below). We have never known much about the avian mesopallium, and understand even less about the obscure reptilian one. It is not a markedly thalamo-recipient pallium region (but see Guirado et al., 2005), and, therefore, it is likely to sustain pallio-pallial connections (as expected in a claustro-insular homolog). The lateropallial mesopallium region of birds has some superficial corticoid structures (Puelles et al., 2007) that might or might not represent a variant of an ancestral insular/rhinal cortex, but we should recall that neuronal stratification in the avian pallium is largely outside-in (review in Puelles et al., 2007), so that the more precocious claustrum homolog should be expected to be superficial. Eventually it should be possible to identify with appropriate markers a "cortex + nucleus" homolog entity in reptiles and birds, irrespective of variant histogenetic and differentiative properties.

Interestingly, although attempts have been made to discern how the LPall, as defined by Puelles et al. (2000), participates in the amygdalar region of sauropsida, as expected in the tetrapartite pallial model (Martínez-García et al., 2007, 2012; Puelles et al., 1999, 2000, 2007; Redies et al., 2001), a convincing demonstration of this point with LPall molecular markers has not been forthcoming (note *Emx1* is not specific for LPall). The avian mesopallium as presently understood apparently evades continuity with the avian amygdalar region (Puelles et al., 2007). The presently redefined LPall notion, which regards the avian mesopallium as the possible field-homolog of a claustro-insular complex that is also unrelated to the amygdala, suggests a reason why this may be so.

On the other hand, if we wholly exclude the huge sauropsidian thalamo-recipient VPall (the nidopallium of birds; Puelles et al., 2007; Reiner et al., 2004; see Figure 4.4) from our claustrum homolog candidates (consistently with the present model, but departing from the Holmgren-based line of thought followed by Puelles et al., 1999, 2000, and other recent sources in the literature), then the nidopallium becomes the hugely enlarged field homolog of the reservoir (the mammalian diminutive BEC), plus parts possibly corresponding to the ventral endopiriform nucleus and the olfactory cortex (see Puelles et al., 2007 on this). In addition, some large and topologically caudal nidopallial region may represent the homolog of some part of the mammalian pallial amygdala, as previously suggested by Bruce and Neary (1995) and other authors, including myself (reviews in Bruce, 2012; Puelles et al., 2007; Striedter, 1997). The avian and reptilian olfactory cortex, which also partially expresses *Emx1*, can probably be likewise ascribed wholly to the VPall subpial stratum (Puelles et al., 2000; review in Puelles et al., 2007). This is a possible sign that Hirata cells (see Figure 4.16) also occur in sauropsida.

The candidate locus for the DPall homolog in birds continues to be, in my opinion, the hyperpallium (a mystifying structure, because it is very differently organized than the mammalian model), possibly with the addition of the corticoid dorsolateral area, as suggested by Redies et al. (2001). This issue continues to be highly controversial, even with the advent of some modern studies with molecular markers (Dugas-Ford et al., 2012; Suzuki et al., 2012). Rather than clarify anything, these recent authors have added much confusion to the field by not bothering to select neocortical layer markers that are not simultaneously expressed in other pallial domains, such as the claustrum, the pallial amygdala, or even the hippocampus. Their rationale in practice was that the sauropsidian pallium only needs to be compared with the mammalian neocortex (they apparently did not expect any amygdala, hippocampus or claustrum). Any pallial spot that expressed what the authors wrongly held to be a specific mammalian cortical marker was automatically believed to be homologous to a mammalian neocortex layer, irrespective of topologic position, developmental pattern, hodology or whatever other comparative criteria are available (an exercise of molecular tabula rasa). Typically, these studies did not illustrate the entire expression pattern of the chosen markers across the whole avian pallium; the authors apparently picked out those labeled areas that agreed with some preconceived notions (theirs, or of others). Under these conditions, these molecular era reports unfortunately do not throw much light upon the evolution and comparative anatomy of the claustrum. More sophisticated efforts are expected to come out shortly, hopefully consistently with present results.

References

Abbie, A.A., 1940. Cortical lamination in the monotremata. J. Comp. Neurol. 72, 429–467.

Addens, J.L., Kurotsu, T., 1936. Die Pyramidenbahn von Echidna. Proc. Kon. Akad. Wetensch. Amsterdam, vol. 39, pp. 1143–1151.

Anderson, S.A., Eisenstat, D.D., Shi, L., Rubenstein, J.L., 1997. Interneuron migration from basal forebrain to neocortex: dependence on Dlx genes. Science 278, 474–476.

Ariens Kappers, C.U., 1921. Die vergleichende anatomie des nervensystems der wirbeltiere und des menschen. Bd.2. Haarlem, Erven Bohn.

Arimatsu, Y., Ishida, M., 2002. Distinct neuronal populations specified to form cortico-cortical and corticothalamic projections from layer VI of developing cerebral cortex. Neuroscience 114, 1033–1045.

Arimatsu, Y., Nihonmatsu, I., Hirata, K., Takiguchi-Hayashi, K., 1994. Cogeneration of neurons with a unique molecular phenotype in layers V and VI of widespread lateral neocortical areas in the rat. J. Neurosci. 14, 2020–2031.

Arimatsu, Y., Ishida, M., Sato, M., Kojima, M., 1999a. Corticocortical associative neurons expressing latexin: specific cortical connectivity formed in vivo and in vitro. Cereb. Cortex 9, 569–576.

Arimatsu, Y., Ishida, M., Takiguchi-Hayashi, K., Uratani, Y., 1999b. Cerebral cortical specification by early potential restriction of progenitor cells and later phenotype control of postmitotic neurons. Development 126, 629–638.

Arimatsu, Y., Kojima, M., Ishida, M., 1999c. Area- and lamina-specific organization of a neuronal subpopulation defined by expression of latexin in the rat cerebral cortex. Neuroscience 88, 93–105.

Arimatsu, Y., Ishida, M., Kaneko, T., Ichinose, S., Omori, A., 2003. Organization and development of corticocortical associative neurons expressing the orphan nuclear receptor. Nurr1. J. Comp. Neurol. 466, 180–196.

Arimatsu, Y., Nihonmatsu, I., Hatanaka, Y., 2009. Localization of latexin-immunoreactive neurons in the adult cat cerebral cortex and claustrum/endopiriform formation. Neuroscience 162, 1398–1410.

Ashwell, K.W.S., Hardman, C., Paxinos, G., 2004. The claustrum is not missing from all monotreme brains. Brain Behav. Evol. 64, 223–241.

Assimacopoulos, S., Grove, E.A., Ragsdale, C.W., 2003. Identification of a Pax6-dependent epidermal growth factor family signaling source at the lateral edge of the embryonic cerebral cortex. J. Neurosci. 23, 6399–6403.

Bai, W.Z., Ishida, M., Arimatsu, Y., 2004. Chemically defined feedback connections from infragranular layers of sensory association cortices in the rat. Neuroscience 123, 257–267.

Bayer, S.A., 1986. Neurogenesis in the rat primary olfactory cortex. Int. J. Dev. Neurosci. 4, 251–271.

Bayer, S.A., Altman, J., 1991a. Neocortical Development. Raven Press, New York.

Bayer, S.A., Altman, J., 1991b. Development of the endopiriform nucleus and the claustrum in the rat brain. Neuroscience 45, 391–412.

Bayer, S.A., Altman, J., 2006. The Human Brain during the Late First Trimester. CRC Press, Boca Raton.

Bayer, S.A., Altman, J., Russo, R.J., Dai, X.F., Simmons, J.A., 1991. Cell migration in the rat embryonic neocortex. J. Comp. Neurol. 307, 499–516.

Bielle, F., Griveau, A., Narboux-Nême, N., Vigneau, S., Sigrist, M., Arber, S., et al., 2005. Multiple origins of Cajal-Retzius cells at the borders of the developing pallium. Nat. Neurosci. 8, 1002–1012.

Bisconte, J.-C., Marty, R., 1975. Analyse chronoarchitectonique du cerveau de rat par radio-autographie. I. Histogenèse du télencéphale. J. Hirnforsch. 16, 55–74.

Braak, H., Braak, E., 1982. Neuronal types in the claustrum of man. Anat. Embryol. 163, 447–460.

Brand, S., 1981. A serial section Golgi analysis of the primate claustrum. Anat. Embryol. 162, 475–488.

Brodmann, K., 1909. Vergleichende lokalisationslehre der grosshirnrinde. Barth, Leipzig.

Bruce, L.L., 2012. The puzzle of forebrain evolution. Brain Behav. Evol. 79, 141–143.

Bruce, L.L., Neary, T.J., 1995. The limbic system of tetrapods: a comparative analysis of cortical and amygdalar populations. Brain Behav. Evol. 46, 224–234.

Burwell, R.D., Witter, M.P., Amaral, D.G., 1995. Perirhinal and postrhinal cortices of the rat: a review of the neuroanatomical literature and comparison with findings from the monkey brain. Hippocampus 5, 390–408.

Butler, A.B., Molnar, Z., Manger, P.R., 2002. Apparent absence of claustrum in monotremes: implications for forebrain evolution in amniotes. Brain Behav. Evol. 60, 230–240.

Dávila, J.C., Real, M.A., Olmos, L., Legaz, I., Medina, L., Guirado, S., 2005. Embryonic and postnatal development of GABA, calbindin, calretinin, and parvalbumin in the mouse claustral complex. J. Comp. Neurol. 481, 42–57.

De Vries, E., 1910. Bemerkungen zur Ontogenie und vergleichenden Anatomie des Claustrums. Folia Neurobiol. 4, 481–513.

Dinopoulos, A., Papadopoulos, G.C., Michaloudi, H., Parnavelas, J.G., Uylings, H.B.M., Karamanlidis, A.N., 1992. Claustrum in the hedgehog (*Erinaceus europeus*) brain: cytoarchitecture and connections with cortical and subcortical structures. J. Comp. Neurol. 316, 187–205.

Divac, I., Holst, M.C., Nelson, J., Mckenzie, J.S., 1987. Afferents of the frontal cortex in the echidna (*Tachyglossus aculeatus*). Brain Behav. Evol. 30, 303–320.

Druga, R., Chen, S., Bentivoglio, M., 1993. Parvalbumin and calbindin in the rat claustrum: an immunocytochemical study combined with retrograde tracing ftom frontoparietal cortex. J. Chem. Neuroanat. 6, 399–406.

Dugas-Ford, J., Rowell, J.J., Ragsdale, C.W., 2012. Cell-type homologies and the origins of the neocortex. Proc. Natl. Acad. Sci. U.S.A. 109, 16974–16979.

Edinger, L., 1908. Vorlesungen über den bau der nervösen zentralorgane der menschen und der tiere band ii. vergleichende anatomie des gehirns, seventh ed. F.C.W.Vogel Verlag, Leipzig.

Elliot, S.G., 1919. A preliminary note on the morphology of the corpus striatum and the origin of the neopallium. J. Anat. 53, 271–291.

Filimonoff, N., 1966. The claustrum, its origin and development. J. Hirnforsch. 8, 503–528.

Flames, N., Pla, R., Gelman, D.M., Rubenstein, J.L., Puelles, L., Marín, O., 2007. Delineation of multiple subpallial progenitor domains by the combinatorial expression of transcriptional codes. J. Neurosci. 27, 9682–9695.

Gadisseux, J.F., Evrard, P., Misson, J.P., Caviness, V.S., 1989. Dynamic structure of the radial glial fiber system of the developing murine cerebral wall. An immunocytochemical analysis. Brain Res. Dev. Brain Res. 50, 55–67.

Gorski, J.A., Talley, T., Qiu, M., Puelles, L., Rubenstein, J.L., Jones, K.R., 2002. Cortical excitatory neurons and glia, but not GABAergic neurons, are produced in the Emx1-expressing lineage. J. Neurosci. 22, 6309–6314.

Guirado, S., Real, M.A., Dávila, J.C., 2005. The ascending tectofugal visual system in amniotes: new insights. Brain Res. Bull. 66, 290–396.

Haller, B., 1908. Die phyletische Entfaltung der Grosshirnrinde. Arch. Mikr. Anat. 71, 350–466.

Haller, B., 1909. Die phyletische Stellung des Grosshirnrinde der Insektivoren. Jena Zschr. Naturwiss. 45 (NF38), 279–298. +plate XXVI.

Herrick, C.J., 1910. The morphology of the forebrain in amphibia and reptilia. J. Comp. Neurol. 35, 413–547.

Hinds, J.W., Angevine, J.B., 1965. Autoradiographic study of histogenesis in the area piriformis and claustrum in the mouse. Anat. Rec. 151, 456–457.

Hines, M., 1929. The brain of Ornithorhynchus anatinus. Phil. Trans. Roy. Soc./London Series B. 201, 155–287.

Hirata, T., Nomura, T., Takagi, Y., Sato, Y., Tomioka, N., Fujisawa, H., et al., 2002. Mosaic development of the olfactory cortex with Pax6-dependent and –independent components. Dev. Brain Res. 136, 17–26.

Hirata, T., Li, P., Lanuza, G.M., Cocas, L.A., Huntsman, M.M., Corbin, J.G., 2009. Identification of distinct telencephalic progenitor pools for neuronal diversity in the amygdala. Nat. Neurosci. 12, 141–149.

His, W., 1904. Die entwicklung des menschlichen gehirns während des ersten monate. S.Hirzel Verlag, Leipzig.

Hochstetter, F., 1919. Beiträge zur entwicklungsgeschicte des menschlichen gehirns I teil. Franz Deuticke, Wien/Leipzig.

Holmgren, N., 1922. Points of view concerning forebrain morphology in lower vertebrates. J. Comp. Neurol. 34, 391–459.

Holmgren, N., 1925. Points of view concerning forebrain morphology in higher vertebrates. Acta Zool. 6, 414–477.

Humphrey, T., 1936. The telencephalon of the bat. J. Comp. Neurol. 65, 603–711.

Johnston, J.B., 1908. The Nervous System of Vertebrates. John Murray, London.

Johnston, J.B., 1916. The development of the dorsal ventricularridge in turtles. J. Comp. Neurol. 26, 481–505.

Johnston, J.B., 1923. Further contributions to the study of the evolution of the forebrain. J. Comp. Neurol. 35, 337–481.

Källén, B., 1951. The nuclear development in the mammalian forebrain with special regard to the subpallium. Kgl. Fysiogr. Sällsk. Lund. Handl. 46, 3–43.

Kim, A.S., Anderson, S.A., Rubenstein, J.L., Lowenstein, D.H., Pleasure, S.J., 2001. Pax-6 regulates expression of SFRP-2 and Wnt-7b in the developing CNS. J. Neurosci. 21, RC132.

Kodama, S., 1927. Über die entwicklung des striären systems beim menschen. neurol. psychiatr. abhandl. aus dem schweizer arch. neurol. u. psychiatr., heft 5. Zürich/Leipzig/Berlin, Orell Füssli Verlag.

Korematsu, K., Redies, C., 1997. Expression of cadherin-8 mRNA in the developing mouse central nervous system. J. Comp. Neurol. 387, 291–306.

Kowiański, P., Moryś, J.M., Wojcik, S., Dziewiatkowski, J., Luczynska, A., Spodnik, E., et al., 2004. Neuropeptide-containing neurons in the endopiriform region of the rat: morphology and colocalization with calcium-binding proteins and nitric oxyde synthase. Brain Res. 996, 97–110.

Kuhlenbeck, H., 1924. Über den Ursprung der Basalganglien des Grosshirns. Anat. Anz. 58, 49–74.

Kuhlenbeck, H., 1927. Vorlesungen über das zentralnervensystem der wirbeltiere. Gustav Fischer Verlag, Jena.

Kuhlenbeck, H., 1973. The Central Nervous System of Vertebrates (vol. 3) Part II: Overall Morphological Pattern. S. Karger, Basel.

Kuhlenbeck, H., 1977. The Central Nervous System of Vertebrates (vol. 5), Part I: Derivatives of the Prosencephalon: Diencephalon and Telencephalon. S. Karger, Basel.

Künzle, H., Radtke-Schuller, S., 2000. Multiarchitectonic characterization of insular, perirhinal and related regions in a basal mammal, Echinops telfairi. Anat. Embryol. (Berlin) 202, 507–522.

Künzle, H., Radtke-Schuller, S., 2001. Cortical connections of the claustrum and subjacent cell groups in the hedgehog tenrec. Anat. Embryol. (Berlin) 203, 403–415.

Künzle, H., Radtke-Schuller, von Stebut, B., 2002. Parabrachio-cortical connections with the lateral hemisphere in the Madagascan hedgehog tenrec: prominent projections to layer 1, weak projections from layer 6. Brain Res. Bull. 57, 705–719.

Landau, E., 1919. The comparative anatomy of the nucleus amygdalae, the claustrum and the insular cortex. J. Anat. physiol. 53, 351–360.

Landau, E., 1923a. Anatomie des grosshirns. Bern.

Landau, E., 1923b. Zur Kenntnis des Beziehungen des Claustrums zum Nucleus amygdalae and zur Area piriformis im speziellen zum Tractus olfactorius. Arch. Neurol. Psych. 8, 391–400.

Landau, E., 1936. Quelques nouvelles considerations sur l'avant-mur. Arch. Anat. Strassburg 23, 165–181.

Legaz, I., García-López, M., Medina, L., 2005. Subpallial origin of part of the calbindin-positive neurons of the claustral complex and piriform cortex. Brain Res. Bull. 66, 470–474.

Lindsay, S., Sarma, S., Martínez-de-la-Torre, M., Kerwin, J., Scott, M., Luis Ferran, J., et al., 2005. Anatomical and gene expression mapping of the ventral pallium in a three-dimensional model of developing human brain. Neuroscience 136, 625–632.

Loo, Y.T., 1931. The forebrain of the opossum Didelphis virginiana. J. Comp. Neurol. 52, 1–148.

MacLean, P.D., 1990. The Triune Brain in Evolution. Plenum Press, New York/London.

Macchi, G., 1951. The ontogenetic development of the olfactory telencephalon in man. J. Comp. Neurol. 95, 245–305.

Mamos, L., Narkiewicz, O., Morys, J., 1986. Neurons of the claustrum in the cat: a Golgi study. Acta Neurobiol. 46, 171–178.

Manocha, S.L., Shantha, T.R., Bourne, G.H., 1967. Histochemical mapping of the distribution of monoamine oxidase in the diencephalon and basal telencephalic centers of the brain of the squirrel monkey (Saimiri sciureus). Brain Res. 6, 570–586.

Marín, O., Rubenstein, J.L., 2003. Cell migration in the forebrain. Annu. Rev. Neurosci. 26, 441–483.

Marin-Padilla, M., 1971. Early prenatal ontogenesis of the cerebral cortex (neocortex) of the cat (Felis domestica). A Golgi study. Part I. The primordial neocortical organization. Z. Anat. Entwickl-gesch. 134, 117–145.

Marin-Padilla, M., 1978. Dual origin of the mammalian neocortex and evolution of the cortical plate. Anat. Embryol. 152, 109–126.

Marin-Padilla, M., 2011. The Human Brain: Prenatal Development and Structure. Springer, Heidelberg.

Martínez-García, F., Novejarque, A., Lanuza, E., 2007. Evolution of the amygdala in vertebrates. In: Kaas, J.H., Bullock, T., Krubitzer, L.A., Preuss, T.M., Rubenstein, J.L.R., Striedter, G.F. (Eds.), Evolution of Nervous Systems. Vol.2. Non-Mammalian Vertebrates Academic Press/Elsevier, London, pp. 255–334. Chief ed.

Martínez-García, F., Novejarque, A., Gutiérrez-Castellanos, Lanuza, E., 2012. Piriform cortex and amygdala. In: Watson, C., Paxinos, G., Puelles, L. (Eds.), The Mouse Nervous System Academic Press/Elsevier, London, pp. 140–172.

Mathur, B.N., Caprioli, R.M., Deutch, A.Y., 2009. Proteomic analysis illuminates a novel structural definition of the claustrum and insula. Cereb. Cortex 19, 2372–2379.

Medina, L., Abellán, A., 2012. Subpallial structures. In: Watson, C., Paxinos, G., Puelles, L. (Eds.), The Mouse Nervous System Academic Press/Elsevier, London, pp. 173–220.

Medina, L., Legaz, I., González, G., De Castro, F., Rubenstein, J.L., Puelles, L., 2004. Expression of Dbx1, Neurogenin 2, Semaphorin 5A, Cadherin 8, and Emx1 distinguish ventral and lateral pallial histogenetic divisions in the developing mouse claustroamygdaloid complex. J. Comp. Neurol. 474, 504–523.

Meynert, T., 1868. Neue Untersuchungen über den Bau der Grosshirnrinde und ihre örtliche Verschiedenheiten. Allg. Wien med. Ztg. 13, 419–428.

Meynert, T., 1872. The brain of mammals (J.J. Putnam, Trans.). In: Stricker, S., A manual of histology (American Trans. by) Buck, A.H. (Ed.), W.Wood, New York, pp. 650–766.

Misson, J.P., Edwards, M.A., Yamamoto, M., Caviness Jr., V.S., 1988. Identification of radial glial cells within the developing murine central nervous system: studies based upon a new immunohistochemical marker. Brain Res. Dev. Brain Res. 44, 95–108.

Misson, J.P., Austin, C.P., Takahashi, T., Cepko, C.L., Caviness Jr., V.S., 1991. The alignment of migrating neural cells in relation to the murine neopallial radial glial fiber system. Cereb. Cortex 1, 221–229.

Miura, E., Iijima, T., Yuzaki, M., Watanabe, M., 2006. Distinct expression of Cbln family mRNAs in developing and adult mouse brains. Eur. J. Neurosci. 24, 750–760.

Narkiewicz, O., Mamos, L., 1990. Relation of the insular claustrum to the neocortex in insectivora. J. Hirnforsch. 5, 623–633.

Nieuwenhuys, R., 2009. The structural organization of the forebrain: a commentary on the papers presented at the 20th Annual Karger Workshop Forebrain evolution in fishes. Brain Behav. Evol. 74, 77–85.

Obst-Pernberg, K., Medina, L., Redies, C., 2001. Expression of R-cadherin and N-cadherin by cell groups and fiber tracts in the developing mouse forebrain: relation to the formation of functional circuits. Neuroscience 106, 505–533.

Perlmann, T., Wallén-Mackenzie, Å., 2004. Nurr1, an orphan nuclear receptor with essential functions in developing dopamine cells. Cell Tissue Res. 318, 45–52.

Pirone, A., Cozzi, B., Edelstein, L., Peruffo, A., Lenzi, C., Quilici, F., et al., 2012. Topography of Gng2- and NetrinG2-expression suggests an insular origin of the human claustrum. PLoS One 7 (9), e44745. doi: 10.1371/journal.pone.0044745 Published online 2012 September 5.

Pombal, M.A., Megías, M., Bardet, S.M., Puelles, L., 2009. New and old thoughts on the segmental organization of the forebrain in lampreys. Brain Behav. Evol. 74, 7–19.

Puelles, L., 2001. Thoughts on the development, structure and evolution of the mammalian and avian telencephalic pallium. Philos. Trans. R. Soc. Lond. B. Biol. Sci. 356, 1583–1598.

Puelles, L., 2011. Pallio-pallial tangential migrations and growth signaling: new scenario for cortical evolution? Brain Behav. Evol. 78, 108–127.

Puelles, L., Kuwana, E., Puelles, E., Rubenstein, J.L., 1999. Comparison of the mammalian and avian telencephalon from the perspective of gene expression data. Eur. J. Morphol. 37, 139–150.

Puelles, L., Kuwana, E., Puelles, E., Bulfone, A., Shimamura, K., Keleher, J., et al., 2000. Pallial and subpallial derivatives in the embryonic chick and mouse telencephalon, traced by the expression of the genes *Dlx-2, Emx-1, Nkx-2.1, Pax-6,* and *Tbr-1*. J. Comp. Neurol. 424, 409–438.

Puelles, L., Martinez, S., Martinez-de-la-Torre, M., Rubenstein, J.L.R., 2004. Gene maps and related histogenetic domains in the forebrain and midbrain. In: Paxinos, G. (Ed.), The Rat Nervous System, third ed. Academic Press (Elsevier), San Diego, pp. 3–25.

Puelles, L., Martinez-de-la-Torre, M., Paxinos, G., Watson, C., 2007. The Chick Brain in Stereotaxic Coordinates. An Atlas Featuring Neuromeric Subdivisions and Mammalian Homologies. Academic Press (Elsevier), San Diego.

Ramon y Cajal, S., 1902. Studien über die Hirnrinde des Menschen. III. Heft: Die Hörrinde NB, this contains also the insular cortex; Translated into German by Bresler, J, pp. 63, Leipzig, Barth.

Real, M.A., Dávila, J.C., Guirado, S., 2003. Expression of calcium-binding proteins in the mouse claustrum. J. Chem. Neuroanat. 25, 151–160.

Redies, C., Medina, L., Puelles, L., 2001. Cadherin expression by embryonic divisions and derived gray matter structures in the telencephalon of the chicken. J. Comp. Neurol. 438, 253–285.

Reiner, A., Perkel, D.J., Bruce, L.L., Butler, A.B., Csillag, A., Kuenzel, W., Avian Brain Nomenclature Forum, 2004. Revised nomenclature for avian telencephalon and some related brainstem nuclei. J. Comp. Neurol. 473, 377–414.

Reiner, A., Yang, M., Cagle, M.C., Honig, M.G., 2011. Localization of Cerebellin-2 in late embryonic chicken brain: implications for a role in synapse formation and for brain evolution. J. Comp. Neurol. 519, 2225–2251.

Remedios, R., Subramanian, L., Tole, S., 2004. LIM genes parcellate the embryonic amygdala and regulate its development. J. Neurosci. 24, 6986–6990.

Rose, M., 1927. Der Allocortex bei Tier and Mensch. II. Die sogenante Riechrinde beim Menschen und beim Affen. J. Psychol. Neurol. 34, 261–401.

Rose, M., 1928. Die Ontogenie der Inselrinde; zugleich ein Beitrag zur histogenetischen Rindeneinteilung. J. Psychol. Neurol. 36, 182–209.

Sanides, F., 1969. Comparative architectonics of the neocortex of mammals and their evolutionary interpretation. Ann. N. Y. Acad. Sci. 167, 404–423. doi: 10.1111/j.1749-6632.1969.tb20459.x.

Sato, Y., Hirata, T., Ogawa, M., Fujisawa, H., 1998. Requirement for early-generated neurons recognized by monoclonal antibody lot1 in the formation of lateral olfactory tract. J. Neurosci. 18, 7800–7810.

Schneider, R., 1966. Das Gehirn von *Rousettus aegypticus* (E. Geoffroy 1810) (*Megachiroptera, Chiroptera, Mammalia*). Senkenberg Abh. 513, 1–159.

Schuster, E., 1910. Cerebral cortex of *Echidna*. Proc. R. Soc. London Series B. Biol. 82, 113–123. + plates IV and V.

Smart, I.H., Smart, M., 1977. The location of nuclei of different labeling intensities in autoradiographs of the anterior forebrain of postnatal mice injected with [3H]thymidine on the eleventh and twelfth days post-conception. J. Anat. 123, 515–525.

Smart, I.H., Smart, M., 1982. Growth patterns in the lateral wall of the mouse telencephalon: I. Autoradiographic studies of the histogenesis of the isocortex and adjacent areas. J. Anat. 134, 273–298.

Smith-Fernandez, A.S., Pieau, C., Repérant, J., Boncinelli, E., Wassef, M., 1998. Expression of the Emx-1 and Dlx-1 homeobox genes define three molecularly distinct domains in the telencephalon of mouse, chick, turtle and frog embryos: implications for the evolution of telencephalic subdivisions in amniotes. Development 125, 2099–2111.

Spatz, H., 1921. Zur Anatomie der Zentren des Streifenhügels. Münch.Med.Wsch. 45, 1441–1445.

Spiegel, E., 1919. Die Kerne im Vorderhirn der Säuger. Arb. Neurol. Inst. Wiener Univ. 22, 418–497.

Stenman, J.M., Wang, B., Campbell, K., 2003. *Tlx* controls proliferation and patterning of lateral telencephalic progenitor domains. J. Neurosci. 23, 10568–10576.

Striedter, G.F., 1997. The telencephalon of tetrapods in evolution. Brain Behav. Evol. 49, 179–213.

Suzuki, I.K., Kawasaki, T., Gojobori, T., Hirata, T., 2012. The temporal sequence of the mammalian neocortical neurogenetic program drives mediolateral pattern in the chick pallium. Dev. Cell 22, 1–8.

Swanson, L.W., Petrovich, G.D., 1998. What is the amygdala? Trends Neurosci. 21, 323–331.

Takiguchi-Hayashi, K., 2001. In vitro clonal analysis of rat cerebral cortical neurons expressing latexin, a subtype-specific molecular marker of glutamatergic neurons. Dev. Brain Res. 132, 87–90.

Teissier, A., Griveau, A., Vigier, L., Piolot, T., Borello, U., Pierani, A., 2010. A novel transient glutamatergic population migrating from the pallial-subpallial boundary contributes to neocortical development. J. Neurosci. 30, 10563–10574.

Teissier, A., Waclaw, R.R., Griveau, A., Campbell, K., Pierani, A., 2012. Tangentially migrating transient glutamatergic neurons control neurogenesis and maintenance of cerebral cortical progenitor pools. Cereb. Cortex 22, 403–416.

Thomson, A.M., 2010. Neocortical layer 6, a review. Front. Neuroanat. 4, 13. doi: 10.3389/fnana.2010.00013.

Tigges, J., Shantha, T.R., 1969. *A Stereotaxic Brain Atlas of the Tree Shrew* (Tupaia glis). Williams & Wilkins, Baltimore.

Tole, S., Remedios, R., Saha, B., Stoykova, A., 2005. Selective requirement of Pax6, but not Emx2, in the specification and development of several nuclei of the amygdaloid complex. J. Neurosci. 25, 2753–2760.

Tomioka, N., Osumi, N., Sato, Y., Inoue, T., Nakamura, S., Fujisawa, H., et al., 2000. Neocortical origin and tangential migration of guidepost neurons in the lateral olfactory tract. J. Neurosci. 20, 5802–5812.

Urban, I., Richard, P., 1972. A Stereotaxic Atlas of the New Zealand Rabbit's Brain. Charles C. Thomas, Springfield.

Valverde, F., Santacana, M., 1994. Development and early postnatal maturation of the primary olfactory cortex. Brain Res. Dev. Brain Res. 80, 96–114.

Valverde, F., Facal-Valverde, M.V., Santacana, M., Heredia, M., 1989. Development and differentiation of early generated cells of sublayer VIb in the somatosensory cortex of the rat: a correlated Golgi and autoradiographic study. J. Comp. Neurol. 290, 118–140.

Valverde, F., López-Mascaraque, L., Santacana, M., De Carlos, J.A., 1995. Persistence of early-generated neurons in the rodent subplate: assessment of cell death in neocortex during the early postnatal period. J. Neurosci. 7, 5014–5024.

Vandevelde, L.L., Duckworth, E., Reep, R.L., 1996. Layer VII and the gray matter trajectories of corticocortical axons in rats. Anat. Embryol. 194, 581–593.

Von Economo, C., Koskinas, G.N., 1925. Die Zytoarchitektonik der Hirnrinde des erwachsenen Menschen. Atlas mit 112 mikrophotographishen Tafeln in besonderer Mappe. Julius Springer Verlag, Wien.

Witschi, E., 1956. Development of Vertebrates. W.B. Saunders, Philadelphia.

Xu, Q., Cobos, I., De La Cruz, E., Rubenstein, J.L., Anderson, S.A., 2004. Origins of cortical interneuron subtypes. J. Neurosci. 24, 2612–2622.

Yun, K., Potter, S., Rubenstein, J.L., 2001. *Gsh2* and *Pax6* play complementary roles in dorsoventral patterning of the mammalian telencephalon. Development 128, 193–205.

Ziehen, T., 1908. Das Centralnervensystem der Monotremen und Marsupialier. II. Mikroskopische Anatomie. Part 2. Der Faserverlauf im Gehirn von Echidna und Ornithorhynchus nebst vergleichenden Angaben über den Faserverlauf des Gehirns von Perameles und Macropus. Semon's Zoologische Forschungreisen in Australien und dem Malayischen Archipel. vol.III, part 2, 790–921.

5

Physiology of the Claustrum

Helen Sherk

Department of Biological Structure, University of Washington Seattle,
WA, USA

INTRODUCTION

Although physiological investigation of the claustrum began in the 1970s, the number of such studies to date has been small. The location and, in larger brains, the shape of the claustrum make it difficult to target with microelectrodes. Functional imaging methods are poorly suited to investigating the human claustrum because of its narrow, sheet-like configuration. Physiological studies have been undertaken in only two species, the cat and the macaque monkey, and therefore this chapter will be devoted to these animals. The sections at the end of the chapter deal with the physiological action of claustral efferents to cortex and hypotheses about claustral function.

CAT

Visual Zone

The cat's claustrum can be divided into three zones. Its wide dorsal "head" contains a somatosensory zone and a visual zone (shown as *hatched* and *black stipple*, respectively, in Figure 5.1A). The remainder of the nucleus will be referred to as the ventral zone.

The visual zone in the cat's claustrum has a single, orderly representation of the contralateral visual hemifield, with a small inclusion of the central ipsilateral visual field (LeVay and Sherk, 1981b; Olson and Graybiel, 1980). Each visual field point is represented along a line through the claustrum, and each iso-azimuth line in the visual field is represented along a curved sheet (Figure 5.2A and C). The vertical

J. Smythies, L. Edelstein and V. Ramachandran (Eds):
The Claustrum.

DOI: http://dx.doi.org/10.1016/B978-0-12-404566-8.00005-2

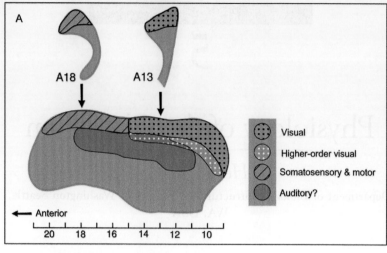

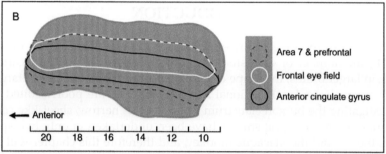

FIGURE 5.1   Diagrams of the cat's claustrum, in a lateral view. The scale shows antero-posterior location in Horsley-Clarke coordinates. A: Sensory zones, two of which (visual and somatosensory) have been defined by neural responses as well as anatomical connections (Olson and Graybiel, 1980; LeVay and Sherk, 1981a). The "higher-order visual" zone and the putative auditory zone have been identified only by their connections with identified cortical areas (Sherk, 1986; Updyke, 1993; Beneyto and Prieto, 2001). Above the parasagittal drawing are shown two coronal sections at two locations (A18 and A13). B: Outlines of claustral regions connected to four cortical regions. These outlines are based on anatomical data, in most cases by labeling claustral neurons that project to the cortex.

meridian is the core of this three-dimensional map. The upper visual field is represented posteriorly, and the lower visual field, anteriorly. What, if anything, is mapped along the mediolateral axis of the visual zone is unknown. One possibility is preferred orientation, as suggested in the diagram in Figure 5.2D. Domains of iso-orientation preference appear to run in an anteroposterior direction in the visual zone (Sherk and LeVay, 1981).

Four areas of visual cortex (areas 17, 18, 19, and the lateral suprasylvian visual area) send retinotopically superimposed inputs to the visual

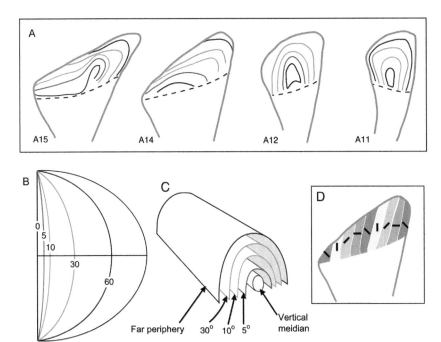

FIGURE 5.2   Schematic representations of the retinotopic map in the cat's visual zone. A: Coronal sections through the left claustrum, with each line representing a visual field point at the azimuth indicated in the visual hemifield drawing (B). Lower visual field is represented in the more anterior sections (A15, A14), the horizontal meridian coincides approximately with section A12, and the upper field is represented more posteriorly (A11). The *dashed lines* shows the approximate ventral limit of the visual zone. B: Diagram of the right visual hemifield. C: Cartoon of the retinotopic organization in the claustrum. Iso-azimuth "sheets" are curved, while the vertical meridian occupies the central core. D: Hypothetical organization of iso-orientation domains in one coronal section.

zone (Jayaraman and Updyke, 1979; LeVay and Sherk, 1981a; Olson and Graybiel, 1980; Sanides and Buchholtz, 1979; Squatrito et al., 1980). An interesting feature of the claustral visual map is its emphasis on the visual periphery compared to maps in its cortical sources. Although the claustral representation of central visual field is expanded compared to the actual size of central field, this expansion is notably less pronounced than in any of its sources. The claustral map may be more similar to the retinotopic map in the superior colliculus, which also de-emphasizes central visual field (McIlwain, 1975).

Claustral neurons have response properties that are similar to those of neurons in area 17. They are orientation selective, and respond well to elongated bars or edges moving at moderate speeds (Sherk and LeVay, 1981). Their most distinctive feature is the length of their receptive fields: they may show response summation for stimuli of 30 degrees or more.

In other respects, they are nonselective, being indifferent to direction of motion and showing no strong preference for one eye over the other.

The neurons in layer 6 of area 17 that provide the major input to the visual zone also have strikingly long receptive fields, though not as extreme as those in the claustrum (Grieve and Sillito, 1996). Corticoclaustral neurons tend to be direction selective, and to be strongly dominated by input from one eye, properties not typical of neurons in the claustrum. Probably each claustral neuron receives convergent input from more than one cortical neuron, but this convergence maintains orientation selectivity in the target neurons.

An early study of the claustrum reported that neurons in at least part of the visual zone had somatosensory and/or auditory responses, and rarely responded to visual stimuli (Spector et al., 1974). Similarly, Clarey and Irvine (1986) reported that most neurons in the visual zone responded not to visual but to auditory stimuli (clicks). These responses were "commonly rather weak and labile" (p 435), and had latencies considerably longer than those in auditory cortex. Because the visual zone does not receive input from either auditory or somatosensory cortex, the origin of these responses is puzzling. Notably, both studies were undertaken using chlorolose anesthesia, which in studies of cat cortex has led to problematic results. For example, area 7 was thought to be multimodal because in cats anesthetized with chlorolose, a large majority of neurons had brief, nonspecific, habituating responses to both auditory and somatosensory (paw-shock) stimulation (Dow and Dubner, 1969; Robertson et al., 1975). However, in alert cats, neurons in area 7 had only visual responses (Olson and Lawler, 1987; Pigarev and Rodionova, 1998).

Below the visual zone is a region that is reciprocally connected with several areas of "higher-order" visual cortex such as PLLS (the middle lateral bank of the suprasylvian sulcus), AMLS (the anterior medial bank of the suprasylvian sulcus) and area 20 (Sherk, 1986; Updyke, 1993) (see *white stipple* in Figure 5.1A). All of these cortical areas have quite large receptive fields, crude or no retinotopic organization, and weak or no visual responsiveness in anesthetized cats. Not surprisingly, Olson and Graybiel (1980) and Sherk and LeVay (1981) failed to detect visual responses in this region.

Somatosensory and Motor Zone

The somatosensory zone lies anterior to the visual zone (Figure 5.1A, *hatched*). It is strongly connected with primary motor cortex (M1; area 4) and with area 5 (parietal motor cortex) (Clascá et al., 1992) as well as with the primary somatosensory cortex (S1) (Minciacchi et al., 1995; Olson and Graybiel, 1980), but no information is available about this zone's motor-related activity. The somatosensory zone contains a

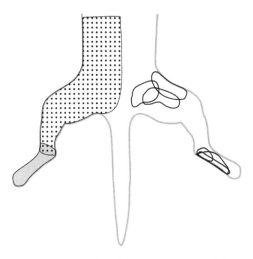

FIGURE 5.3   Somatosensory receptive fields on hind legs of the cat. On the right are outlined six receptive fields belonging to neurons in S1 (Dykes et al., 1980; Minciacchi et al., 1995). On the left, shown by *stippling* and *shading*, are two receptive fields belonging to neurons in the somatosensory zone of the claustrum (Olson and Graybiel, 1980).

somatotopic map. Starting at the dorsal surface, receptive fields were found on the hind leg; proceeding ventrally, on the body, then on the foreleg, and finally on the face (Olson and Graybiel, 1980). Based on labeling of claustral neurons after tracer injections in S1, Minciacchi et al. (1995) concluded that the foreleg occupies the largest region in the somatosensory zone. The map may expand the legs' representation at the expense of the face (akin to the visual map's expansion of the visual periphery), at least as compared to the map in S1. Neither anatomical nor physiological mapping has shown what somatotopic dimension is mapped along the anteroposterior axis of the claustrum, or along its mediolateral axis.

Olson and Graybiel (1980) found that receptive fields in the somatosensory zone are larger than in S1 (Figure 5.3). Neurons responded to "firm pressure or sharp taps" (p. 480) applied to the receptive field, with some neurons also responsive to light touch. No somatosensory neuron had a visual or auditory response. The absence of visual responses is consistent with the apparent lack of overlap between visual and somatosensory zones defined anatomically (Minciacchi et al., 1995).

The absence of auditory responses, on the other hand, is not altogether consistent with anatomical data. Beneyto and Prieto (2001) reported that neurons in some parts of the somatosensory zone send output to auditory cortex. This overlap, if it exists, is interesting because hypotheses about the claustrum's function emphasize cross-modal processing in the claustrum. Three possibilities might reconcile the physiological and anatomical observations:

1. Auditory responses by somatosensory neurons might be suppressed by barbiturate anesthesia.
2. Neurons may project to auditory cortex, but not have auditory responses.
3. The apparent overlap between somatosensory and auditory zones in the claustrum may be an artifact of pooling data across different cats. There is variability among cats in the cross-sectional shape of the claustrum, so attempts to superimpose data across animals may obscure boundaries that are actually distinct.

Auditory Zone

Anatomical experiments have shown that a zone of claustrum below the somatosensory and visual zones contains numerous neurons that project to auditory cortex (*dark gray* in Figure 5.1A). Although input to the claustrum from primary auditory cortex has been somewhat controversial (discussed in Sherk, 1986), input from some higher-order auditory areas is well documented (Beneyto and Prieto, 2001; Irvine and Brugge, 1980; Olson and Graybiel, 1980).

Olson and Graybiel (1980) found brisk auditory responses in the putative auditory zone just deep to the somatosensory zone. Neurons responded via either ear to clicks and tone bursts, as well as to a variety of sounds at different locations. Notably, auditory-responsive cells were confined to a thin layer, spanning only 100–200 μm below the somatosensory zone, and thus much more limited than the anatomically defined auditory region shown in Figure 5.1A.

Ventral Zone

The region ventral to the visual, somatosensory, and putative auditory zones is connected with a diverse array of cortical areas (see Figure 5.1B), but not with primary motor or sensory areas, possibly excepting A1. Investigators have failed to find neurons with sensory responses in the ventral zone, with the exception of the narrow layer of auditory responses noted above.

The regions illustrated in Figure 5.1B have been defined mostly on the basis of retrograde tracers injected into the cortex (Beneyto and Prieto, 2001; Clascá et al., 1992; Irvine and Brugge, 1980; Musil and Olson, 1988a,b; Olson and Jeffers, 1987; Olson and Lawler, 1987; Updyke, 1993). Thus although we know that claustral neurons project to a particular cortical area, we do not always know whether the same cortical area sends input to the claustrum. Typically, the population of claustral cells projecting to one cortical area extends along the full rostrocaudal extent of the nucleus. Inevitably, these different populations overlap extensively. Indeed, Figure 5.1B underestimates the overlap, because it leaves out the large population projecting to entorhinal cortex and the hippocampal subiculum (Room and Groenewegen, 1986; Witter and Groenewegen, 1986).

MACAQUE

Visual Zone

Compared to the cat, the macaque monkey's claustrum is difficult to divide into discrete zones based on connections with the cortex. A visual zone has been identified in the ventroposterior claustrum (Figure 5.4A), though its limits are uncertain. In Figure 5.4A, this zone has been divided into two regions. The posterior region (*black stipple*) sends output to areas V1, V2, V4, MT, MST, and FST, which are the early and middle stages of visual cortex (Boussaoud et al., 1992; Dineen and Hendrickson, 1982; Kennedy and Bullier, 1985; Lysakowski et al., 1988; Mizuno et al., 1981; Perkel et al., 1986; Ungerleider et al., 1984; Weller et al., 2002). The anterior region ("higher-order", *white stipple*) sends output to higher stages of visual cortex, that is the inferotemporal and intraparietal areas (Baizer et al., 1993; Cheng et al., 1997).

The posterior visual region in the macaque differs from the cat's visual zone in three key ways. First, the macaque's visual zone may lack input from area V1, although it does send axons to this cortical area. Secondly its retinotopic map may be rather crude (Ungerleider et al., 1984; Kennedy and Bullier, 1985), though there appears to be some separation between neurons projecting to the central visual field in V1, and those projecting to the periphery (Perkel et al., 1986). Thirdly, the macaque's visual zone contains neurons projecting to non-visual areas of cortex (compare Figures 5.4A and B). Two of these areas are involved in eye movement control: the frontal eye field (FEF; area 8) and supplementary eye field (SEF) (Huerta and Kaas, 1990; Künzle and Akert, 1977; Pearson et al., 1982; Shook et al., 1988; Stanton et al., 1988).

Another population that overlaps the claustral visual zone sends axons to the superior temporal gyrus, a region of higher-order auditory cortex (Pearson et al., 1982). Scattered throughout most of the visual zone are neurons that send axons to prefrontal cortex (area 46) (Mufson and Mesulam, 1982; Tanné-Gariépy et al., 2002). A large extent of claustrum (not shown in Figure 5.4B) contains neurons that project to the cingulate and retrosplenial gyri (Mesulam and Mufson, 1982; Parvizi et al., 2006). Note that these various claustral territories do not necessarily have

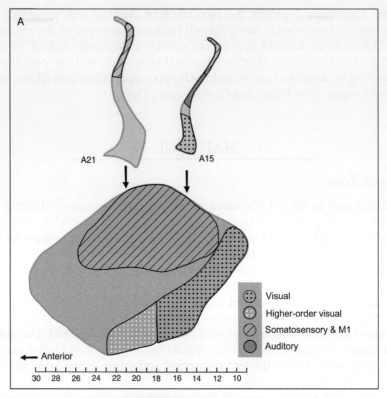

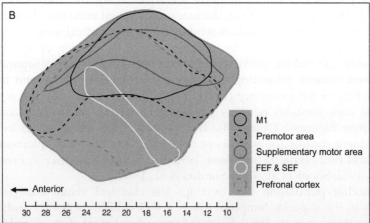

FIGURE 5.4 Diagrams of the macaque monkey's claustrum, in a lateral view. The curvature visible in coronal sections has been flattened out. A: Sensory zones identified by anatomical data. As in the cat, the somatosensory zone appears to coincide with the region connected to M1 (shown again in lower diagram). The visual zone has been divided into a posterior region connected with early to mid-level visual areas (V1, V2, MT, MST, FST,

strictly reciprocal connections with cortex. Prefrontal area 46, for example, appears to send input to a limited claustral territory that avoids the visual zone (Selemon and Goldman-Rakic, 1988).

There has been only one study of visual responses in the primate claustrum. In alert macaque monkeys, Remedios et al. (2010) recorded from visually responsive neurons located within the anatomically defined visual zone. These authors did not plot cells' receptive fields or test their response properties, but instead used movies of jungle scenes as stimuli. Neurons had transient responses to movie onset, followed by modest activity for the 1-second duration of the stimulus. Notably, neurons did not respond to the audio track of the movie when it was played without the video, nor were their visual responses affected when the audio track was combined with the video.

Auditory Responses

In the same study, Remedios et al. (2010) also encountered neurons with responses to the audio track of their stimulus movies. These neurons were located in a region about 6 mm above the visual zone (in Figure 5.4A, see *dark gray* region in coronal section at A15). Like the visual neurons, they fired briefly to onset of the audio; they had weak sustained activity that tended to fall to baseline within about 500 msec. These responses were neither increased nor diminished when the video track was paired with the audio.

This observation of auditory responses in the primate claustrum is both intriguing and puzzling. The only connections that have been reported between auditory cortex and the macaque's claustrum are those noted above, with the superior temporal gyrus. However, this putative auditory region is in the extreme ventral claustrum, below the location of neurons with auditory responses.

Somatosensory and Motor Zone

A large dorsal zone of the macaque's claustrum is connected to somatosensory cortex and to M1 (*hatching*, Figure 5.4A; Minciacchi

---

and V4), and an anterior region connected with "higher level" visual areas (inferotemporal cortex and areas located in the intraparietal sulcus). At top are two coronal sections through the left claustrum showing the locations of neurons studied electrophysiologically. Motor-related neurons (*hatching*) were found by Shima et al. (1996) in a zone that extended from A14 to A24. Neurons with auditory responses were found in the *dark gray* region, and neurons with visual responses were found in the *black stippled* region (Remedios et al., 2010). B: Motor-related zones, based mostly on populations of claustral neurons projecting to identified cortical areas (Künzle and Akert, 1977; Leichnetz, 1986; Stanton et al., 1988; Shook et al., 1991; Tanné-Gariépy et al., 2002).

et al., 1991; Tanné-Gariépy et al., 2002). Shima et al. (1996) described the movement-related activity of neurons in this zone in active macaque monkeys. The animal grasped a manipulandum and performed three movements, rotate, push, and pull, in varying sequences. Neurons had robust activity accompanying each movement. Timing of neural activity was indistinguishable from that in M1, starting well before the movement and continuing throughout it. Interestingly, a large majority of claustral neurons responded well with every kind of movement, and they were unaffected by the order of these movements. Their activity was the same whether visual cues (colored lights) accompanied the movements, or whether the monkey relied on memory to perform the sequence. However, neurons were specific for arm/hand movements: none were active when the monkey moved its leg, foot, mouth, tongue, facial muscles, or eyes. Whether neurons in some other part of the claustral motor zone fire in conjunction with movement of body parts other than the hand and arm remains unknown.

ACTION OF CLAUSTRAL EFFERENTS ON CORTEX

Claustrocortical axons are believed to be excitatory (da Costa et al., 2010; LeVay, 1986), but in cats electrical stimulation of the claustrum has been found to suppress neural firing in several cortical areas, including area 4 (M1) and area 6 (accessory motor area) (Salerno et al., 1984). Typically, spontaneous activity was silenced for about 200 msec starting on average 20 msec after electrical stimulation of the claustrum. Some neurons in the cat's FEF, on the other hand, were excited by electrical stimulation of the claustrum, with a brief burst of firing commonly followed by a longer suppression (Cortimiglia et al., 1991; Salerno et al., 1989).

In area 17, many neurons showed suppression of on-going activity after electrical stimulation of the claustrum (Ptito and Lassonde, 1981; Tsumoto and Suda, 1982). It seems probable, however, that the claustrum also has an excitatory influence on visual cortex, given that its neurons synapse primarily on dendritic spines, and these belong to excitatory neurons (LeVay, 1973). Indeed, Tsumoto and Suda (1982) observed that many neurons in area 17 were briefly excited by stimulation of the claustrum.

Only one study has investigated the consequences of lesioning the claustrum. When the visual zone of the cat's claustrum was destroyed, neurons in area 17 showed a modest reduction in end-stopping (suppression of responses by long stimuli) (Sherk and LeVay, 1983). This result should be regarded somewhat cautiously. Subtle changes can easily be missed in lesion studies, and changes in motor or cognitive behavior cannot be detected in anesthetized animals. A more informative approach might make use of reversible inactivation of the claustrum in behaving animals.

HYPOTHESES OF CLAUSTRAL FUNCTION

Several hypotheses have been proposed regarding the function of the claustrum. Theories tend to focus on the fact that information from different sensory modalities converges on one nucleus, albeit apparently to separate zones within it. Claustral output to most of the cortex might in theory integrate information across modalities.

Crick and Koch (2005) suggested that the claustrum binds together different modalities of a real-world event to create a highly salient percept. Thus, for example, a glimpse of a pattern of black spots paired with a rustling sound at the same location might evoke a monkey's percept of a leopard hidden in foliage. Notably, in a direct test of this hypothesis, neither visual nor auditory responses in the claustrum were enhanced when visual and auditory stimuli were paired together (Remedios et al., 2010). However, the outcome of this study is not easy to interpret. The movies used as stimuli may not have been perceived by the monkeys as behaviorally significant: the strongest responses were brief, nonspecific onset responses, with much weaker or no subsequent response to movie content.

A greater difficulty with this theory concerns the known anatomy within the claustrum. The hypothesis requires integration between zones processing different sensory modalities (in this example, visual and auditory); thus, intraclaustral axons must link together different sensory zones of the claustrum, as Crick and Koch (2005) point out. However, Rahman and Baizer (2007), after exhaustively searching the literature, concluded that: "Despite numerous anatomical studies of the claustrum… there is surprisingly little information available about connections within it" (p. 107). Golgi studies have revealed rich but quite local axonal arbors within the primate and cat claustrums, arising from what are most likely inhibitory neurons (Braak and Braak, 1982; Brand, 1981; LeVay and Sherk, 1981a). Their probable suppressive effect, as well as their short range, makes these axons unsuitable for extensive intraclaustral linkage.

A second detailed hypothesis (Smythies et al., 2012) proposes that a "mismatch" signal is sent to the claustrum from the thalamus when sensory information arrives that fails to match "the brain's expectation of what that information should be" (p. 2). The claustrum then sends a burst of spikes at gamma frequency (approximately 40 Hz) to visual cortex at the appropriate retinotopic location. Furthermore, this signal is propagated within the claustrum and subsequently fans out to other cortical areas, both nonvisual sensory and motor.

Current understanding of physiological properties of claustral neurons is insufficient to evaluate this hypothesis. To test for a mismatch response, one must record from claustral neurons in an alert animal that has expectations about future events, which has not been attempted. One prediction of the hypothesis which can be evaluated, however, is that claustral neurons will have multimodal responses. Setting aside early experiments performed on cats under chlorolose anesthesia, which appear to poorly predict response properties in alert animals, multimodal responses in the claustrum have not been found (Olson and Graybiel, 1980; Sherk and LeVay, 1981).

At present, anatomy presents a greater obstacle to the mismatch hypothesis than the physiological evidence. The mismatch signal depends on retinotopically precise activity relayed from the lateral geniculate nucleus (LGN) via the thalamic reticular nucleus to the intralaminar nuclei, and finally to the claustrum (Kolmac and Mitrofanis, 1997). There is, however, no retinotopic organization in the intralaminar nuclei (Salin et al., 1989). Furthermore, within the claustrum, the hypothesis envisions activity propagated from the visual zone to the somatosensory and ventral zones, but, as noted above, there is no known anatomical substrate for such propagation. Finally, a role for the claustrum in synchronizing activity across widely separated cortical areas seems doubtful. Claustral axons to the cortex are of small diameter and conduct slowly (Cortimiglia et al., 1991; da Costa et al., 2010; LeVay, 1986; Salerno et al., 1989; Tsumoto and Suda, 1982). Following a putative burst of claustral firing, the arrival times of this activity in different cortical areas would be far from synchronous.

The two hypotheses discussed above are the rational outcome of available physiological information, almost all of which concerns sensory processing. However, physiological studies have given us a lopsided picture of the claustrum. They have focused on sensory responses because these can be investigated in anesthetized animals, or in unanesthetized but passive and immobile animals. It is far more challenging to study motor or cognitive functions, since animals must be alert and engaged in active behavior. Yet even a cursory look at the claustrum's cortical connections suggests that it may be more concerned with motor or cognitive processing (see Figures 5.1B and 5.4B). Motor-related connectivity is particularly striking in the macaque: almost the entire nucleus is interconnected with motor areas. In both cat and macaque, regions of cortex that are likely to be involved in cognitive activities, for example prefrontal cortex and cingulate gyrus, have extremely broad connections with the claustrum. If the only information available were anatomical, we would probably conclude that the claustrum's function involves motor control and/or cognitive processes such as planning and memory.

References

Baizer, J.S., Desimone, R., Ungerleider, L.G., 1993. Comparison of subcortical connections of inferior temporal and posterior parietal cortex in monkeys. Vis. Neurosci. 10, 59–72.

Beneyto, M., Prieto, J.J., 2001. Connections of the auditory cortex with the claustrum and the endopiriform nucleus in the cat. Brain Res. Bull. 54, 485–498.

Boussaoud, D., Desimone, R., Ungerleider, L.G., 1992. Subcortical connections of visual areas MST and FST in macaques. Vis. Neurosci. 9, 291–302.

Braak, H., Braak, E., 1982. Neuronal types in the claustrum of man. Anat. Embryol. 163, 447–460.

Brand, S., 1981. A serial section Golgi analysis of the primate claustrum. Anat. Embryol. 162, 475–488.

Cheng, K., Saleem, K.S., Tanaka, K., 1997. Organization of corticostriatal and corticoamygdalar projections arising from the anterior inferotemporal area TE of the macaque monkey: a Phaseolus vulgaris leucoagglutinin study. J. Neurosci. 17, 7902–7925.

Clarey, J.C., Irvine, D.R.F., 1986. Auditory response properties of neurons in the claustrum and putamen in the cat. Exp. Brain Res. 61, 432–437.

Clascá, F., Avendaáo, C., Román-Guindo, A., Llamas, A., Reinoso-Suárez, F., 1992. Innervation from the claustrum of the frontal association and motor areas: axonal transport studies in the cat. J. Comp. Neurol. 326, 402–422.

Cortimiglia, R., Crescimanno, G., Salerno, M.T., Amato, G., 1991. The role of the claustrum in the bilateral control of frontal-oculomotor neurons in the cat. Exp. Brain Res. 84, 471–477.

Crick, F.C., Koch, C., 2005. What is the function of the claustrum? Philos. Trans. R. Soc. Lond. B Biol. Sci. 360, 1271–1279.

da Costa, N.M., Fursinger, D., Martin, K.A.C., 2010. The synaptic organization of the claustral projection to the cat's visual cortex. J. Neurosci. 30, 13166–13170.

Dineen, J.T., Hendrickson, A., 1982. Cortical and subcortical pathways of the geniculate-recipient prestriate cortex in the Macaca monkey. Anat. Rec. 202, 45A.

Dow, B.M., Dubner, R., 1969. Visual receptive fields and responses to movement in an association area of cat cerebral cortex. J. Neurophysiol. 32, 773–784.

Dykes, R.W., Rasmusson, D.D., Hoeltzell, P.B., 1980. Organization of primary somatosensory cortex in the cat. J. Neurophysiol. 43, 1527–1546.

Grieve, K.L., Sillito, A.M., 1996. Differential properties of cells in the feline primary visual cortex providing the corticofugal feedback to the lateral geniculate nucleus and visual claustrum. J. Neurosci. 15, 4868–4874.

Huerta, M.F., Kaas, J.H., 1990. Supplementary eye field as defined by intracortical microstimulation: connections in macaques. J. Comp. Neurol. 293, 299–330.

Irvine, D.R.F., Brugge, J.F., 1980. Afferent and efferent connections between the claustrum and parietal association cortex in cat: a horseradish peroxidase and autoradiographic study. Neurosci. Lett. 20, 5–10.

Jayaraman, A., Updyke, B.V., 1979. Organization of visual cortical projections to the claustrum in the cat. Brain Res. 178, 107–115.

Kennedy, H., Bullier, J., 1985. A double-labeling investigation of the afferent connectivity to cortical areas V1 and V2 of the macaque monkey. J. Neurosci. 5, 2815–2830.

Kolmac, C.I., Mitrofanis, J., 1997. Organisation of the reticular thalamic projection to the intralaminar and midline nuclei in rats. J. Comp. Neurol. 377, 165–178.

Künzle, H., Akert, K., 1977. Efferent connections of cortical area 8 (frontal eye field) in Macaca fascicularis. A reinvestigation using the autoradiographic technique. J. Comp. Neurol. 173, 147–164.

LeVay, S., 1973. Synaptic patterns in the visual cortex of the cat and monkey: electron microscopy of Golgi preparations. J. Comp. Neurol. 150, 53–86.

LeVay, S., 1986. Synaptic organization of claustral and geniculate afferents to the visual cortex of the cat. J. Neurosci. 6, 3564–3575.

LeVay, S., Sherk, H., 1981a. The visual claustrum of the cat. I. Structure and connections. J. Neurosci. 1, 956–980.

LeVay, S., Sherk, H., 1981b. The visual claustrum of the cat. II. The visual field map. J. Neurosci. 1, 981–992.

Leichnetz, G.R., 1986. Afferent and efferent connections of the dorsolateral precentral gyrus (area 4, hand/arm region) in the macaque monkey, with comparisons to area 8. J. Comp. Neurol. 254, 450–492.

Lysakowski, A., Standage, G.P., Benevento, L.A., 1988. An investigation of collateral projections of the dorsal lateral geniculate nucleus and other subcortical structures to cortical areas V1 and V4 in the macaque monkey: a double label retrograde tracer study. Exp. Brain Res. 69, 651–661.

McIlwain, J.T., 1975. Visual receptive fields and their images in the superior colliculus of the cat. J. Neurophysiol. 38, 219–230.

Mesulam, M.-M., Mufson, E.J., 1982. Insula of the old world monkey. III: efferent cortical outputs and comments on function. J. Comp. Neurol. 212, 38–52.

Minciacchi, D., Granato, A., Barbaresi, P., 1991. Organization of claustro-cortical projections to the primary somatosensory area of primates. Brain Res. 553, 309–312.

Minciacchi, D., Granato, A., Antonini, A., Tassinari, G., Satarelli, M., Zanolli, L., et al., 1995. Mapping subcortical extrarelay afferents onto primary somatosensory and visual areas in cats. J. Comp. Neurol. 362, 46–70.

Mizuno, N., Uchida, K., Nomura, S., Nakamura, Y., Sugimoto, T., Uemura-Sumi, M., 1981. Extrageniculate projections to the visual cortex in the macaque monkey: an HRP study. Brain Res. 212, 454–459.

Mufson, E.J., Mesulam, M.-M., 1982. Insula of the old world monkey. II. Afferent cortical input and comments on the claustrum. J. Comp. Neurol. 212, 23–37.

Musil, S.Y., Olson, C.R., 1988a. Organization of cortical and subcortical projections to anterior cingulate cortex in the cat. J. Comp. Neurol. 272, 203–218.

Musil, S.Y., Olson, C.R., 1988b. Organization of cortical and subcortical projections to medial prefrontal cortex in the cat. J. Comp. Neurol. 272, 219–241.

Olson, C.R., Graybiel, A.M., 1980. Sensory maps in the claustrum of the cat. Nature 288, 479–481.

Olson, C.R., Jeffers, I., 1987. Organization of cortical and subcortical projections to area 6 m of the cat. J. Comp. Neurol. 266, 73–94.

Olson, C.R., Lawler, K., 1987. Cortical and subcortical afferent connections of a posterior division of feline area 7 (area 7p). J. Comp. Neurol. 259, 13–30.

Parvizi, J., Van Hoesen, G.W., Buckwalter, J., Damasio, A., 2006. Neural connections of the posteromedial cortex in the macaque. Proc. Natl. Acad. Sci. U.S.A. 103, 1563–1568.

Pearson, R.C.A., Brodal, P., Gatter, K.C., Powell, T.P.S., 1982. The organization of the connections between the cortex and the claustrum in the monkey. Brain Res. 234, 435–441.

Perkel, D.J., Bullier, J., Kennedy, H., 1986. Topography of the afferent connectivity of area 17 in the macaque monkey: a double-labeling study. J. Comp. Neurol. 253, 374–402.

Pigarev, I.N., Rodionova, E.I., 1998. Two visual areas located in the middle suprasylvian gyrus (cytoarchitectonic field 7) of the cat's cortex. Neurosci. 85, 717–732.

Ptito, M., Lassonde, M.C., 1981. Effects of claustral stimulation on the properties of visual cortex neurons in the cat. Exp. Neurol. 73, 315–320.

Rahman, E., Baizer, J.S., 2007. Neurochemically defined cell types in the claustrum of the cat. Brain Res. 1159, 94–111.

Remedios, R., Logothetis, N.K., Kayser, C., 2010. Unimodal responses prevail within the multisensory claustrum. J. Neurosci. 29, 12902–12907.

Robertson, R.T., Mayers, K.S., Teyler, T.J., Bettinger, L.A., Birch, H., Davis, J.L., et al., 1975. Unit activity in posterior association cortex of cat. J. Neurophysiol. 38, 780–794.

Room, P., Groenewegen, H.J., 1986. Connections of the parahippocampal cortex in the cat. II. Subcortical afferents. J. Comp. Neurol. 251, 451–473.

Salerno, M.T., Cortimiglia, R., Crescimanno, G., Amato, G., Infantellina, F., 1984. Effects of claustrum stimulation on spontaneous bioelectric activity of motor cortex neurons in the cat. Exp. Neurol. 86, 227–239.

Salerno, M.T., Cortimiglia, R., Crescimanno, G., Amato, G., 1989. Effect of claustrum activation on the spontaneous unitary activity of frontal eye field neurons in the cat. Neurosci. Lett. 98, 299–304.

Salin, P.A., Bullier, J., Kennedy, H., 1989. Convergence and divergence in the afferent projections to cat area 17. J. Comp. Neurol. 283, 486–512.

Sanides, D., Buchholtz, C.S., 1979. Identification of the projection from the visual cortex to the claustrum by anterograde axonal transport in the cat. Exp. Brain Res. 34, 197–200.

Selemon, L.D., Goldman-Rakic, P.S., 1988. Common cortical and subcortical targets of the dorsolateral prefrontal and posterior parietal cortices in the rhesus monkey: evidence for a distributed neural network subserving spatially guided behavior. J. Neurosci. 8, 4049–4068.

Sherk, H., 1986. The claustrum and the cerebral cortex. In: Jones, E.G. Peters, A. (Eds.), Cerebral Cortex, vol. 5. Plenum Publishing Corp.

Sherk, H., LeVay, S., 1981. The visual claustrum of the cat. III. Receptive field properties. J. Neurosci. 1, 993–1002.

Sherk, H., LeVay, S., 1983. Contribution of the cortico-claustral loop to receptive field properties in area 17 of the cat. J. Neurosci. 3, 2121–2127.

Shima, K., Hoshi, E., Tanji, J., 1996. Neuronal activity in the claustrum of the monkey during performance of multiple movements. J. Neurophysiol. 76, 2115–2119.

Shook, B.L., Schlag-Rey, M., Schlag, J., 1991. Primate supplementary eye field. II. Comparative aspects of connections with the thalamus, corpus striatum, and related forebrain nuclei. J. Comp. Neurol. 317, 562–583.

Smythies, J., Edelstein, L., Ramachandran, V., 2012. Hypotheses relating to the function of the claustrum. Front. Int. Neurosci. 6, 1–16.

Spector, I., Hassmannova, J., Albe-Fessard, D., 1974. Sensory properties of single neurons of cat's claustrum. Brain Res. 66, 39–65.

Squatrito, S., Battaglini, P.P., Galletti, C., Riva Sanseverino, E., 1980. Autoradiographic evidence for projections from cortical visual areas 17, 18, 19, and the Clare-Bishop area to the ipsilateral claustrum in the cat. Neurosci. Lett. 19, 265–269.

Stanton, G.B., Goldberg, M.E., Bruce, C.J., 1988. Frontal eye field efferents in the macaque monkey: I. Subcortical pathways and topography of striatal and thalamic terminal fields. J. Comp. Neurol. 271, 473–492.

Tanné-Gariépy, J., Boussaud, D., Bouiller, E.M., 2002. Projections of the claustrum to the primary motor, premotor, and prefrontal cortices in the macaque monkey. J. Comp. Neurol. 454, 140–157.

Tsumoto, T., Suda, K., 1982. Effects of stimulation of the dorsocaudal claustrum on activities of striate cortex neurons in the cat. Brain Res. 240, 345–359.

Ungerleider, L.G., Desimone, R., Galkin, T.W., Mishkin, M., 1984. Subcortical projections of area MT in the macaque. J. Comp. Neurol. 223, 368–386.

Updyke, B.V., 1993. Organization of visual corticostriatal projections in the cat, with observations on visual projections to claustrum and amygdala. J. Comp. Neurol. 327, 159–193.

Weller, R.E., Steele, G.E., Kaas, J.H., 2002. Pulvinar and other subcortical connections of dorsolateral visual cortex in monkeys. J. Comp. Neurol. 450, 215–240.

Witter, M.P., Groenewegen, H.J., 1986. Connections of the parahippocampal cortex in the cat. IV. Subcortical efferents. J. Comp. Neurol. 254, 51–77.

Neurocomputation and Coding in the Claustrum: Comparisons with the Pulvinar

Klaus M. Stiefel[1] and Alex O. Holcombe[2]

[1]University of Western Sydney, MARCS Institute, Bioelectronics and Neuroscience, University of Western Sydney, NSW, Australia
[2]School of Psychology, University of Sydney, NSW, Australia

THE BINDING PROBLEM IN PERCEPTION

Picking up one's cup of coffee evokes several sensations. We perceive the color and texture of the cup, the shape of its handle, the sound it makes as it moves off the table, the smoothness of its surface, and its mass. Some of these features are initially analyzed by very distinct populations of neurons, responsible for visual signals, auditory signals, tactile signals, and proprioceptive signals. Yet in our perceptual experience, these independent signals are somehow linked to their common object, the coffee cup.

This impressive feat of neural systems integration may be particularly challenging in everyday scenes, as they are often filled with objects emitting competing sensory signals. The brain must have some way of determining which features go together and linking them into a representation of a common object.

An obvious cue to whether features belong to the same object is their location. Features originating from a common location and at a particular time typically belong to the same object. In the case of visual features, like color and motion, that may be processed in different brain regions, a convenient location index is available from the retinotopic maps that represent both motion and color location. Features sharing a location can be linked together. Retinotopic location cannot be relied on exclusively, witness the existence of motion transparency (where two objects are perceived in the

DOI: http://dx.doi.org/10.1016/B978-0-12-404566-8.00006-4

same location) and illusions where motion causes features to be attributed to the wrong location (Cai and Schlag, 2001). However, retinotopic location narrows down the solution.

Across sensory cortices, however, the retinotopic code does not fly. Somatosensory and auditory cortex represent features in skin-based and head-centered coordinates, respectively. Binding therefore requires that the brain calculate the mapping between these coordinate systems. This computation is thought to be done by parietal cortex.

POSSIBLE COMPUTATIONAL SOLUTIONS FOR THE BINDING PROBLEM

How can a neural structure successfully master the binding problem? Several solutions have been proposed, among them neurons with highly convergent inputs, the "grandmother neurons," and self-organizing temporal assemblies of neurons coding for the percepts of a particular object. We will now briefly review these computational strategies, and their occurrence or absence in the cortex.

Grandmother Neurons

What happens in the brain when someone encounters their grandmother? The person will visually perceive her glasses and her gray hair, hear her voice and feel the skin of her hand. The percepts from the visual, auditory and somatosensory modalities will be represented in different structures of the person's brain. Additionally, different aspects of each perceptual modality, like the visual shape and motion, will be represented in different brain structures, many of them being cortical sensory areas (shape being in the "what" pathway, motion in the "where" pathway, Mishkin et al., 1983). A neuron receiving convergent input from all of these regions when, and only when, they are activated by a grandmother stimulus, would be active when, and only when, grandmother is encountered. Such a neuron would be highly specialized for recognizing grandma, and would provide a convenient way of coding for the convergence of all stimuli relevant in this situation (Figure 6.1).

Grandmother-neuron-like cells have been found in the human cortex. Koch and co-workers recorded from neurons in awake patients awaiting antiepileptic surgery that fired very specifically in response to being presented with faces of US celebrities. These neurons were located in the higher visual areas, specifically the medial temporal lobe (Quiroga et al., 2005).

However, serious conceptual issues with coding via grandmother neurons can emerge. At the heart of these issues is a combinatorial explosion.

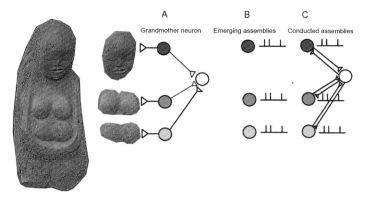

FIGURE 6.1   Coding schemes. Several neurons code for different components of the sculpture of a mother holding her child (at the Nimbin Bush Theater, Australia). A: (Grand) mother neuron. All component neurons converge on one neuron. B: Emergent temporal assemblies. Self-organizing cortical dynamics synchronize the activity of all neurons encoding components of the sculpture. C: Conducted temporal assemblies. One or more central conductors enhance the self-organization of assemblies via reciprocal connections. We suggest that the claustrum is one of the brain regions conducting temporal assemblies.

We can perceive and distinguish a large number of individual humans and other objects, and it is impossible that for each and every feature combination a grandmother neuron exists. The celebrity-specific neurons recorded by Quiroga and colleagues are likely the result of learning after repeated perceptions of the faces of these people. Grandmother-neuron coding will inevitably fail when it comes to novel objects where no such learning has yet taken place.

Emerging Temporal Assemblies

An alternative proposed to grandmother-neuron coding is coding via self-organizing, or emerging, assemblies. The mammalian cortex is a layered structure with rich lateral connections of different types, and lateral spread of excitation is well known to occur. Furthermore, the cortex shows a variety of ongoing dynamics, such as oscillations of different frequencies (Buzsaki, 2011). Several frequency bands seem to be both mechanistically distinct and separate in their behavioral correlates. Slow wave, delta, oscillations (0.5–2 Hz) are large in amplitude and synchronous over large cortical areas. They occur during deep sleep. Delta oscillations are thought to be caused by the interaction of excitatory connections between pyramidal neurons and their slow after-hyperpolarization K^+ conductances (Steriade et al., 1991, 1993b, Tiesinga, 2001).

Theta oscillations (1–5 Hz) are correlated with resting awake states. Gamma oscillations (30–80 Hz) occur during vigilant wakefulness and rapid-eye movement sleep. These oscillations are spatially and temporally restricted and are thought to be generated by an interneuronal network, linked by inhibitory synapses and gap-junctions. The neuromodulator acteylcholine plays a role in switching the brain to a gamma-oscillatory state (Steriade et al., 1993a).

All these types of oscillations themselves are complex phenomena, and they also interact in sophisticated ways. One example is mammalian sleep (Buzsaki, 2011). Typically, episodes of deep sleep, with delta oscillations, are succeeded with great regularity by rapid-eye movement sleep with gamma oscillations. During the course of a night's sleep, the duration of the delta to gamma successions increasingly shortens. Thus, in sleep, a long time-scale structure (hours: long to short delta − gamma successions) is superimposed on short time-scale structures (milliseconds: delta or gamma oscillations).

Another example of the structured interaction of brain oscillations is the nesting of gamma oscillations in theta oscillations in the hippocampal cortex. The rodent hippocampus contains "place cells," neurons which fire when the animal is in a specific location in its environment, the place fields. In these neurons, short bursts of gamma activity occur during theta oscillations. The further advanced the animal is on its path through the place field, the further advanced in the theta cycle the gamma activity occurs (Burgess et al., 2002). Again we see a short time-scale structure (10s of milliseconds, gamma oscillations) interlaced within a long-time-scale structure (100s of milliseconds, theta oscillations).

These oscillatory dynamics are a rich substrate for the formation of spatiotemporal assemblies. According to the emerging-assembly coding hypothesis, such assemblies are triggered and molded by sensory input; binding is achieved by bundling the activity corresponding to each feature of a perceived object into one spatiotemporal assembly (Singer, 1999).

There is intriguing evidence for the emerging-assembly coding hypothesis. Castelo-Branco and colleagues (2002) found evidence that neural responses in the cat cortex to different parts of moving gratings are synchronous when, and only when, they are configured to be perceived as part of one object. Another study supporting the assembly idea is the work by Iurilli et al. (2012), showing evidence for cross-modal activity from one to another modality of early sensory cortex. However, in contrast, Thiele and Stoner (2003) found no support for this hypothesis in recordings from macaque mediotemporal (MT) cortex, and emerging assemblies have remained difficult to study or find consistently.

Also supporting the emerging-assembly coding hypothesis is the large-scale structure of the cortex. Van Essen and colleagues assembled global wiring diagrams of the cortex, and no apparent cortical top level is obvious from these diagrams (Felleman and Van Essen, 1991; Van Essen et al., 1992). There is no cortical area into which all sensory information converges; the cortical grandmother area does not seem to exist. This connectivity argues for a distributed coding scheme by way of exclusion, and the emerging-assembly coding scheme is a good candidate, but not the only one.

Conducted Temporal Assemblies

The fact that the cortex most likely encodes sensory information in a distributed manner does not exclude the existence of a central coordinating structure. An orchestra can serve as an analogy (Crick and Koch, 2005): the individual musicians each perform (synchronously) a part of a symphony, but their performance is greatly aided by the presence of a skilled conductor. Equally, while the different cortical areas code for sensory information via the activity of spatiotemporal assemblies, a structure coordinating these assemblies would still improve the coding performance.

While the rich dynamics in the cortex are potentially a good substrate for the emergence of assemblies, there are also limitations. The binding of adjacent parts of a visual stimulus into one whole object could be facilitated by the interaction between adjacent patches in one cortical area. Equally, the binding of visual features between different visual regions (like those coding for shape and motion) could be brought about by topologically organized inter-area connections. But the binding between visual and auditory components of a perceived object could be more problematic.

While temporal assemblies may arise entirely through self-organization, a brain structure conducting the emergence of these assemblies could make this process faster, more reliable and less error-prone. Hence, several functional arguments point to the usefulness of a conductor of the cortex. Which brain structures are candidates for this conducting role?

POSSIBLE NEURAL SUBSTRATES FOR AN ASSEMBLY CONDUCTOR

Several brain structures have been proposed as higher-order integration centers involved in binding and conscious perception. These are areas of parietal cortex, the intralaminar thalamic nuclei, the pulvinar, and the claustrum. We will now discuss each of these areas in the context of higher-order integration in the brain.

The Pulvinar

The pulvinar is located at the dorsal end of the thalamus. The lateral geniculate nucleus (visual thalamus) and medial geniculate nucleus (auditory) look like small bumps stuck on the much larger pulvinar, which contains about 14 million neurons (Highley et al., 2003). The pulvinar is undoubtedly an integration center of the brain, and is believed to be driven primarily by cortical areas rather than the sensory organs (Shipp, 2003). It is reciprocally connected with many cortical areas. In some animals at least it also receives retinal input via the superior colliculus (Beckstead, 1984; Lin et al., 1984) and projects to the amygdala (Linke et al., 1999) and striatum (Harting and Updyke, 2006; Künzle, 2006). Interestingly, the pulvinar appears to connect cortical areas that are themselves connected directly to each other. This and the rhythmicity of pulvinar neural activity (like that of other areas in the thalamus) makes it a prime candidate for a conductor of synchronized neural assemblies (Shipp, 2003).

One aspect of this neural conductor role appears to be to regulate information transmission among cortical areas to favor the focus of visuospatial attention. In a recent physiological study, when a monkey attended to the location corresponding to a pulvinar receptive field, synchrony between its activity and those of cortical areas increased (Saalmann et al., 2012).

Maintaining a focus of attention that selects processing of the same external stimulus across cortices may be crucial for a unified sensory experience. Focusing attention on a particular object appears to allow binding of the features of that object (Treisman and Gelade, 1980). Furthermore, the capacity for conscious access to bound objects appears to be severely limited to one or a few objects at any one time (Treisman and Gelade, 1980), and thus may run afoul of a binding bottleneck. The difficult problem of coordinating widespread brain activity may be the reason for this bottleneck. The pulvinar and parietal cortices appear to participate in processing the coordinate frames (such as retinotopy) that facilitate linkage across sensory cortices.

Evidence from behavioral studies of patients strongly implicates the pulvinar in binding object features. Patients with such lesions make striking errors in determining which colors belong to which shapes, even though their perception of the individual shapes and colors seems intact (Arend et al., 2008; Ward et al., 2002). The neural coding of such links may have the same possibilities as those schematized in Figure 6.1, in the example of the sculpture. The deficits caused by pulvinar damage can be specific to the part of the visual field corresponding to the area of the pulvinar retinotopic map that was damaged.

This evidence indicates that the pulvinar is a conductor of neural assemblies. In particular, the pulvinar facilitates linking information that belongs together by virtue of originating at an attended location in the visual field. It may also serve to link other neural populations into an assembly, but this is harder to study than location-based assemblies. Location-based linking is feasible to study thanks to the retinotopic receptive fields of cells that can be assessed. These provide a sensible target for assessing synchrony, as was done in the study by Saalmann et al. (2012). As the claustrum does not have clear retinotopy, it may be harder to establish its role as a conductor, purely for methodological reasons.

While much of sensory binding is likely to be based on retinotopic location, integration of more abstract information is based on more complex representations. Objects move, so features originating at disparate locations across time should be represented as belonging to the same object (Cavanagh et al., 2008; Kahneman et al., 1992; Nishida et al., 2007). In addition, a single object or surface is typically spatially extended and all the features that belong to it should be represented together. This aspect of neural representation is often referred to as "object-based" representation.

Neuropsychological evidence suggests that the pulvinar may also be involved in object-based representation. In a patient with a unilateral lesion, visual targets were poorly localized contralateral to the lesion (Ward and Arend, 2007). At the same time, for a group of targets entirely in the ipsilateral field, those on the side of the array closer to the contralateral field were less likely to be found than those on the other side of the array.

A further important type of coordinate representation is known as "spatiotopic" representation. Every time we move our eyes, the retinal representation changes dramatically, yet we perceive the world as stable. To achieve this stability of conscious experience, the brain infers the location of objects relative to our body or to the scene rather than relative to our eyes. Achieving this spatiotopic coordinate frame requires combining eye-centered retinotopic information with eye position and head position and possibly scene landmarks. The parietal cortex seems to be heavily involved in this computation (Batista et al., 1999; Buneo and Anderson, 2012).

A fully spatiotopic map in the brain has not been found. Rather than creating a fully spatiotopic representation, the brain may instead use information from a weakly retinotopic cortical area such as the MT, in combination with attentional pointers and representations in memory (Burr and Morrone, 2011; Cavanagh et al., 2010). The interpretation of activity in a particular brain area may change over time, with the array of area MT for example being used to code retinal locations at one time and spatiotopic locations at another. This requires sophisticated dynamic routing in order to achieve the right population activity at the right time.

Parietal Cortex

The parietal cortex, particularly inferior parts, are at the crossroads for various sensory and motor information. Besides the pulvinar, the parietal cortex is the only brain area that, when lesioned, has been shown to lead to profound binding deficits. Specifically, patients presented with multiple colored letters frequently report the wrong pairings of color and letter, while perception of individual features is relatively intact (Arguin et al., 1994; Friedman-Hill et al., 1995; Robertson et al., 1997). The intraparietal sulcus may be particularly important for feature binding (Esterman et al., 2007). Another parietal area, the angular gyrus, is important for body image, with electrical stimulation sometimes yielding out-of body experiences (Blanke et al., 2002). This, together with its involvement in motor selection may give it a critical role in feelings of agency (Chambon et al., 2013). Thus the angular gyrus appears to link vestibular, somatosensory, and action plan information, whereas the role of the intraparietal sulcus is in linking sensory information, possibly visual features only. Regarding its structure, as an area of cortex with typical laminar structure it does not appear to be specialized for connections among brain areas. These parietal areas seem to process information relevant to particular functions, such as visual feature binding and body image and agency. The pulvinar and claustrum may link complex neural assemblies more broadly.

Intralaminar Thalamic Nuclei

The intralaminar thalamic nuclei are a part of the thalamus, the relay station for visual, somatosensory, and auditory information to the cortex. These nuclei are also involved in attention and arousal (Kinomura et al., 1996). As in the case of the pulvinar, we believe that the intralaminar thalamic nuclei are crucial to higher-order sensory processing, but are not the brain region conducting sensory binding.

The Claustrum

The claustrum lies medial of the insular cortex and consists of a thin sheet of neurons, which are of uncertain ontogenetic origin.

Different cortical areas are connected to different parts of the claustrum (Figure 6.2). Typically, topological relationships such as retinotopy

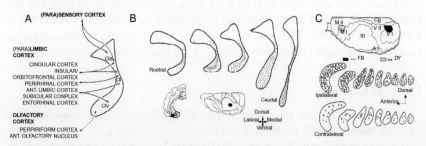

FIGURE 6.2   Cortex–claustrum connections. A: The claustrum is extensively reciprocally connected to the cortex. B: Connections with olfactory and non-sensory cortices in the ventral claustrum. *From Witter et al., 1988.* C: Connections with sensory cortices in the dorsal claustrum. *From Sadowski et al., 1997.*

are maintained (Olson and Graybiel, 1980), and the input from each modality spans a large fraction of the claustrum. Claustro-cortical projections target layer IV of the cortex (LeVay and Sherk, 1981).

The ventral part of the claustrum is connected to the olfactory cortex and to non-sensory cortices, such as the cingulate, insular, orbitofrontal, perirhinal, anterior limbic, subicular and entorhinal cortex (Witter et al., 1988). The dorsal part of the claustrum is reciprocally connected to sensory cortices. It is further divided into an audiovisual and a somatomotor part (Sadowski et al., 1997), bundling modalities which are likely to co-occur in a stimulus. Vision and audition are more likely to represent objects at a distance, while somatosensation and motor output deal with events at the person's body.

This anatomical pattern of connectivity supports the hypothesis that the claustrum is a conductor of cortical assemblies (Crick and Koch, 2005). Whereas the connections of the pulvinar include thalamic and subcortical areas, the claustrum is focused on the cortex. The claustrum may link together higher-order cortical areas while the pulvinar's role includes more the linkage between areas of primary sensory activity.

As we review below, the electrophysiological properties are consistent with the cortical conductor hypothesis. The (limited) psychophysical and psychopharmacological data available also support this proposition.

RELEVANT EMPIRICAL DATA FROM ELECTROPHYSIOLOGY

The intrinsic properties and connections of the neurons in the claustrum are crucial for its suggested role as a high-level cortical integrator. Together, they must generate dynamics capable of conducting cortical assemblies. How exactly this coding function would emerge from the computational function and the physiological function (for a discussion of these concepts see Torben-Nielsen and Stiefel, 2010) is not clear; however, we need to assume that a certain amount of lateral synaptic interaction between the neurons of the claustrum is taking place.

This seems to be the case, given that the neurons in the claustrum are connected with excitatory synapses. Electron microscope studies have demonstrated the presence of asymmetric (usually excitatory) and symmetric (usually inhibitory) synapses (Juraniec et al., 1974; Norita and Hirata, 1976). Synaptic connections exist directly between neurons in the claustrum, and axon collaterals of claustro-cortical projections also synapse onto other claustral neurons (LeVay and Sherk, 1981). The firing dynamics of principal claustral neurons are similar to those of regular firing pyramidal neurons in the cortex (Shibuya and Yamamoto, 1998).

Stimulating claustral neurons in the anesthetized cat in vivo caused a marked reduction of firing rates in the primary visual cortex, during both spontaneous firing and stimulation with stimuli matching the neurons' orientation preference (Ptito and Lassonde, 1981). This is intriguing, since it suggests that cortico-claustral connections target inhibitory interneurons in the cortex. Such interneurons are believed to generate the gamma oscillations in the cortex (Buhl et al., 1998).

In vivo recordings from awake macaques have shown that while the claustrum as a whole is multimodal, individual neurons in the claustrum respond to one modality (visual or auditory in this study) only (Remedios et al., 2010). The authors conclude that this speaks against the claustrum's role as an integrator of sensory information. However, we believe that a structure with tightly interleaved and laterally connected neurons responsive to different modalities can serve as an integrator of these modalities. This highlights the mystery of what computation the within-claustrum connections compute. More specifically, if the conductor hypothesis is correct, what criteria does the claustrum use to determine what cortical areas to link together? In any case, we believe that the known anatomy and electrophysiology of the claustrum generally supports its suggested role as a cortical conductor.

RELEVANT EMPIRICAL DATA FROM PSYCHOPHARMACOLOGY

We have recently found evidence for a role of the claustrum in integration and binding in subjective reports of pharmacological effects (Stiefel and Holcombe, in preparation). The mint relative *Salvia divinorum* contains the kappa-opiate agonist Salvitorin A, a substance with strong psychoactive effects. Immunohistochemical studies show that while they are expressed in several brain regions, kappa-opiate receptors are most densely present in the claustrum (Peckys and Landwehrmeyer, 1999). Hence, any psychological effects of kappa-opiate receptor agonists like Salvitorin A are likely to reflect mainly effects on the claustrum. Human subjects exposed to strong doses of *Salvia divinorum* report experiences consistent with a disruption of the conductor of consciousness or a structure involved in binding. Reports include a loss of the sense of existing in three-dimensional space and the impression of being in places unlike the current location of physical existence. These effects point to a disruption of high-level cortical binding.

In contrast, classical serotonergic hallucinogens mainly lead to distortions of visual and auditory percepts, neuropharmacological effects quite distinct from those elicited by *Salvia divinorum*.

Hence, a substance which seems mainly to lead to a perturbation of the claustrum (by kappa-opiate receptor agonists) leads to a significant perturbation of binding.

TESTABLE PREDICTIONS AND FUTURE DIRECTIONS

Patients with unilateral ablations of the claustrum exist, and after recovery patients did not show any noticeable abnormalities and returned to their normal social and professional lives (Duffau et al., 2007). Duffau and colleagues raised the possibility that brain restructuring during the development of the gliomas led to the absence of noticeable effects on the patients' behavior. Additionally, the remaining claustrum in the other hemisphere could take over some function. In this study, only basic neurological tests were conducted. We speculate that more in-depth examinations of such patients would reveal subtle changes in:

- the binding of percepts from the hemisphere contralateral to the removed claustrum. Psychophysically difficult binding tasks will probably be executed with more errors and more slowly than tasks allocated to the intact hemisphere. Candidate tasks would be those used to reveal visual feature binding deficits in parietal and pulvinar patients, and more sensitive psychophysical tests that have been used in normals but not yet in patients (Clifford et al., 2004; Holcombe, 2009; Holcombe and Cavanagh, 2001; Holcombe et al., 2011; Seymour et al., 2009).
- the EEG of the affected hemisphere. A reduction in task-related gamma oscillations would be expected. This will be the case if the hypotheses are correct that: 1) gamma oscillations are involved in binding, and 2) the claustrum conducts the formation of assemblies bound by gamma oscillations.

Ultimately, the computational hypotheses laid out in this chapter can only be supported by multiple coincident recordings from the claustrum and sensory cortical areas during perception. If the hypothesis presented here – that the claustrum is a conductor of binding – is correct, then claustral activity would precede episodes of cortical activity synchronous between two or more cortical areas. Recordings from the claustrum in awake monkeys have already been conducted (Remedios et al., 2010).

Furthermore, a thorough characterization of the connectivity and intrinsic electrophysiology of the neurons in the claustrum will aid our understanding of this brain structure and allow the construction of detailed computational models of the claustrum. We speculate that the dynamics of selecting a "spotlight of consciousness" will follow rapid winner-takes-all dynamics.

We believe that all of these suggestions are well within the technical capabilities of modern neuroscience and hope that they will be taken up by experimentalists and modelers to elucidate the function of this fascinating brain structure.

CONCLUSIONS

We argue here that a small number of brain areas are particularly important for high-level coordination of brain activity. The main areas appear to be parts of parietal cortex, the intralaminar thalamic nuclei, the pulvinar and the claustrum. The literature provides anatomical, physiological and clinical/neuropharmacological support for this proposition.

Furthermore, different neuromodulatory responsiveness in these highest-level coordination areas will be indicative of their function. The claustrum is known to contain S2 serotonin receptors (Altar et al., 1985) and respond to both acetylcholine and dopamine (Salerno et al., 1981). The pulvinar equally responds to acetylcholine (Godfraind, 1975), serotonin (Monckton and McCormick, 2002) and dopamine (García-Cabezas et al., 2007). The angular gyrus as a part of the cortex also receives input from all three of these neuromodulatory systems. It seems likely that subtle differences in the responses of these brain regions to these neuromodulators will determine their state-dependent activation and precise role in conducting activity in the rest of the brain.

In the light of the probability that there are several high-order integration areas, we suggest a new analogy. Although a classical music orchestra has only a single conductor, this is not true of the brain, and it is probably not true even of narrower functions such as consciousness. Rather than the claustrum being a lone conductor, it co-exists with other conductors that have somewhat different roles.

A better analogy may be that of a group of American football coaches, whose hand signals all influence the action on the field simultaneously. Different coaches have different responsibilities, such as for the defensive line or for the defensive backfield. The activity of the line and backfield are partly distinct but do overlap and interact. The anatomy provides some clues to how our neural "coaches" differ in their roles. Physiological evidence indicates that one role of the pulvinar is to synchronize visual neurons across brain areas responsive to stimuli at the focus of attention. Areas of parietal cortex are likely involved in the coordination of these brain regions. Parietal areas seem to be involved in bringing together sensory signals and combining their differing reference frames. The claustrum, with exclusively cortical connections, probably acts as a more general coordinator of the cortex.

References

Altar, C.A., O'Neil, S., Walter, R.J., Marshall, J.F., 1985. Brain dopamine and serotonin receptor sites revealed by digital subtraction autoradiography. Science 228, 597–600.

Arend, I., Rafal, R., Ward, R., 2008. Spatial and temporal deficits are regionally dissociable in patients with pulvinar lesions. Brain 131 (8), 2140–2152.

Arguin, M., Cavanagh, P., Joanette, Y., 1994. Visual feature integration with an attentional deficit. Brain Cogn. 24, 44–56.

Batista, A.P., Buneo, C.A., Snyder, L.H., Andersen, R.A., 1999. Reach plans in eye-centered coordinates. Science 285, 257–260.

Beckstead, R.M., 1984. The thalamostriatal projection in the cat. J. Comp. Neurol. 223, 313–346.

Blanke, O., Ortigue, S., Landis, T., Seeck, M., 2002. Stimulating illusory own-body perceptions. Nature 419 (6904), 269–270.

Buhl, E.H., Tamás, G., Fisahn, A., 1998. Cholinergic activation and tonic excitation induce persistent gamma oscillations in mouse somatosensory cortex in vitro. J. Physiol. 513, 117–126.

Buneo, C.A., Andersen, R.A., 2012. Integration of target and hand position signals in the posterior parietal cortex: effects of workspace and hand vision. J. Neurophysiol. 108, 187–199.

Burgess, N., Maguire, E.A., O'Keefe, J., 2002. The human hippocampus and spatial and episodic memory. Neuron 35, 625–641.

Burr, D.C., Morrone, M.C., 2011. Spatiotopic coding and remapping in humans. Phil. Trans. R. Soc. B 366 (1564), 504–515.

Buzsaki, G., 2011. Rhythms of the Brain, first ed. Oxford University Press, USA.

Cai, R.H., Schlag, J., 2001. Asynchronous feature binding and the flash-lag illusion. Invest. Ophth. & Vis. Sci. 42, S711

Cavanagh, P., Holcombe, A.O., Chou, W., 2008. Mobile computation: spatiotemporal integration of the properties of objects in motion. J. Vis. 8 (12), 1–23.

Cavanagh, P., Hunt, A.R., Afraz, A., Rolfs, M., 2010. Visual stability based on remapping of attention pointers. Trends Cogn. Sci. 14 (4), 147–153.

Chambon, V., Wenke, D., Fleming, S.M., Prinz, W., Haggard, P., 2013. An online neural substrate for sense of agency. Cereb. Cortex 23 (5), 1031–1037.

Clifford, C.W.G., Holcombe, A.O., Pearson, J., 2004. Rapid global form binding with loss of associated colors. J. Vis. 4 (12), 1090–1101.

Crick, F.C., Koch, C., 2005. What is the function of the claustrum? Phil. Trans. R. Soc. B 360, 1271–1279.

Duffau, H., Mandonnet, E., Gatignol, P., Capelle, L., 2007. Functional compensation of the claustrum: lessons from low-grade glioma surgery. J. Neurooncol. 81, 327–329.

Esterman, M., Verstynen, T., Robertson, L.C., 2007. Attenuating illusory binding with TMS of the right parietal cortex. NeuroImage 35 (3), 1247–1255.

Felleman, D.J., Van Essen, D.C., 1991. Distributed hierarchical processing in the primate cerebral cortex. Cereb Cortex 1, 1–47.

Friedman-Hill, S.R., Robertson, L.C., Treisman, A., 1995. Parietal contributions to visual feature binding: evidence from a patient with bilateral lesions. Science 269, 853–856.

García-Cabezas, M.Á., Rico, B., Sánchez-González, M.Á., Cavada, C., 2007. Distribution of the dopamine innervation in the macaque and human thalamus. NeuroImage 34, 965–984.

Godfraind, J.M., 1975. Micro-electrophoretic studies in the cat pulvinar region: effect of acetylcholine. Exp. Brain Res. 22, 243–254.

Harting, J.K., Updyke, B.V., 2006. Oculomotor-related pathways of the basal ganglia. Prog. Brain Res. 151, 441–460.

Highley, J.R., Walker, M.A., Crow, T.J., Esiri, M.M., Harrison, P.J., 2003. Low medial and lateral right pulvinar volumes in schizophrenia: a postmortem study. Am. J. Psych. 160 (6), 1177–1179.

Holcombe, A.O., 2009. Temporal binding favors the early phase of color changes, but not of motion changes, yielding the color-motion asynchrony illusion. Vis. Cogn. 17 (1-2), 232–253. [Special issue on feature binding].

Holcombe, A.O., Cavanagh, P., 2001. Early binding of feature pairs for visual perception. Nat. Neurosci. 4 (2), 127–128.

Holcombe, A.O., Linares, D., Vaziri-Pashkam, M., 2011. Perceiving spatial relations via attentional tracking and shifting. Curr. Biol. 21 (13), 1135–1139.

Iurilli, G., Ghezzi, D., Olcese, U., Lassi, G., Nazzaro, C., Tonini, R., et al., 2012. Sound-driven synaptic inhibition in primary visual cortex. Neuron 73 (4), 814–828.

Juraniec, J., Wrzolkowa, T., Narkiewicz, O., 1974. Types of synapses in the claustrum of the cat. Acta Neurobiol. Exp. (Wars) 34, 233–252.

Kahneman, D., Treisman, A., Gibbs, B.J., 1992. The reviewing of object files: object-specific integration of information. Cogn. Psychol. 24, 175–219.

Kinomura, S., Larsson, J., Gulyás, B., Roland, P.E., 1996. Activation by attention of the human reticular formation and thalamic intralaminar nuclei. Science 271, 512–515.

Künzle, H., 2006. Thalamo-striatal projections in the hedgehog tenrec. Brain Res. 1100, 78–92.

LeVay, S., Sherk, H., 1981. The visual claustrum of the cat. I. Structure and connections. J. Neurosci. 1, 956–980.

Lin, C.-S., May, P.J., Hall, W.C., 1984. Nonintralaminar thalamostriatal projections in the gray squirrel (Sciurus carolinensis) and tree shrew (Tupaia glis). J. Comp. Neurol. 230, 33–46.

Linke, R., De Lima, A.D., Schwegler, H., Pape, H.C., 1999. Direct synaptic connections of axons from superior colliculus with identified thalamo-amygdaloid projection neurons in the rat: possible substrates of a subcortical visual pathway to the amygdala. J. Comp. Neurol. 403, 158–170.

Mishkin, M., Ungerleider, L.G., Macko, K.A., 1983. Object vision and spatial vision: two cortical pathways. Trends Neurosci. 6, 414–417.

Monckton, J.E., McCormick, D.A., 2002. Neuromodulatory role of serotonin in the ferret thalamus. J. Neurophysiol. 87, 2124–2136.

Nishida, S., Watanabe, J., Kuriki, I., Tokimoto, T., 2007. Human visual system integrates color signals along a motion trajectory. Curr. Biol. 17 (4), 366–372.

Norita, M., Hirata, Y., 1976. Some electron microscope findings of the claustrum of the cat. Arch. Histol. Jpn. 39, 33–49.

Olson, C.R., Graybiel, A.M., 1980. Sensory maps in the claustrum of the cat. Nature 288, 479–481.

Peckys, D., Landwehrmeyer, G., 1999. Expression of mu, kappa, and delta opioid receptor messenger RNA in the human CNS: a 33 P in situ hybridization study. Neuroscience 88, 1093–1135.

Ptito, M., Lassonde, M.C., 1981. Effects of claustral stimulation on the properties of visual cortex neurons in the cat. Exp. Neurol. 73, 315–320.

Quiroga, R.Q., Reddy, L., Kreiman, G., Koch, C., Fried, I., 2005. Invariant visual representation by single neurons in the human brain. Nature 435, 1102–1107.

Remedios, R., Logothetis, N.K., Kayser, C., 2010. Unimodal responses prevail within the multisensory claustrum. J. Neurosci. 30, 12902–12907.

Robertson, L., Treisman, A., Friedman-Hill, S., Grabowecky, M., 1997. The interaction of spatial and object pathways: evidence from Balint's syndrome. J. Cogn. Neurosci. 9, 295–317.

Saalmann, Y.B., Pinsk, M.A., Wang, L., Li, X., Kastner, S., 2012. The pulvinar regulates information transmission between cortical areas based on attention demands. Science 337, 753–756.

Sadowski, M., Moryś, J., Jakubowska-Sadowska, K., Narkiewicz, O., 1997. Rat's claustrum shows two main cortico-related zones. Brain Res. 756, 147–152.

Salerno, M.T., Zagami, M.T., Cortimiglia, R., Infantellina, F., 1981. The action of iontophoretically applied acetylcholine and dopamine on single claustrum neurones in the cat. Neuropharmacology 20, 895–899.

Seymour, K.J., Scott McDonald, J., Clifford, C.W., 2009. Failure of colour and contrast polarity identification at threshold for detection of motion and global form. Vis. Res. 49 (12), 1592–1598.

Shibuya, H., Yamamoto, T., 1998. Electrophysiological and morphological features of rat claustral neurons: an intracellular staining study. Neuroscience 85, 1037–1049.

Shipp, S., 2003. The functional logic of cortico-pulvinar connections. Phil. Trans. R. Soc. B. 358, 1605–1624.

Singer, W., 1999. Time as coding space? Curr. Opin. Neurobiol. 9, 189–194.

Steriade, M., Dossi, R.C., Paré, D., Oakson, G., 1991. Fast oscillations (20–40 Hz) in thalamocortical systems and their potentiation by mesopontine cholinergic nuclei in the cat. Proc. Natl. Acad. Sci. USA 88, 4396–4400.

Steriade, M., McCormick, D.A., Sejnowski, T.J., 1993a. Thalamocortical oscillations in the sleeping and aroused brain. Science 262, 679–685.

Steriade, M., Nunez, A., Amzica, F., 1993b. Intracellular analysis of relations between the slow (<1 Hz) neocortical oscillation and other sleep rhythms of the electroencephalogram. J. Neurosci. 13, 3266–3283.

Thiele, A., Stoner, G., 2003. Neuronal synchrony does not correlate with motion coherence in cortical area MT. Nature 421, 366–370.

Tiesinga, P.H.E., Fellous, J.-M., José, J.V., Sejnowski, T.J., 2001. Computational model of carbachol-induced delta, theta, and gamma oscillations in the hippocampus. Hippocampus 11, 251–274.

Torben-Nielsen, B., Stiefel, K.M., 2010. An inverse approach for elucidating dendritic function. Front. Comput. Neurosci. 4, 128.

Treisman, A.M., Gelade, G., 1980. A feature-integration theory of attention. Cogn. Psychol. 12, 97–136.

Van Essen, D.V., Anderson, C.H., Felleman, D.J., 1992. Information processing in the primate visual system: an integrated systems perspective. Science 255, 419–423.

Ward, R., Arend, I., 2007. An object-based frame of reference within the human pulvinar. Brain 130 (9), 2462–2469.

Ward, R., Danziger, S., Owen, V., Rafal, R., 2002. Deficits in spatial coding and feature binding following damage to spatiotopic maps in the human pulvinar. Nat. Neurosci. 5 (2), 99–100.

Witter, M.P., Room, P., Groenewegen, H.J., Lohman, A.H.M., 1988. Reciprocal connections of the insular and piriform claustrum with limbic cortex: an anatomical study in the cat. Neuroscience 24, 519–539.

Structural and Functional Connectivity of the Claustrum in the Human Brain

Sudhir Pathak[1] and Juan C. Fernandez-Miranda[2]

[1]Department of Psychology, Learning Research and Development Center, University of Pittsburgh, PA, USA [2]Department of Neurological Surgery, University of Pittsburgh Medical Center, Pittsburgh, PA, USA

INTRODUCTION

The claustrum is a telencephalic and subcortical structure found in all mammalian brains. In the human brain, it is located deep inside the insula and superficial to the basal ganglia. Multiple animal studies have demonstrated that the claustrum shows a topologically organized reciprocal connectivity with the cerebral cortex through claustrocortical and corticoclaustral projection fibers (Carman et al., 1964; Druga, 1966, 1968, 1972, 1982; Kemp and Powell, 1970; Narkiewicz, 1964, 1966, 1972; Pearson et al., 1982). Based on this widespread connectivity, Crick and Koch (2005) theorized that the key functional role of the claustrum is to integrate cortical information that might give rise to conscious percept.

Most investigations on the connectivity of the claustrum are histological studies with injected tracers and neurophysiological recording in animals (mouse, cat, monkey). These studies provided the basis for understanding the connectivity and function of the claustrum but are, however, difficult to extrapolate to the human brain. Consequently, the function and connectivity of the claustrum in the human brain is currently largely unknown (Edelstein and Denaro, 2004). Recent advancements and improvements in MRI techniques, such as functional MRI (fMRI; Buxton and Frank, 1997) and diffusion MRI (dMRI; Basser, et al.,

J. Smythies, L. Edelstein and V. Ramachandran (Eds):
The Claustrum.

DOI: http://dx.doi.org/10.1016/B978-0-12-404566-8.00007-6

1994), can provide new insights into its functional role and underlying structural connectivity. Importantly, these advanced MRI techniques can be applied "in vivo" to study the connectivity of claustral fibers and may facilitate better understanding of its role in human cognition (Fernández-Miranda et al., 2008a).

In this chapter, we aim to briefly review the anatomy of the human claustrum and the connectivity of the claustrum in animals, summarize the main fMRI studies on this topic, and expand on our structural human brain connectivity studies, which employ advanced diffusion imaging MR-based techniques.

MACROSCOPIC AND WHITE MATTER ANATOMY OF THE HUMAN CLAUSTRUM

The anatomy of the claustrum in the human brain can be studied by examining axial, coronal, and sagittal slices of post-mortem brains in order to understand its topology and macroscopic configuration. Based on these studies, the claustrum is divided into two parts: the dorsal claustrum, also referred to as compact (Rae, 1954a, 1954b) or insular claustrum (Brand, 1981; Filimonoff, 1966; Morys et al., 1993); and the ventral claustrum, also named fragmented (Rae, 1954a, 1954b) prepiriform, amygdalar, or temporal claustrum (Filimonoff, 1966; Morys et al., 1993).

The *dorsal claustrum* is a continuous irregular lamina of gray matter lying between the putamen (from which it is separated by the external capsule) and the insular cortex (from which it is separated by the extreme capsule) (Figure 7.1). It has the form of a plate, narrowing in the superior direction and widening in the inferior direction, giving it a triangular form in the coronal cross-section. This contrasts with the external capsule, which is wider in the superior direction and then becomes thin and narrow, or even non-existent, in the area adjacent to the lower part of the dorsal claustrum.

The *ventral claustrum* consists of a group of diffuse or island-like gray masses fragmented by the uncinate and the inferior occipitofrontal fascicles. We divide the ventral claustrum into two parts: superior and inferior. The superior part of the ventral claustrum is continuous with the anteroinferior pole of the dorsal claustrum and extends downward toward the base of the frontal lobe below the putamen, adjoining the prepiriform cortex. The inferior part of the ventral claustrum is continuous with the posteroinferior pole of the dorsal claustrum and is directed toward the amygdalar region (Figures 7.1–7.4). In this area, the relationship between the ventral claustrum and the amygdala is so close that it is sometimes difficult to delineate one from the other clearly. This close

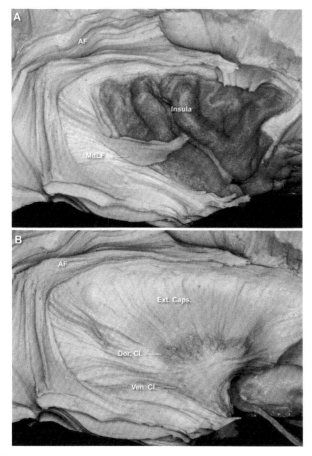

FIGURE 7.1   Fiber microdissection of post-mortem human brain (right hemisphere). A: the frontal, parietal, and temporal opercula have been removed to expose the insula. B: the insular cortex and extreme capsule have been dissected to expose the claustrum and external capsule. The fragmented or ventral claustrum can be identified at the periphery of the compact or dorsal claustrum. *MdLF, middle longitudinal fascicle; AF, arcuate fascicle; Ext. Caps, external capsule; Dor. Cl., dorsal claustrum; Ven Cl., ventral claustrum.*

anatomical relationship has been found in other ontogenetic and phylogenetic studies (Filimonoff, 1964, 1966).

In order to study the white-matter anatomy of the claustrum we employed Klingler's microsurgical fiber-dissection technique in prepared, post-mortem human brains (Fernández-Miranda et al., 2008b). After removal or dissection of the frontal, parietal, and temporal operculae, the cortical surface of the insula is exposed (see Figure 1A). The gray matter of the short and long insular gyri is carefully removed to expose the

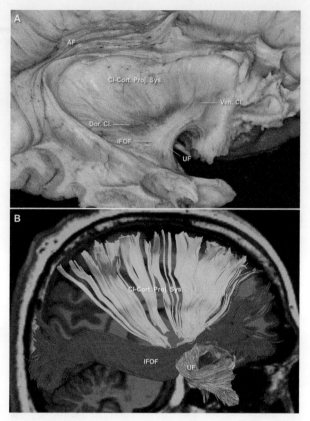

FIGURE 7.2   A: fiber microdissection of post-mortem human brain (right hemisphere, different specimen). The ventral and dorsal claustrum are identified. The dorsal portion of the external capsule is mainly formed by the radiating fibers from the dorsal claustrum, which form the claustrocortical projection system (Cl-Cor. Proj. Sys.). The ventral portion of the external capsule contains the inferior fronto-occipital fascicle (IFOF) and uncinate fascicle (UF), which cross the ventral claustrum. B: High-definition fiber tractography (HDFT) reconstruction of the claustrocortical system, IFOF, and UF in a normal subject (35 years-old male, healthy volunteer). Note the similarities with the fiber microdissection specimen. AF, arcuate fascicle; Dor. Cl., dorsal claustrum; Ven Cl., ventral claustrum.

extreme capsule and the claustrum. The extreme capsule, like the claustrum, has a ventral part composed of the lateral fibers of the uncinate and inferior occipitofrontal fascicles, and a thinner dorsal part formed by multiple short association fibers which connect the different insular gyri with each other and with the adjacent frontal, parietal, and temporal operculae. After removal of the dorsal part of the extreme capsule, the fibers of the dorsal external capsule converge in and merge with the gray matter of the dorsal claustrum, forming a characteristic spoke-and-wheel

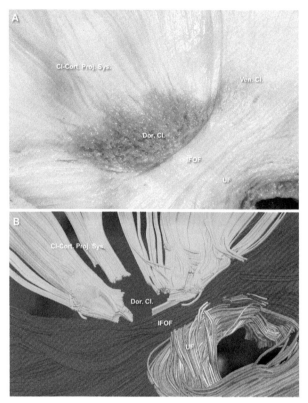

FIGURE 7.3   A: fiber microdissection of post-mortem human brain (right hemisphere). Close-up view of the claustrum region. B: High-definition fiber tractography (HDFT) reconstruction (same as 2B), close-up view. Cl-Cor. Proj. Sys., claustrocortical projection system; IFOF, inferior fronto-occipital fascicle; Dor. Cl., dorsal claustrum; UF, uncinate fascicle; Ven Cl., ventral claustrum.

pattern with its center at the dorsal claustrum (see Figures 7.1B–7.4). The fibers of the uncinate and inferior occipitofrontal fascicles traverse the most anterior and inferior parts of the claustrum to create the gray-matter islands that form the ventral claustrum (see Figures 7.2 and 7.3). Our fiber dissection studies clearly established that the so-called dorsal external capsule consists mainly of the *claustrocortical projection system*, which interconnects the claustrum with multiple cortical areas (see Figures 7.1–7.4). Therefore, we challenged the conventional "static" conceptualization of the external capsule as a group of fibers located deep to the claustrum and propose a more dynamic view of the external capsule as, indeed, the claustrocortical system (Fernández-Miranda et al., 2008a,b).

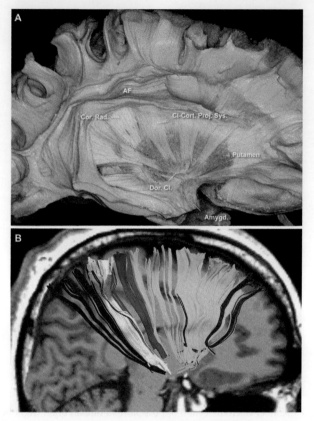

FIGURE 7.4   A: fiber microdissection of post-mortem human brain (right hemisphere). The claustrocortical fibers have been partially dissected to expose the underlying corona radiata (Cor. Rad.) and putamen. The ventral claustrum is continuous with the amygdala (amyg). B: High-definition fiber tractography (HDFT) reconstruction of the claustrocortical projection system. Each color represents a different cortical termination region (see Figure 7.5 for further details). Note the somatotopic organization of the projection fibers. AF, arcuate fascicle; Cl-Cor. Proj. Sys., claustrocortical projection system; Dor. Cl., dorsal claustrum.

CONNECTIVITY OF THE CLAUSTRUM: ANIMAL STUDIES

The location and phylogenetic origin of the claustrum suggest that it might be part of the basal ganglia (Andersen, 1968; Cortimiglia et al., 1991; Druga, 1972; Künzle, 1977; Sloniewski et al., 1986; Stanton et al., 1988). Several animal studies have shown that the claustrum's strongest connectivity is with the caudate nucleus, and some authors have suggested that the claustrum is part of the basal ganglia, which is directly connected to the cerebral cortex, as opposed to the caudate nucleus, which has indirect connections through the thalamus (Arikuni and Kubota, 1985).

Multiple studies in the rat, cat, and monkey have demonstrated that the claustrum connectivity shows topological organization with different neural systems and is reciprocally connected to the cerebral cortex and several subcortical regions. These features make the claustrum resemble the thalamus more than the basal ganglia. The claustrum shows a non-overlapping and distinct connectivity with functionally different neural circuits: the dorsal claustrum is connected to motor and somatosensory cortices and the ventral claustrum is connected to visual areas with topo-logical organization (Druga et al., 1990; George, 1986; Minciacchi et al., 1991; Selemon and Goldman-Rakic, 1988; Stanton et al., 1988; Tokuno and Tanji, 1993). The connectivity of the claustrum with the cerebral cortex follows a particular property: the claustral projections of two distinct cortical regions overlap if both regions are interconnected. For example, the projection fibers from motor areas to the claustrum do not overlap with projections from the visual areas, hippocampus or amygdala, while projections from the primary motor, premotor, and supplementary motor cortices overlap with projections from the parietal cortex, somatosensory cortex, cingulate cortex, and supplementary eye field (Tanné-Gariépy et al., 2002).

The rostral and caudal parts of the dorsal claustrum in animals also have some topological organized connectivity with the cerebral cortex (George, 1986; Pearson et al., 1982). The rostral part of the dorsal claustrum is connected with premotor, supplementary motor area, and Brodmann area (BA) 46 (Druga et al., 1990) but not with the primary motor cortex, while the caudal region is connected with the primary motor and posterior parietal cortices. Functionally, it has been suggested that the rostral part of the claustrum is involved in attention, working memory, higher order cognitive function, goal directed action, planning and temporal organization of tasks, and the caudal region of the dorsal claustrum is involved in preparation and execution of movements (Tanné-Gariépy et al., 2002). Stanton et al. (1988) showed by intracortical microstimulation of the frontal eye field in monkeys that the size of saccade depends on the topological organization of efferent projections on the claustrum, and concluded that it might be involved in integration of planned saccade and skeletomotor control.

FUNCTIONAL MRI STUDIES OF THE HUMAN CLAUSTRUM

Functional MRI measures neural activity by probing changes in MR signal of oxygenated (oxygen rich) and de-oxygenated (oxygen poor) blood flow (Buxton and Frank, 1997). The basic assumption in all fMRI-based studies is that cerebral blood flow is directly related to neural activity, and statistical methods are used to find activation areas (increased blood flow) when a particular functional task is undertaken while in a scanner. These studies provide better spatial localization of neural activity when compared to other functional techniques, such as electro- and magnetoencephalography (EEG, MEG), and PET (positron emission tomography), but has poorer time resolution (a few seconds). The main limitation of using the fMRI technique to study the functional role of the claustrum in the living human brain is the current voxel size (2–3 mm^3). Given the thin sheet-like structure of the claustrum, it is challenging to definitely identify clusters of activation within the claustrum and separate them from neighboring functional areas such as insula, basal ganglia (putamen), or amygdala.

With that limitation in mind, there have been several fMRI studies showing selective activation of the claustrum when performing a functional task. Most studies showed that the claustrum is activated during tasks that require integration of multimodal sensory information, suggesting that the claustrum is a multimodal convergence zone. Recently, Baugh et al. (2011) showed claustrum activation during a visual-motor adaptation task, which required integration of proprioceptive and visual information. Naghavi et al. (2007) demonstrated selective claustrum activation during an event-related fMRI study which included conceptually related and unrelated paired audiovisual stimuli of objects, concluding that the claustrum (and insular region) integrate conceptually related sounds and pictures. Rene Vohn and colleagues (Vohn et al., 2007) showed claustrum activation during two within-modal (auditory/auditory, visual/visual) and one cross-modal (auditory/visual) divided-attention, fMRI tasks. An investigation into how muscle proprioception and tactile messages can result in a unified percept of one's own body movement was conducted by Kavounoudias et al. (2008); in this study, insula-claustrum regions were active during the fMRI tasks that involved a congruent proprio-tactile costimulation condition. Studies like these provide stronger evidence that the human claustrum is involved in general *multimodal integration of information* from multiple neural circuits. Finally, a PET study (Qurrat-ul-Ain and Abidi, 2005) showed that the claustrum was active during a task involving visual sexual stimuli, suggesting that the claustrum might also be implicated in motivational behaviors.

DIFFUSION MRI STUDIES OF THE HUMAN CLAUSTRUM

Traditional structural MRI (T1 and T2 weighted imaging techniques) is unable to study fiber tracts in living human brains; however, dMRI is a unique technique to study the microstructure of anisotropic tissue

(Basser et al., 1994); methods based on this provide diffusion characteristics of intra- and extracellular water (Assaf and Basser, 2005; Assaf et al., 2004). These directional-dependent diffusion characteristics of water can be used to create fiber pathways that resemble axonal bundles, which makes dMRI a robust tool to study connectivity between functional areas in human brains (Mori and van Zijl, 2002).

The initial and most popular method in the literature, diffusion tensor imaging (DTI) (Basser et al., 1994), can describe anisotropy and microstructure of biological tissues. DTI-based tractography has been extensively employed to study the fiber tracts in the living human brain (Fernández-Miranda et al., 2008a). In 2008, Fernández-Miranda et al. completed the first DTI study on the claustrocortical projection system in the human brain (Fernández-Miranda et al., 2008b). Our DTI-based tractography studies showed that the external capsule is composed of multiple fiber bundles coming from the superior frontal, precentral, postcentral, superior parietal, and parieto-occipital regions, converging in the area of the dorsal claustrum. Thus, we concluded that the dorsal external capsule was mainly composed of projection (intrahemispheric corticosubcortical) and not association (intrahemispheric interlobar corticocortical) fibers. Furthermore, our tractography studies revealed a topographical organization in the dorsal claustrum and external capsule, where posterior cortical areas project into the posterior part of the dorsal claustrum, and more anterior cortical areas converge in the anterior part. These exciting results, however, had to be interpreted with caution. DTI has limited capability in terms of fiber tracking, as it fails to model complex fiber geometry in crossing regions and is poor in showing endpoint connectivity information between cortical and subcortical regions. In fact, the claustrocortical fibers in our study were not truly ending in the claustrum but converging in it, and there was evident false continuation in between the fiber tracts at the level of the claustrum. Also, the cortical regions of termination of the claustrocortical fibers were approximated by using an overlaid T1 structural sequence but we were not able to accurately demonstrate their precise cortical termination or even less quantify the volume of fibers terminating at each cortical region.

More advanced dMRI methods such as Q-ball imaging (Descoteaux et al., 2007; Tuch, 2004) and constrained spherical deconvolution (Lenglet et al., 2009; Tournier et al., 2007; Tristán-Vega, Westin, & Aja-Fernández) can resolve fiber crossings and may provide a more accurate representation of the anatomy. *High-definition fiber tractography* (HDFT) (Fernandez-Miranda et al., 2010, 2012; Wang et al., 2012) is a novel method that is based on optimization of each stage in diffusion analysis (scanning, reconstruction, fiber tracking, validation). HDFT uses diffusion spectrum imaging (DSI) scanning protocol (Wedeen et al., 2005), generalized q-sampling imaging (GQI) as reconstruction (Yeh et al., 2010), and the multi-FACT algorithm with trilinear interpolation of diffusion orientations for fiber tractography. Our human connectivity studies using HDFT (Fernández-Miranda et al., 2012; Jarbo et al., 2012; Pyles et al., 2013; Shin et al, 2012; Verstynen et al., 2011, 2012; Wang et al., 2012) have shown accurate fiber tracking in crossing regions and robust end-point connectivity between cortical and subcortical regions. We have completed validation studies that have demonstrated that, with the judicious use of HDFT, we can accurately replicate known neural circuits and fiber tracts (as compared with the fiber microdissection technique) with unprecedented detail (Fernandez-Miranda et al., 2012; Jarbo et al., 2012; Pyles et al., 2013; Shin et al, 2012; Verstynen, et al., 2011; Wang et al., 2012; Yeh et al., 2010). Importantly, we have effectively employed HDFT to map the claustrocortical projection system in multiple control subjects. In the following sections we will describe the HDFT imaging methodology and fiber tracking results.

Imaging Protocol

Data from DSI were acquired using 257 unique non-collinear gradient directions with a twice-refocused, spin-echo EPI (echo planar imaging) sequence. Multiple b-values were used to sample the q-space with TR = 9916 ms, TE = 157 ms, FOV = 231 × 231 mm, voxel size = 2.4 × 2.4 × 2.4 mm, and b-max = 7000 s/mm^2.

The total MR acquisition time for the DSI sequence was 43 minutes. A high resolution T1 weighted structural image was acquired with TR = 2110 ms, TE = 2.63 ms, flip angle = 8°, 176 slices, FOV = 256 × 256 − mm^2, and voxel size = 1.0 × 1.0 × 1.0 mm^3.

The total MR acquisition time for the T1 weighted image was 9 minutes.

Freesurfer Reconstruction

Freesurfer analysis is performed on MP-RAGE images to create the gray-white matter surfaces. Freesurfer segments gray-white matter into different cortical regions based on sulci and gyri pattern, and also segments major subcortical regions in the brain. All cortical and subcortical regions of interest (ROIs) are then transformed to the diffusion space using FSL's Flirt package (Jenkinson et al., 2002; Jenkinson and Smith, 2001). An affine linear registration is performed with 12 degrees of freedom. An inverse transform is also created to transform fiber tracks into MP-RAGE using the TrackVis (www.Trackvis.org) track_transform function. After transformation takes place in diffusion space these ROIs are then used to seed for fiber tracking. Since Freesurfer does not provide automatic claustrum segmentation, the claustrum is manually traced on MP-RAGE space.

Diffusion Analysis

The DSI (Wedeen et al., 2005) dataset is reconstructed using the GQI method to create the oriented distributed functions (ODF) (Descoteaux et al., 2007; Lenglet et al., 2009; Tristán-Vega et al., 2009; Tuch, 2004). ODF values are defined on a discrete unit sphere with 362 points. A mean diffusion length of 1.2 is used to reconstruct the ODFs (Yeh et al., 2010). The local maxima of the ODF is calculated on the unit sphere for each voxel. The local maxima for each voxel provide an estimate for major fiber orientations, and these in turn are used to create streamlines using the multi-FACT algorithm. A trilinear interpolation is used to create fiber orientations at subvoxel resolution for each step in the fiber tracking algorithm. All analysis described above was performed in DSI Studio (http://dsi-studio.labsolver.org).

The local seeding method is used to create all the claustrum projection fibers, in contrast to whole brain seeding, which involves first generating seed points in the whole brain and then using ROIs to segment fiber tracts of interest. The local seeding approach uses known ROIs to generate seeds for creating fiber tracts and this provides much denser fiber tracts and more detailed end-point cortical projections. Fiber tracking is performed for each seed point by randomly choosing an initial fiber orientation for each voxel in order to reduce bias for major fiber bundles. The claustrum ROI is manually traced in diffusion space. Fiber tracking is then performed using the local seeding approach (claustrum ROI as the seed ROI). Fibers that end in the claustrum region are then selected and saved in the TrackVis file format. All fibers along with T1 underlay are then rendered using the TrackVis visualization program (http://trackvis.org).

HDFT Results: The Claustrocortical Projection System

Our HDFT studies have shown the claustrum connectivity in great detail, including the end-point projection pattern onto the cerebral cortex. The claustrum fiber bundles show a ribbon-like pattern that highly resembles the fiber microdissection findings (see Figures 7.2 and 7.4). In contrast to our previous DTI studies, these show that the claustrocortical fibers are clearly terminating at the level of the dorsal claustrum with minimal degree of false continuation in between the fibers. Importantly, the fibers of the uncinate fascicle (UF) and inferior fronto-occipital fascicle (IFOF) pass just ventral to the dorsal claustrum without any of their fiber showing a termination at that point. The IFOF runs from the prefrontal to the parieto-occipital regions, and the UF interconnects the orbital-frontal cortex with the medial temporal lobe and temporal pole. Both the IFOF and UF are supposed to cross the scattered gray matter of

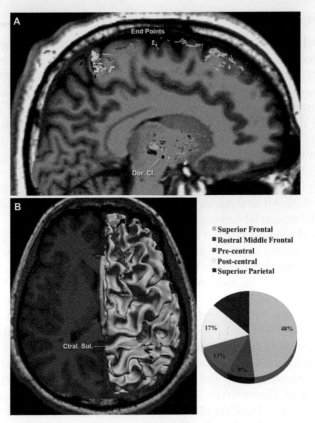

FIGURE 7.5   A: HDFT reconstruction of the end points of the claustrocortical projection system. Note the clustering of the fiber end points at the level of the claustrum, and the somatotopic organization of the cortical termination end points. B: axial view; the cortical surface of the right hemisphere has been reconstructed to allow for identification of the cortical region of termination. The sectorial diagram shows the distribution of the fiber termination onto different cortical regions and the color coding for each region. Ctral. Sul, central sulcus; Dor. Cl., dorsal claustrum.

the ventral claustrum, as shown in our fiber dissection studies, but with fiber tracking techniques we cannot investigate these complex structural (and functional) relationships (see Figures 7.2 and 7.3).

The detailed analysis of the cortical termination of the claustrum fibers revealed interconnections with the superior and rostral middle frontal gyri, premotor, somatosensory, and posterior parietal cortex areas (Figure 7.5). It is important to note that with HDFT we did not find any connection from dorsal claustrum to the occipital cortex. The claustrum is connected to both primary motor and sensory areas, but the predominant connectivity is with associative cortical areas (superior frontal gyrus or supplementary motor region and posterior parietal cortex) (Table 7.1) (see

TABLE 7.1   Fiber Volume for Each Segment of the Claustrocortical Fiber Pathways and their Relative Percentage Volume

Connection Area	Volume (ml)	Length (mm)	Std Length	Voxel Count	Track Count	% Volume
Superior Frontal	21.28	70.22	4.19	1531	10134	48.25
Rostral Middle Frontal	4.07	68.41	6.35	293	189	9.23
Pre-central	5.57	65.57	2.05	401	2366	12.64
Post-central	7.31	69.33	4.46	526	17831	16.58
Superior Parietal	5.86	68.99	5.17	422	5494	13.30
Total	44.09	342.52		3173	36014	100

Figures 7.4 and 7.5). In particular, the frontal cortex shows very high connectivity when compared with the remaining cortical areas. This structural connectivity pattern definitely supports the notion of claustrum involvement in multimodal integration of sensory and motor information.

SUMMARY

The connectivity of the claustrum has been extensively studied in animals, but there is very limited knowledge on the structural connectivity of the human claustrum. Fiber microdissection techniques in post-mortem brains allow for an initial approximation to the complex architecture of the fiber tract systems closely related to the human claustrum. Diffusion MRI techniques are rapidly evolving to offer unique insights into the structural connectivity of the brain in general, and the claustrum in particular. HDFT is an optimized dMRI technique that overcomes many of the limitations of DTI. Our HDFT studies have shown claustrum connectivity in great detail, including the end-point projection pattern onto the cerebral cortex. Next steps should include the combination of high quality fiber-tracking studies, such as HDFT, with advanced functional imaging methods, such as MEG, task-based fMRI, and resting state fMRI, in order to bring new light onto the functional role of the human claustrum.

References

Andersen, D., 1968. Some striatal connections to the claustrum. Exp. Neurol. 20 (2), 261–267.
Arikuni, T., Kubota, K., 1985. Claustral and amygdaloid afferents to the head of the caudate nucleus in macaque monkeys. Neurosci. Res. 2 (4), 239–254.

Assaf, Y., Basser, P., 2005. Composite hindered and restricted model of diffusion (CHARMED) MR imaging of the human brain. Neuroimage 27 (1), 48–58.

Assaf, Y., Freidlin, R.Z., Rohde, G.K., Basser, P.J., 2004. New modeling and experimental framework to characterize hindered and restricted water diffusion in brain white matter. Magn. Reson. Med. 52 (5), 965–978.

Basser, P., Mattiello, J., LeBihan, D., 1994. Estimation of the effective self-diffusion tensor from the NMR spin echo. J. Magn. Reson. B 103 (3), 247–254.

Baugh, L., Lawrence, J., Marotta, J., 2011. Novel claustrum activation observed during a visuomotor adaptation task using a viewing window paradigm. Behav. Brain Res. 223 (2), 395–402.

Brand, S., 1981. A serial section Golgi analysis of the primate claustrum. Anat. Embryol. 162 (4), 475–488.

Buxton, R., Frank, L., 1997. A model for the coupling between cerebral blood flow and oxygen metabolism during neural stimulation. J. Cereb. Blood Flow Metab. 17 (1), 64–72.

Carman, J., Cowan, W., Powell, T., 1964. The cortical projection upon the claustrum. J. Neurol. Neurosurg. Psychiatr. 27, 46–51.

Cortimiglia, R., Crescimanno, G., Salerno, M.T., Amato, G., 1991. The role of the claustrum in the bilateral control of frontal oculomotor neurons in the cat. Exp. Brain Res. 84 (3), 471–477.

Crick, F., Koch, C., 2005. What is the function of the claustrum? Philos. Trans. R. Soc. Lond., B, Biol. Sci. 360 (1458), 1271–1279.

Descoteaux, M., Angelino, E., Fitzgibbons, S., Deriche, R., 2007. Regularized, fast, and robust analytical Q-ball imaging. Magn. Reson. Med. 58 (3), 497–510.

Druga, R., 1966. Cortico-claustral connections. I. Fronto-claustral connections. Folia Morphol. (Praha) 14 (4), 391–399.

Druga, R., 1968. Cortico-claustral connections. II. Connections from the parietal, temporal and occipital cortex to the claustrum. Folia Morphol. (Praha) 16 (2), 142–149.

Druga, R., 1972. Efferent projections from the claustrum (an experimental study using Nauta's method). Folia Morphol. (Praha) 20 (2), 163–165.

Druga, R., 1982. Claustro-neocortical connections in the cat and rat demonstrated by HRP tracing technique. J. Hirnforsch. 23 (2), 191–202.

Druga, R., Rokyta, R., Benes, V., 1990. Claustro-neocortical projections in the rhesus monkey (projections to area 6). J. Hirnforsch. 31 (4), 487–494.

Edelstein, L., Denaro, F., 2004. The claustrum: a historical review of its anatomy, physiology, cytochemistry and functional significance. Cell. Mol. Biol. 50 (6), 675–702.

Fernández-Miranda, J., Engh, J.A., Pathak, S.K., et al., 2010. High-definition fiber tracking guidance for intraparenchymal endoscopic port surgery. J. Neurosurg. 113 (5), 990–999.

Fernández-Miranda, J., Pathak, S., Engh, J., et al., 2012. High-definition fiber tractography of the human brain: neuroanatomical validation and neurosurgical applications. Neurosurgery 71 (2), 430–453.

Fernández-Miranda, J., Rhoton Jr., A.L., Álvarez-Linera, J., et al., 2008a. Three-dimensional microsurgical and tractographic anatomy of the white matter of the human brain. Neurosurgery 62 (6 Suppl. 3), 989–1028.

Fernández-Miranda, J., Rhoton Jr., A.L., Kakizawa, Y., et al., 2008b. The claustrum and its projection system in the human brain: a microsurgical and tractographic anatomical study. J. Neurosurg. 108 (4), 764–774.

Filimonoff, I., 1964. Homologies of the cerebral formations of mammals and reptiles. J. Hirnforsch. 7 (2), 229–251.

Filimonoff, I., 1966. The claustrum, its origin and development. J. Hirnforsch. 8 (5), 503–528.

George, R.L., 1986. Afferent and efferent connections of the dorsolateral precentral gyrus (area 4, hand/arm region) in the macaque monkey, with comparisons to area 8. J. Comp. Neurol., 254.

Jarbo, K., Verstynen, T., Schneider, W., 2012. In vivo quantification of global connectivity in the human corpus callosum. Neuroimage 59 (3), 1988–1996.

Jenkinson, M., Bannister, P.R., Brady, J.M., Smith, S.M., 2002. Improved optimisation for the robust and accurate linear registration and motion correction of brain images. Neuroimage 17 (2), 825–841.

Jenkinson, M., Smith, S.M., 2001. A global optimisation method for robust affine registration of brain images. Med. Image Anal. 5 (2), 143–156.

Kavounoudias, A., Roll, J.P., Anton, J.L., et al., 2008. Proprio-tactile integration for kinesthetic perception: an fMRI study. Neuropsychologia 46 (2), 567–575.

Kemp, J., Powell, T., 1970. The cortico-striate projection in the monkey. Brain 93 (3), 525–546.

Künzle, H., 1977. Projections from the primary somatosensory cortex to basal ganglia and thalamus in the monkey. Exp. Brain Res. 30 (4), 481–492.

Lenglet, C., Campbell, J.S.W., Descoteaux, M., et al., 2009. Mathematical methods for diffusion MRI processing. Neuroimage 45 (Suppl. 1), 22.

Minciacchi, D., Granato, A., Barbaresi, P., 1991. Organization of claustro-cortical projections to the primary somatosensory area of primates. Brain Res. 553 (2), 309–312.

Mori, S., van Zijl, P., 2002. Fiber tracking: principles and strategies - a technical review. NMR Biomed. 15 (7–8), 468–480.

Morys, J., Narkiewicz, O., Wisniewski, H., 1993. Neuronal loss in the human claustrum following ulegyria. Brain Res. 616 (1-2), 176–180.

Naghavi, H., Eriksson, J., Larsson, A., Nyberg, L., 2007. The claustrum/insula region integrates conceptually related sounds and pictures. Neurosci. Lett. 422 (1), 77–80.

Narkiewicz, O., 1964. Degenerations in the claustrum after regional neocortical ablations in the cat. J. Comp. Neurol. 123, 335–355.

Narkiewicz, O., 1966. [Connections of the claustrum with the cerebral cortex]. Folia Morphol. (Praha) 25 (4), 555–561.

Narkiewicz, O., 1972. Frontoclaustral interrelations in cats and dogs. Acta Neurobiol. Exp. (Wars) 32 (2), 141–150.

Pearson, R., Brodal, P., Gatter, K.C., Powell, T.P., 1982. The organization of the connections between the cortex and the claustrum in the monkey. Brain Res. 234 (2), 435–441.

Pyles, J.A., Verstynen, T.D., Schneider, W., Tarr, M.J., 2013. Explicating the Face Perception Network with White Matter Connectivity. PLoS ONE, 8.

Qurrat-ul-Ain, Abidi, T., 2005. Unraveling the function of claustrum. JPMA. J. Pak. Med. Assoc. 55 (3), 123–125.

Rae, A., 1954a. The connections of the claustrum. Confin. Neurol. 14 (4), 211–219.

Rae, A., 1954b. The form and structure of the human claustrum. J. Comp. Neurol. 100 (1), 15–39.

Selemon, L., Goldman-Rakic, P., 1988. Common cortical and subcortical targets of the dorsolateral prefrontal and posterior parietal cortices in the rhesus monkey: evidence for a distributed neural network subserving spatially guided behavior. J. Neurosci. 8 (11), 4049–4068.

Shin, S., Verstynen, T., Pathak, S., et al., 2012. High-definition fiber tracking for assessment of neurological deficit in a case of traumatic brain injury: finding, visualizing, and interpreting small sites of damage. J. Neurosurg. 116 (5), 1062–1069.

Sloniewski, P., Usunoff, K., Pilgrim, C., 1986. Diencephalic and mesencephalic afferents of the rat claustrum. Anat. Embryol. 173 (3), 401–411.

Stanton, G., Goldberg, M., Bruce, C., 1988. Frontal eye field efferents in the macaque monkey: I. Subcortical pathways and topography of striatal and thalamic terminal fields. J. Comp. Neurol. 271 (4), 473–492.

Tanné-Gariépy, J., Boussaoud, D., Rouiller, E., 2002. Projections of the claustrum to the primary motor, premotor, and prefrontal cortices in the macaque monkey. J. Comp. Neurol. 454 (2), 140–157.

Tokuno, H., Tanji, J., 1993. Input organization of distal and proximal forelimb areas in the monkey primary motor cortex: a retrograde double labeling study. J. Comp. Neurol. 333 (2), 199–209.

Tournier, J.D., Calamante, F., Connelly, A., 2007. Robust determination of the fibre orientation distribution in diffusion MRI: non-negativity constrained super-resolved spherical deconvolution. Neuroimage 35 (4), 1459–1472.

Tristán-Vega, A., Westin, C.-F., Aja-Fernández, S., 2009. Estimation of fiber orientation probability density functions in high angular resolution diffusion imaging. Neuroimage 47 (2), 638–650.

Tuch, D., 2004. Q-ball imaging. Magn. Reson. Med. 52 (6), 1358–1372.

Verstynen, T., Badre, D., Jarbo, K., Schneider, S., 2012. Microstructural organizational patterns in the human corticostriatal system. J. Neurophysiol. 107 (11), 2984–2995.

Verstynen, T., Jarbo, K., Pathak, S., Schneider, W., 2011. In vivo mapping of microstructural somatotopies in the human corticospinal pathways. J. Neurophysiol. 105 (1), 336–346.

Vohn, R., Fimm, B., Weber, J., et al., 2007. Management of attentional resources in within-modal and cross-modal divided attention tasks: an fMRI study. Hum. Brain Mapp. 28 (12), 1267–1275.

Wang, Y., Fernández-Miranda, J.C., Verstynen, T., et al., 2012. Rethinking the Role of the Middle Longitudinal Fascicle in Language and Auditory Pathways. Cereb. Cortex August 8.

Wedeen, V., Hagmann, P., Tseng, W.-Y., et al., 2005. Mapping complex tissue architecture with diffusion spectrum magnetic resonance imaging. Magn. Reson. Med. 54 (6), 1377–1386.

Yeh, F.-C., Wedeen, V., Tseng, W.-Y.I., 2010. Generalized q-sampling imaging. IEEE Trans. Med. Imaging 29 (9), 1626–1635.

Delayed Development of the Claustrum in Autism

Jerzy Wegiel[1], Janusz Moryś[2], Przemysław Kowiański[2], Shuang Yong Ma[1], Izabela Kuchna[1], Krzysztof Nowicki[1], Humi Imaki[1], Jarek Wegiel[1], Michael Flory[3], W. Ted Brown[4], and Thomas Wisniewski[5]

[1]Department of Developmental Neurobiology, NYS Institute for Basic Research in Developmental Disabilities, Staten Island, NY, USA [2]Department of Anatomy and Neurobiology, Medical University of Gdansk, Poland [3]Department of Infant Development, NYS Institute for Basic Research in Developmental Disabilities, Staten Island, NY, USA [4]Department of Human Genetics, NYS Institute for Basic Research in Developmental Disabilities, Staten Island, NY, USA [5]Department of Psychiatry, New York University School of Medicine, New York, NY, USA

AUTISM CLINICAL DIAGNOSIS AND PREVALENCE

Autism is a severe developmental disorder that is characterized by:

- qualitative impairments in reciprocal social interactions,
- qualitative impairments in verbal and nonverbal communication,
- restricted repetitive and stereotyped patterns of behavior, interests and activities, and
- onset prior to the age of 3 years (American Psychiatric Association, 2000).

The 2008 study of 8.4 percent of the US population of 8-year-old children reported a 1 in 88 overall prevalence of autism spectrum disorder (ASD), with a 1 in 54 prevalence for males and a 1 in 252 prevalence for females (male to female ratio of 4.7:1) (US Department of Health and Human Services, 2012). Of children diagnosed with ASD, 38 percent

J. Smythies, L. Edelstein and V. Ramachandran (Eds):
The Claustrum.

DOI: http://dx.doi.org/10.1016/B978-0-12-404566-8.00008-8

are classified as having intellectual disability, having an IQ of less than 70, and 24 percent were in the borderline range with an IQ of 71–85. Approximately 33 percent of individuals with autism are diagnosed with epilepsy (Tuchman and Rapin, 2002).

AUTISM NEUROPATHOLOGY

The clinical understanding of autism is based on the examination of thousands of patients, but post-mortem neuropathological studies have only been undertaken on a small number of brains. Between 1980 and 2003 only 58 brains were examined (Palmen et al., 2004). Recent studies expanded this number by more than 30 subjects with idiopathic autism (Casanova et al., 2006; Courchesne et al., 2011; Jacot-Descombes et al., 2012; Santos et al., 2011; Schuman and Amaral, 2006, Van Kooten et al., 2008; Wegiel et al., 2010, 2012; Whitney et al., 2008, 2009; Yip et al., 2007, 2008) and 10 with autism associated with duplications 15q11.2-q13 (Wegiel et al., 2012). However, morphological markers and neuropathological diagnostic criteria of autism are not established. The concept that autism is associated with neuropathological changes was explored in the first studies reported between 1980 and 1989 (Damasio et al.,1980; Bauman and Kemper, 1985; Ritvo E.R. et al., 1986; Courchesne et al., 1987; Gaffney et al., 1987; Hashimoto et al., 1989; Murakami et al., 1989). This view was expanded by detection of global markers of abnormal brain development, including accelerated brain growth in the first year of life (Courchesne et al., 2003), a slower rate of brain growth between 2 and 4 years of age (Carper et al., 2002; Courchesne et al., 2001), and the link between an altered trajectory of brain development and severity of autism (Courchesne et al., 2003). Overgrowth of the frontal and temporal lobes and the amygdala, which are involved in cognitive, social and emotional functions, and language development, suggests contribution of these developmental alterations to the clinical features of autism (Carper et al., 2002, Courchesne et al., 2001, Sparks et al., 2002).

The diversity of neuropathological findings corresponds to developmental impairments in many interacting brain networks and to an expansion of autism pathology from "local" abnormalities to global defects of the cognitive system. Localizing models are still the main tools for identification of pathological changes as a component of the networks of structural and functional abnormalities (Müller, 2007).

CLAUSTRUM CONNECTIVITY

In 2005 Crick and Koch proposed the claustrum as the ideal candidate for the brain's "consciousness" center because it acts like a conductor in an orchestra to integrate information from various modalities. The claustrum has extensive reciprocal connections to and from almost all brain regions (Fernández-Miranda et al., 2008). The information from the various brain regions is segmented topographically in the claustrum in a partially overlapping manner. The anterior claustrum projects to and receives signals from the frontal cortex. The middle claustrum is associated with the parietal cortex. The posterior/inferior claustrum is associated with the temporal and occipital cortices. The dorsal claustrum forms a visual loop with Brodmann area (BA) 17 (Fernández-Miranda et al., 2008, Mory et al., 1993, Pearson et al., 1982). The claustrum also projects to the amygdala (Amaral and Insausti, 1992), the hippocampus (Amaral and Cowan, 1980), and the striatum (Arikuni and Kubota, 1985, Crick and Koch, 2005). Social brain circuitry (orbitofrontal cortex, superior temporal gyrus and amygdala) (Brothers et al., 1990) projects to the claustrum and receives claustral connections (Fernández-Miranda et al., 2008).

CLAUSTRUM FUNCTIONING IN NORMAL AND PATHOLOGICAL CONDITIONS

The claustrum is involved in:

- high-order cognitive functions such as fear recognition (Stein et al., 2007),
- experiential dread (Berns et al., 2006),
- cognitive impairment (Dubroff et al., 2008),
- memory storage (Moryś et al., 1996),
- associative learning (Chachich and Powell, 2004),
- repetitive behaviors and addiction (Moryś et al., 1996; Naqvi et al., 2007),
- multimodal processing of olfactory, auditory, visual, and tactile information, and of emotional and behavioral responses (Bennet and Baird, 2006),
- suppression of natural urges (Lerner et al., 2009),
- seizures (Zhang et al., 2001) and
- psychoses (Sperner et al., 1996).

Almost all of these functions are modified in autistic subjects and all of these alterations are present in autism.

The claustrum receives inputs from many cortical areas, integrates multiple inputs into a new signal, and redirects sensory information throughout the striatum and thalamus. Interconnectivity with subcortical nuclei and sensory cortical areas indicates the claustrum's involvement in sensorimotor integration, and potentially in the most complex human brain function – consciousness – as well as in higher orders of functionality that enable the organism to adapt rapidly to the changing environment

(Edelstein and Denaro, 2004). The attraction to routines and sameness appears to be one of the very striking behavioral alterations characteristic of autism. It appears that claustrum immaturity, reflected in the neuronal soma deficit of 29 percent in children and of 17 percent in adults, and the very striking deficit of neuronal nucleus volume (42% and 22% respectively), may be responsible for the claustrum neurons' functional impairment and deficits in adaptability and consciousness.

VARIATIONS IN THE CLAUSTRUM ASSOCIATED WITH AUTISM

Claustral Volume

The claustrum is a thin ribbon of gray matter located between the insular cortex and the striatum, closely associated with the amygdale, and built of five types of cells of both cortical and subcortical origin. The vast majority of neurons in the claustrum (94%) have numerous coarse pigment granules, long projections and spiny dendrites (Braak and Braak, 1982).

Our study compared the brain hemispheres of autistic and control subjects aged from 4 to 36 years (nine hemispheres in each group). The mean claustral volume of those with autism in the subgroup whose ages ranged from 4 to 8 years (388 mm^3) was 22 percent less than in the control subjects (494 mm^3). In the older subgroup, the volume of the claustrum in the autistic subjects increased by 18% (to 472 mm^3), which was insignificantly more than in age-matched controls (452 mm^3).

This observation is consistent with MRI studies of 16 high-functioning (IQ 85) autistic and 14 control males ranging from 7 to 12 years of age, which revealed that those with autism had a smaller claustral volume (555 mm^3) than the control subjects (701 mm^3) (Davies, 2008). The smaller volumes detected in post-mortem study in comparison to clinical MRI are the result of brain tissue shrinkage during dehydration.

NUMBER OF NEURONS IN THE CLAUSTRUM

The mean numerical density of neurons was almost identical in the autistic (29,156/mm^3; $n = 9$) and control subjects (29,260/mm^3; $n = 9$). The mean total number of neurons in the nine autistic subjects (12.5 million) was insignificantly less (by 8%) than in the control subjects (13.6 million). Brain region-specific modifications in the number of neurons reveal a 67 percent increase in the number of neurons in the prefrontal cortex (Courchesne et al., 2011), and a 53 percent increase in the ratio between von Economo neurons (mirror neurons) and pyramidal neurons in the fronto-insular cortex (Santos et al., 2011), but a reduced number of neurons in the fusiform gyrus involved in face recognition (Van Kooten et al., 2008), a regional decrease of the number of Purkinje cells (Fatemi et al., 2002, Lee et al., 2002, Whitney et al., 2008), and most likely prenatal loss of Purkinje cells (Whitney et al., 2008, 2009). Striking differences between the almost normal number of neurons in the claustrum overall and the reported changes in the number of neurons in cortical projection areas may reflect desynchronized development of structures contributing to a broad spectrum of clinical manifestations of autism.

Mean Neuronal Volume

The mean volume of the neuronal body in the 4- to 8-year-old autistic children was 29 percent less ($1,410\,\mu m^3$, $n = 4$) than in the control group ($1,999\,\mu m^3$, $n = 4$; $P < 0.001$); however, in 13- to 36-year-old autistic subjects it was only 17 percent less ($1,582\,\mu m^3$; $n = 5$) tha$n$ in the controls ($1,904\,\mu m^3$; $n = 5$; $P < 0.001$).

In the 4- to 8-year-old autistic children the mean volume of the neuronal nucleus ($234\,\mu m^3$, $n = 4$), was 42 percent less than in the control group ($400\,\mu m^3$, $n = 4$; $P < 0.001$) whereas, in 13- to 36-year-old autistic subjects, it was 22% less ($266\,\mu m^3$) than in the control group ($342\,\mu m^3$). The initial, very severe, volume deficit in the younger group with autism, and the reduction of this deficit (by 38% in the cell soma volume and by 48% in the neuronal nucleus volume) in the subjects who were more than 8 years old, suggest delay of neuron growth in the younger group and abnormal acceleration in late childhood. The volume deficit in the youngest children indicates altered regulation of neuron growth before the age of 4 years, resulting in lifelong structural and functional abnormalities.

Desynchronized Development of Claustral Circuitry

The study of 16 brain subdivisions in 14 autistic and 14 control subjects revealed: (a) brain-structure-specific delay in the rate of neuronal growth and (b) desynchronization of neuronal growth in the brains of autistic subjects. Desynchronization is reflected in striking differences of neuronal volume deficit in early childhood, including a very severe (>30%) deficit in the nucleus accumbens and Purkinje cells; severe deficit (20% to 30%) in the claustrum, thalamus, caudate nucleus, globus pallidus, dentate nucleus, entorhinal cortex, amygdala, magnocellular basal complex (including in the nucleus basalis of Meynert and the Ammons horn); moderate deficit (10% to 20%) in the putamen, inferior olive and magnocellular lateral geniculate body (LGB); and mild deficit (4% to 5%) in the parvocellular LGB and substantia nigra ($P < 0.001$). The function of the examined structures suggests that delayed and desynchronized development of the affected networks contributes to all three diagnostic domains of autism, including disrupted social and communication development, restricted repetitive and stereotyped patterns of behavior, and intellectual deficits. The conspicuous differences in the developmental volume deficits of neurons in the claustrum, and in brain structures receiving projections from the claustrum and projecting to the claustrum, that were detected in 4- to 8-year-old autistic children are signs of desynchronization of claustral-circuit development. The detected pattern of developmental delays suggests that a number of physiologic processes that collectively ensure the maintenance of undisturbed contact with the environment are altered in early childhood and that the claustrum is one of the most severely modified brain structures in early and late childhood (Wegiel et al., submitted). Executive functions, cognitive processes that allow complex goal-directed behavior including spatial working memory and response inhibition, are impaired in autism (Minshew et al., 1999; Hill, 2004; Ozonoff et al., 2004).

Altered Claustrum Connectivity in Autism

The very noticeable range of developmental delays in the claustral neurons, and in the neurons receiving projections from the claustrum and projecting to the claustrum, suggest that claustral dysfunction is associated with or caused by a desynchronized development of the subcomponents of these multifunctional networks.

The claustrum is involved in long-term response potentiation within the claustral-entorhinal-hippocampal system (Wilhite et al., 1986) and neuronal volume deficits of 29%, 23% and 4%, respectively, are found in these three compartments in people with autism (Wegiel et al., submitted). The frontal, temporal, parietal and occipital cortex project to the claustrum (Druga, 1968; LeVay and Sherk, 1981), whereas the dorsocaudal claustrum (visual claustrum) projects to the visual cortex. The slight reduction of neuronal volume in BA17 in the occipital cortex (Casanova et al., 2006) and a significant reduction in the mean perikaryal volume of the neurons in layers V and VI (by 21.1% and 13.4%, respectively) in the fusiform gyrus (Van Kooten et al., 2008) are indicative of desynchronized development of the claustrocortical networks in autism. It may affect the role of the claustrum as an integrator of input from the somatosensory, auditory and visual cortices, as well as from the respective diencephalic relays (Spector, 1969). The 12% and 5% deficits in the neuronal volume of the magnocellular and parvocellular lateral geniculate nuclei, respectively, and the 27% deficit in neuronal volume in the thalamus (Wegiel et al., submitted) indicate developmental desynchronization of other claustrum-related regions. The rostral portion of the claustrum projects to the caudate nucleus (Arikuni and Kubota, 1985), which is also affected by a 16% deficit in neuronal volume (Wegiel et al., submitted).

THEORIES OF AUTISM AND THE ROLE
OF THE CLAUSTRUM IN AUTISM

Functional MRI and post-mortem studies are the foundation for several theories of autism, including a theory of a deficit of long-range connectivity but increased short-range connectivity, disrupted cortical synchronization (Belmonte et al., 2004; Courchesne et al., 2004; Frith, 2004; Hughes, 2007; Rippon et al., 2007), and a defective theory of mind (Baron-Cohen et al., 1985, 1999; Frith, 2001; Frith et al., 1994; Kobayashi et al., 2007). The hypothesis of a deficit of long-range connectivity is based on detection of modified activity between distant, but functionally and anatomically related, cortical regions (Anderson et al., 2011; Cherkassky et al., 2006; Coben et al., 2008; Dinstein et al., 2011; Just et al., 2007; Kennedy and Courchesne, 2008). Regional increase of the white-matter volume suggests an increase in local short-range connections (Herbert et al., 2003, 2004). The theory of "a developmental disconnection syndrome" (Courchesne and Pierce, 2005; Frith, 2004) is supported by new data indicating a failure of development of connections rather than loss of connection (Geschwind and Levitt, 2007). Concentration of functional MRI studies on the cortex supports a corticocentric theory of autism, but the complexity of the autistic phenotype suggests involvement of the subcortical structures, including the claustrum, interacting with such subcortical subdivisions as the amygdala, which is involved in processing social information, emotional interpretation, fear and anxiety (Amaral et al., 2003; Baron-Cohen et al., 2000; Winston et al., 2002); the thalamus, involved in language functions, attention, and anxiety (Ojemann, 1977; Ojemann and Ward, 1971); and the striatum, involved in repetitive motor behaviors and rituals (Day and Carelli, 2007; Salamone, 1994; Sears et al., 1999). The detected deficits of neuronal growth in 16 subcortical structures, including claustrum, expand the corticocentric model of developmental alterations in autism and support a model of global developmental encephalopathy characterized by delayed and desynchronized development of neurons in interacting structures of the cortex and subcortex (Wegiel et al., submitted).

References

Amaral, D., Insausti, R., 1992. Retrograde transport of D-[3H]-aspartate injected into the monkey amygdaloid complex. Exp. Brain Res. 88, 375–388.

Amaral, D.G., Cowan, W.M., 1980. Subcortical afferents to the hippocampal formation in the monkey. J. Comp. Neurol. 189, 573–591.

Amaral, D.G., Bauman, M.D., Schumann, C.M., 2003. The amygdala and autism: Implications from non-human primate studies. Genes Brain Behav. 2, 295–302.

American Psychiatric Association, 2000. Diagnostic and Statistical Manual of Mental Disorders DSM-IV-TR. American Psychiatric Association, Washington, DC.

Anderson, J.S., Druzgal, T.J., Froehlich, A., Dubray, M.B., Lange, N., Alexander, A.L., Lainhart, J.E., 2011. Decreased interhemispheric functional connectivity in autism. Cereb. Cortex 21, 1134–1146.

Arikuni, T., Kubota, K., 1985. Claustral and amygdaloid afferents to the head of the caudate nucleus in macaque monkeys. Neurosci. Res. 2, 239–254.

Baron-Cohen, S., Leslie, A.M., Frith, U., 1985. Does the autistic child have a "theory of mind"? Cognition 21, 37–46.

Baron-Cohen, S., Campbell, R., Karmiloff-Smith, A., Grant, J., Walker, J., 1999. Are children with autism blind to mentalistic significance of the eyes? Br. J. Dev. Psychol. 13, 379–398.

Baron-Cohen, S., Ring, H.A., Bullmore, E.T., Wheelwright, S., Ashwin, C., Wiliams, S.C., 2000. The amygdala theory of autism. Neurosci. Biobehav. Rev. 24, 355–364.

Bauman, M.L., Kemper, T.L., 1985. Histoanatomic observations of the brain in early infantile autism. Neurology 35, 866–867.

Belmonte, M.K., Allen, G., Beckel-Mitchener, A., Boulanger, L.M., Carper, R.A., Webb, S.J., 2004. Autism and abnormal development of brain connectivity. J. Neurosci. 24, 9228–9231.

Bennett, C.M., Baird, A.A., 2006. Anatomical changes in the emerging adult brain: A voxel-based morphometry study. Hum. Brain Mapp. 27, 766–777.

Berns, G.S., Chappelow, J., Cekic, M., Zink, C.F., Pagnoni, G., Martin-Skurski, M.E., 2006. Neurobiological substrates of dread. Science 312, 754–758.

Braak, H., Braak, E., 1982. Neuronal types in the claustrum of man. Anat. Embryol. 163, 447–460.

Brothers, L., Ring, B., Kling, A., 1990. Response of neurons in the macaque amygdala to complex social stimuli. Behav. Brain Res. 41, 199–213.

Carper, R.A., Moses, P., Tigue, Z.D., Courchesne, E., 2002. Cerebral lobes in autism: Early hyperplasia and abnormal age effects. Neuroimage 16, 1038–1051.

Casanova, M.F., van Kooten, I.A.J., Switala, A.E., Heinsen, H., Steinbusch, W., Hof, P.R., Schmitz, C., 2006. Minicolumnar abnormalities in autism. Acta Neuropathol. 112, 287–303.

Chachich, M.E., Powell, D.A., 2004. The role of claustrum in Pavlovian heart rate conditioning in the rabbit (Oryctolagus cuniculatus): anatomical, electrophysiological, and lesion studies. Behav. Neurosci. 118, 514–525.

Cherkassky, V.L., Kana, R.K., Keller, T.A., Just, M.A., 2006. Functional connectivity in a baseline resting state network in autism. NeuroReport 17, 1687–1690.

Coben, R., Clarke, A.R., Hudspeth, W., Barry, R.J., 2008. EEG power and coherence in autistic spectrum disorder. Clin. Neurophysiol. 119, 1002–1009.

Courchesne, E., Pierce, K., 2005. Why the frontal cortex in autism might be talking only to itself: local over-connectivity but long-distance disconnection. Curr. Opin. Neurobiol. 15, 225–230.

Courchesne, E., Hesselink, J.R., Jernigan, T.L., Yeung-Courchesne, R., 1987. Abnormal neuroanatomy in a nonretarded person with autism. Unusual findings with magnetic resonance imaging. Arch. Neurol. 44, 335–341.

Courchesne, E., Karns, C.M., Davis, H.R., Ziccardi, R., Carper, R.A., Tigue, Z.D., 2001. Unusual brain growth patterns in early life in patients with autistic disorder. An MRI study. Neurology 57, 245–254.

Courchesne, E., Carper, R., Akshoomoff, N., 2003. Evidence of brain overgrowth in the first year of life in autism. J. Am. Med. Assoc. 290, 337–344.

Courchesne, E., Redcay, E., Kennedy, D.P., 2004. The autistic brain: Birth through adulthood. Curr. Opin. Neurobiol. 17, 489–496.

Courchesne, E., Mouton, P.R., Calhoun, M.E., Semendeferi, K., Ahrens-Barbeau, C., Hallet, M.J., et al., 2011. Neuron number and size in prefrontal cortex of children with autism. J. Am. Med. Assoc. 306, 2001–2010.

Crick, F.C., Koch, C., 2005. What is the function of the claustrum? Philos. Trans. R. Soc. Lond., Biol. Sci. 360, 1271–1279.

Damasio, H., Maurer, R.G., Damasio, A.R., Chui, H.C., 1980. Computerized tomographic scan findings in patients with autistic behavior. Arch. Neurol. 37, 504–510.

Davies W.B., 2008. The claustrum in autism and typically developing male children: A quantitative MRI study. [Thesis] Brigham Young University.

Day, J.J., Carelli, R.M., 2007. The nucleus accumbens and Pavlovian reward learning. Neuroscientist 13, 148–159.

Dinstein, I., Pierce, K., Eyler, L., Solso, S., Malach, R., Behrmann, M., Courchesne, E., 2011. Disrupted neural synchronization in toddlers with autism. Neuron 70, 1218–1225.

Druga, R., 1968. Cortico-claustral connections. II. Connections from the parietal, temporal and occipital cortex to the claustrum. Folia Morphol. (Prague) 16, 142–149.

Dubroff, J.G., Ficicioglu, C., Segal, S., Wintering, N., Alavi, A., Newberg, A., 2008. FDG-PET finding of patients with galactosemia. J. Inherit. Metab. Disord. 31, 533–539.

Edelstein, L.R., Denaro, F.J., 2004. The claustrum: A historical review of its anatomy, physiology, cytochemistry and functional significance. Cell. Mol. Biol. 50, 675–702.

Fatemi, S.H., Halt, A.R., Realmuto, G., Earle, J., Kist, D.A., Thuras, P., Merz, A., 2002. Purkinje cell size is reduced in cerebellum of patients with autism. Cell. Mol. Neurobiol. 22, 171–175.

Fernandez-Miranda, J.C., Rhoton, A.L., Kakizawa, Y., Choi, C., Avarez-Linera, J., 2008. The claustrum and its projection system in the human brain: A microsurgical and tractographic anatomical study. J. Neurosurg. 108, 764–774.

Frith, C., 2004. Is autism a disconnection disorder? Lancet Neurol. 3, 577.

Frith, U., 2001. Mind blindness and the brain in autism. Neuron 32, 969–979.

Frith, U., Happe, F., 1994. Autism: beyond "theory of mind". Cognition 50, 115–132.

Gaffney, G.R., Tsai, L.Y., Kuperman, S., Minchin, S., 1987. Cerebellar structure in autism. Am. J. Dis Child 141, 1330–1332.

Geschwind, D.H., Levitt, P., 2007. Autism spectrum disorders: developmental disconnection syndromes. Curr. Opin. Neurobiol. 17, 103–111.

Hashimoto, T., Tayama, M., Mori, K., Fujino, K., Miayzaki, M., Kuroda, Y., 1989. Magnetic resonance imaging in autism: preliminary report. Neuropediatrics 20, 142–146.

Herbert, M.R., Ziegler, D.A., Deutsch, C.K., O'Brien, L.M., Lange, N., Bakardjiev, A., et al., 2003. Dissociations of cerebral cortex, subcortical and cerebral white matter volumes in autistic boys. Brain 126, 1182–1192.

Herbert, M.R., Ziegler, D.A., Makris, N., Filipek, P.A., Kemper, T.L., Normandin, J.J., et al., 2004. Localization of white matter volume increase in autism and developmental language disorder. Ann. Neurol. 55, 530–540.

Hill, E.L., 2004. Executive dysfunction in autism. Trends Cogn. Sci. 8, 26–32.

Hughes, J.R., 2007. Autism: the first firm finding = underconnectivity? Epilepsy Behav. 11, 20–24.

Jacot-Descombes, S., Uppal, N., Wicinski, B., Santos, M., Schmeidler, J., Giannakopoulos, P., et al., 2012. Decreased pyramidal neuron size in Brodmann areas 44 and 45 in patients with autism. Acta Neuropathol. 124, 67–79.

Just, M.A., Cherkassky, V.L., Keller, T.A., Kana, R.K., Minshew, N.J., 2007. Functional and anatomical cortical underconnectivity in autism: evidence from an fMRI study of an executive function task and corpus callosum morphometry. Cereb. Cortex 17, 951–961.

Kennedy, D.P., Courchesne, E., 2008. The intrinsic functional organization of the brain is altered in autism. Neuroimage 39, 1877–1885.

Kobayashi, C., Glover, G.H., Temple, E., 2007. Children's and adults' neural bases of verbal and nonverbal "theory of mind". Neuropsychologia 45, 1522–1532.

LeVay, S., Sherk, H., 1981. The visual claustrum of the cat. I. Structure and connections. J. Neurosci. 1, 956–980.

Lee, M., Martin-Ruiz, C., Graham, A., Court, J., Jaros, E., Perry, R., et al., 2002. Nicotinic receptor abnormalities in the cerebellar cortex in autism. Brain 15, 1483–1495.

Lerner, A., Bagic, A., Janakawa, T., Boudreau, E., Pagan, F., Mari, Z., et al., 2009. Involvement of insula and cingulate cortices in control and suppression of natural urges. Cereb. Cortex 19, 218–223.

Minshew, N.J., Luna, B., Sweeney, J.S., 1999. Oculomotor evidence for neocortical systems but not cerebellar dysfunction in autism. Neurology 52, 917–922.

Moryś, J., Narkiwicz, O., Wisniewski, H.M., 1993. Neuronal loss in the human claustrum following ulegyria. Brain Res. 616, 176–180.

Moryś, J., Bobinski, M., Wegiel, J., Wisniewski, H.M., Narkiewicz, O., 1996. Alzheimer's disease severely affects areas of the claustrum connected with the entorhinal cortex. J. Hirnforsch. 37, 173–180.

Murakami, J.W., Courchesne, E., Press, G.A., Yeung-Courchesne, R., Hesselink, J.R., 1989. Reduced cerebellar hemisphere size and its relationship to vermal hypoplasia in autism. Arch. Neurol. 46, 689–694.

Müller, R.A., 2007. The study of autism as a distributed disorder. Ment. Retard. Dev. Disabi.l Res. Rev. 13, 85–95.

Naqvi, N.H., Rudraunt, D., Damasio, H., Bechara, A., 2007. Damage to the insula disrupts addiction to cigarette smoking. Science 315, 531–534.

Ojemann, G.A., 1977. Asymmetric function of the thalamus in man. Ann. N. Y. Acad. Sci. 299, 380–396.

Ojemann, G.A., Ward, A.A., 1971. Speech representation in ventrolateral thalamus. Brain 94, 669–680.

Ozonoff, S., Cook, I., Coon, H., Dawson, G., Joseph, R.M., Klin, A., et al., 2004. Performance on Cambridge Neuropsychological Test Automated Battery subtests sensitive to frontal lobe function in people with autistic disorder: Evidence from the collaborative programs of excellence in autism network. J. Autism. Dev. Disord. 34, 139–1150.

Palmen, S.J.M.C., van Engeland, H., Hof, P.R., Schmitz, C., 2004. Neuropathological findings in autism. Brain 127, 2572–2583.

Pearson, R.C.A., Brodal, P., Gatter, K.C., Powell, T.P.S., 1982. The organization of the connections between the cortex and the claustrum in the monkey. Brain Res. 234, 435–441.

Rippon, G., Brock, J., Brown, C., Boucher, J., 2007. Disordered connectivity in the autistic brain: Challenges for the "new psychophysiology." J. Psychophysiol. 63, 164–172.

Ritvo, E.R., Freeman, B.J., Scheibel, A.B., Duong, T., Robinson, H., Guthrie, D., Ritvo, A., 1986. Lower Purkinje cell counts in the cerebella of four autistic subjects: Initial findings of the UCLA-NSAC autopsy research report. Am. J. Psychiatry. 143, 862–866.

Salamone, J.D., 1994. The involvement of nucleus accumbens dopamine in appetitive and aversive motivation. Behav. Brain Res. 61, 117–133.

Santos, M., Uppal, N., Butti, C., Wicinski, B., Schmeidler, J., Giannakopoulos, P., et al., 2011. Von Economo neurons in autism: A stereologic study of the frontoinsular cortex in children. Brain Res. 1380, 206–217.

Schumann, C.M., Amaral, D.G., 2006. Stereological analysis of amygdala neuron number in autism. J. Neurosci. 26, 7674–7679.

Sears, L.L., Vest, C., Mohamed, S., Bailey, J., Ranson, B.J., Piven, J., 1999. An MRI study of the basal ganglia in autism. Prog. Neuropsychopharmacology Biol. Psychiatry 23, 613–624.

Sparks, B.F., Friedman, S.D., Shaw, D.W., Aylward, E.H., Echelard, D., Artru, A.A., et al., 2002. Brain structural abnormalities in young children with autism spectrum disorder. Neurology 59, 184–192.

Spector, I., 1969. Functional organization of the claustrum. Anat. Rec. 163, 269.

Sperner, J., Sander, B., Lau, S., Krude, H., Scheffner, D., 1996. Severe transitory encephalopathy with reversible lesions of the claustrum. Pediatr. Radiol. 26, 769–771.

Stein, M.B., Simmons, A.N., Feinstein, J.S., Paulus, M.P., 2007. Increased amygdala and insula activation during emotion processing in anxiety-prone subjects. Am. J. Psychiatry 164, 318–327.

Tuchman, R.F., Rapin, I., 2002. Epilepsy in autism. Lancet Neurol. 1, 352–358.

US Department of Health and Human Services, Centers for Disease Control and Prevention Morbidity and Mortality Weekly Report, 2012. 61, 1–19.

Van Kooten, I.A.J., Palmen, S.J.M.C., von Cappeln, P., Steinbusch, H.W.M., Korr, H., Heinsen, H., et al., 2008. Neurons in the fusiform gyrus are fewer and smaller in autism. Brain 131, 987–999.

Wegiel, J., Kuchna, I., Nowicki, K., Imaki, H., Wegiel, J., Marchi, E., et al., 2010. The neuropathology of autism: defects of neurogenesis and neuronal migration, and dysplastic changes. Acta Neuropathol. 119, 755–770.

Wegiel, J., Schanen, N.C., Cook, E.H., Sigman, M., Brown, W.T., Kuchna, I., et al., 2012. Differences between the pattern of developmental abnormalities in autism associated with duplications 15q11.2-q13 and idiopathic autism. J. Neuropathol. Exp. Neurol. 71, 382–397.

Whitney, E.R., Kemper, T.L., Bauman, M.L., Rosene, D.L., Blatt, G.J., 2008. Cerebellar Purkinje Cells are reduced in a subpopulation of autistic brains: a stereological experiment using Calbindin-D28k. Cerebellum 7, 406–416.

Whitney, E.R., Kemper, T.L., Rosene, D.L., Bauman, M.L., Blatt, G.J., 2009. Density of cerebellar basket and stellate cells in autism: evidence for a late developmental loss of Purkinje cells. J. Neurosci. Res. 87, 2245–2254.

Wilhite, B.L., Teyler, T.I., Hendricks, C., 1986. Functional relations of the rodent claustral-entorhinal-hippocampal system. Brain Res. 365, 54–60.

Winston, J.S., Strange, B.A., O'Doherty, J., Dolan, R.J., 2002. Automatic and intentional brain responses during evaluation of trustworthiness of faces. Nat. Neurosci. 5, 277–283.

Yip, J., Soghomonian, J.J., Nguyen, L., Blatt, G.J., 2007. Decreased *GAD67* mRNA levels in cerebellar Purkinje cells in autism: Pathophysiological implications. Acta Neuropathol. 113, 559–568.

Yip, J., Soghomonian, J.J., Blatt, G.J., 2008. Increased *GAD67* mRNA expression in cerebellar interneurons in autism: implications for Purkinje cell dysfunction. J. Neurosci. Res. 86, 525–530.

Zhang, X., Hannesson, D., Saucier, D., Wallace, A., Howland, J., Corcoran, M.E., 2001. Susceptibility to kindling and neuronal connections of the anterior claustrum. J. Neurosci. 21, 3674–3687.

9

The Claustrum in Schizophrenia

Nicola G. Cascella[1] and Akira Sawa[2,3]

[1]Neuropsychiatry Program, Sheppard Pratt Hospital, Baltimore, MD, USA [2]Department of Psychiatry and Behavioral Sciences, Johns Hopkins University School of Medicine, Baltimore, MD, USA [3]Department of Neuroscience, Johns Hopkins University School of Medicine, Baltimore, Maryland, USA

INTRODUCTION

Investigators have approached the correlation study of brain structure and symptoms of schizophrenia (SZ) using region-of-interest (ROI) measurements and found associations between volume changes in temporal cortex (superior temporal gyrus and middle temporal lobe) and positive symptoms in SZ, including hallucinations, delusions, and thought disorder (Barta et al., 1990, 1997; Bogerts et al., 1993; Flaum et al., 1995; McCarley et al., 1993; Shenton et al., 1992). Applications of voxel based morphometry techniques (VBM) have confirmed correlations of positive symptoms with temporal structures, the prefrontal cortex, and subcortical structures such as the thalamus and the basal ganglia (Gaser et al., 2004; Koutsouleris et al., 2008; Nenadic et al., 2010; Wright et al., 1995).

In contrast, no prior studies have reported finding a relationship between gray-matter (GM) changes in the claustrum and positive symptoms of SZ. Only Shapleske and colleagues (2002) observed a whitematter (WM) excess in the left claustrum among patients with SZ and a predominance of hallucinations. However, Bruen et al. (2008) found that smaller left claustrum volume, along with reductions of other cortical areas, correlated with the severity of delusions in elderly adults with Alzheimer-type dementia.

J. Smythies, L. Edelstein and V. Ramachandran (Eds):
The Claustrum.

DOI: http://dx.doi.org/10.1016/B978-0-12-404566-8.00009-X

We examined the relationship between cerebral GM volume and the severity of delusions and hallucinations in adults with schizophrenia (Cascella et al., 2011).

SEVERITY OF DELUSIONS IN RELATION TO GREY MATTER VOLUME

Our Research

Our study involved 43 participants, aged 19 to 56 years, who met *DSM-IV* (American Psychiatric Association, 1994) criteria for SZ, and who were all receiving medications at the time of the study. Positive and negative symptoms associated with SZ were assessed using the scales for the assessment of positive and negative symptoms (SAPS and SANS; Andreasen and Olsen, 1982). We acquired MRI scans using a Philips Achieva 3.0 Tesla machine and obtained 3D magnetization pre-pared rapid gradient-echo (MP-RAGE) images with sensitive encoding (SENSE). T1-weighted images in contiguous 1.0 mm thick slices were acquired in the coronal plane for each participant. Given variability in head and brain size among participants, the number of slices acquired varied somewhat, but it did not exceed 200 contiguous slices. The acqui-sition parameters were TR/TE = 7.9/3.7 ms, flip angle = 8°, FOV = 200 × 200 mm, voxel size of $1 \times 1 \times 1 \, mm^3$ and in-plane matrix of 256 × 256.

The whole brain images were preprocessed with Statistical Parametric Mapping (SPM5; http://www.fil.ion.ucl.ac.uk/spm/software/spm5/) software. We then implemented voxel-based morphometry (Ashburner and Friston, 2000) with a Dell workstation using Matlab version 7.9 (The MathWorks, Natick, MA), following the unified segmentation method of SPM5 (Ashburner and Friston, 2005). Non-brain regions were removed from the normalized images using MRIcro (http://www.mricro.com). The normalized images were then segmented into GM, WM, and cere-brospinal fluid (CSF) probability maps. Segmentation was based on a cluster analysis to identify each voxel's signal intensity, together with a priori information about the spatial distribution of these tissue segments in normal adults derived from probability maps. The GM images were smoothed using an 8 mm full width at half-maximum isotropic Gaussian kernel. Modulating GM images removes the effect of nonlinear trans-formation of the images on local tissue volumes–applying the Jacobian determinants of the transformation parameters yields corresponding changes of voxel values in areas that were increased or decreased by the transformation. These modulated images were used for statistical analyses of GM volumetric differences. Montreal Neurological Institute (MNI) coordinates were generated by SPM5 and Talairach labels by

Talairach Client (Lancaster et al., 2000) after conversion of MNI coordinates to Talairach coordinates by the mni2tal algorithm (Lancaster et al., 2007) implemented within GingerALE (Laird et al., 2005). The images were analyzed with the general linear model executed in SPM5 to evaluate the relationship between GM volume and ratings on the SAPS Hallucinations and Delusions subscales, as well as SAPS Global Rating. Given the demographics of our sample, we controlled for race, age, and total intracranial volume (TICV) in the analyses. TICV was calculated as the sum of GM, WM, and CSF. Furthermore, TICV was used as a covariate in place of sex, as it allowed us to simultaneously control for individual and sex-related differences in TICV. We did not apply an extent threshold for these analyses, but no significant clusters contained fewer than 30 voxels. A false discovery rate (FDR) error correction method was used to maintain a Type I error rate of $P < 0.05$ for all positive findings relating GM volume to positive symptom ratings. An atlas-based method was utilized in order to identify any specific brain areas that significantly correlated with SAPS scores.

The SAPS Delusion subscale was inversely correlated with left claustrum GM volume after correcting for multiple comparisons ($P < 0.05$). No significant correlations were found between GM volumes and the SAPS Global Rating score or the SAPS Global Rating Hallucinations subscale after correcting for multiple comparisons.

The lack of significant correlations between GM volumes and the SAPS Global Rating score or the SAPS Global Rating Hallucinations subscale was discrepant from previous findings in the literature (Crespo-Facorro et al., 2000; Goldstein et al., 1999; Paillère-Martinot et al., 2001). We explored possible reasons for this discrepancy by conducting some additional analyses to better characterize our sample and further explore our initial findings. First, we conducted a correlation analysis to assess the relationship between the SAPS global rating of hallucinations and the SAPS global rating of delusions in our sample, which revealed a significant correlation ($r = 0.45$, $P = 0.002$) between these two scales on the SAPS. Given this overlap we decided to treat the SAPS global rating of hallucinations as a nuisance variable, and included it as a covariate in our original analysis. Furthermore, we decided to use the SAPS Global Rating Hallucinations subscale as opposed to the Composite score for the SAPS as a covariate because so few participants in this sample demonstrated positive symptoms other than delusions and hallucinations.

After re-running the original analyses and including the SAPS Global Rating of Hallucinations subscales as a covariate, the SAPS Global Rating of Delusions subscale remained significantly and inversely correlated with the left claustrum. The cluster size in the left claustrum region increased in size when we reran these analyses.

Implications of Our Results

Our finding is the first to show a relationship between claustrum structure and a cardinal symptom of SZ like delusion. Delusions of schizophrenia have been thought to represent the result of a disturbance of the corollary discharge systems that integrate the motor systems of thought with the sensory systems of consciousness. Jackson (1958) proposed that thinking is our most complex motor act and, as such, it might conserve and utilize the computational and integrative mechanisms evolved for physical movements. Corollary discharge and feed-forward (CD-FF) are integrative mechanisms that prepare neural systems for the consequences of self-initiated action. In the motor systems of thought, they would act to distinguish self-produced from externally stimulated events in consciousness. Malfunctioning of the CD-FF systems in SZ could explain auditory hallucinations, disruptions of the sense of self and will and the delusions that follow from these disruptions. Whereas the internal feedback associated with simpler motor acts is far below the level of consciousness, one might postulate that corollary discharges accompanying the motor processes of thought themselves affect conscious experience. If this is the case then the subjective experience of these discharges should correspond to nothing less than the experience of will or intention. That this notion is not entirely far-fetched is suggested by the experiments of Penfield (1974). He found that stimulation of the motor cortex of conscious subjects with sufficient intensity to provoke an arm movement would lead them to say: "You caused my arm to move." Moreover, this effect was not limited to motor acts. When stimulation of the temporal lobe brought memories into consciousness, the patient would report: "You caused me to think that." Thus, Penfield's studies support the possibility that a corollary discharge or internal feedback mechanism is normally present in conscious thought.

Patients with SZ often report abnormal sensory perceptions that could reflect a disturbance of the afferent component of the corollary discharge systems. These perceptions represent a disordered internal feedback to the highest sensory centers that Jackson (1958) postulated must represent, in the most complex way, the entire organism or "self."

It is within this framework that we could envision an involvement of the claustrum in delusions of SZ. If the claustrum acts as a cross-modal integration center, binding different sensory input into a unitary perception, or is involved in the timing and coordination of cortical activity resulting ultimately in perceptual experience, then it is possible to speculate that the claustrum could be involved in CD-FF mechanisms similar to the function of the thalamus and basal ganglia.

It is important to note in this regard that in our study we found a significant correlation between severity of delusions and volume changes in the right insula. Significant findings with regards to the insular cortex and individuals with SZ spectrum disorders have been documented across studies. The insular cortex is highly interconnected with somatosensory and other cortical areas and limbic structures (perirhinal and entorhinal cortex, and amygdala). It plays a role in sensory, motor, and language functions and has been described as a limbic-integration area (Bushara et al., 2001; Calvert, 2001). Thus, our finding of an association between reduced right insular volumes and the severity of delusions in SZ might implicate such a disruption in the pathophysiology of this cardinal positive symptom of the disease.

Another aspect of SZ pathophysiology that could relate the disease to the claustrum is the evidence that a set of interneurons may be functionally impaired in SZ brains. Pharmacological agents, such as phencyclidine and ketamine (the NMDA-type glutamate receptor antagonists), can result in SZ-like manifestations. These compounds probably affect glutamatergic neurotransmission in interneurons (Behrens et al., 2007). Electrophysiological abnormalities in SZ patients, such as disturbance in gamma-oscillation associated with cognitive deficits, support this notion. Neuropathological examinations with autopsied brains from SZ patients have reproducibly reported that immunostaining of molecules specifically or preferentially expressed in the interneurons are altered in SZ (Hashimoto et al., 2003). These molecules include an isoform (67-kDa) of the gamma-aminobutyric acid synthesizing enzyme glutamic acid decarboxylase (GAD67) and parvalbumin. Most of the reports have indicated changes in both the frontal cortex and anterior cingulate cortex, and also in the hippocampus. Similar neuropathological change (decreased immunostaining of GAD67 in the cortex) has also been observed in drug-induced animal models treated with phencyclidine and ketamine, and in genetic animal models for SZ. The claustrum is a brain region which is enriched with parvalbumin and GAD67-positive interneurons, and it is plausible that the same changes previously reported in cortical areas and the hippocampus could be found in the claustrum. If so, the claustrum with its integrative function would contribute to the pathophysiology of SZ through a disruption of the gamma oscillations linked to activity of interneurons.

In conclusion, our findings suggest that structural differences in the insula and claustrum might contribute to positive symptoms in schizophrenia, and specifically to delusions. These findings emphasize the need to develop a better understanding of the role and function of the claustrum and insula in healthy and SZ populations.

References

American Psychiatric Association, 1994. Diagnostic and Statistical Manual of Mental Disorders: DSM-IV, fourth ed. American Psychiatric Association, Washington, DC.
Andreasen, N.C., Olsen, S., 1982. Negative v positive schizophrenia. Definition and validation. Arch. Gen. Psychiatry 39 (7), 789–794.

Ashburner, J., Friston, K.J., 2000. Voxel-based morphometry–the methods. Neuroimage 11 (6 Pt 1), 805–821.

Ashburner, J., Friston, K.J., 2005. Unified segmentation. Neuroimage 26 (3), 839–851.

Barta, P.E., Pearlson, G.D., Powers, R.E., Richards, S.S., Tune, L.E., 1990. Auditory hallucinations and smaller superior temporal gyral volume in schizophrenia. Am. J. Psychiatry 147 (11), 1457–1462.

Barta, P.E., Pearlson, G.D., Brill II, L.B., Royall, R., McGilchrist, I.K., Pulver, A.E., et al., 1997. Planum temporale asymmetry reversal in schizophrenia: replication and relationship to gray matter abnormalities. Am. J. Psychiatry 154 (5), 661–667.

Behrens, M.M., Ali, S.S., Dao, D.N., Lucero, J., Shekhtman, G., Quick, K.L., et al., 2007. Ketamine-induced loss of phenotype of fast-spiking interneurons is mediated by NADPH-oxidase. Science 318 (5856), 1645–1647.

Bogerts, B., Lieberman, J.A., Ashtari, M., Bilder, R.M., Degreef, G., Lerner, G., et al., 1993. Hippocampus-amygdala volumes and psychopathology in chronic schizophrenia. Biol. Psychiatry 33 (4), 236–246.

Bruen, P.D., McGeown, W.J., Shanks, M.F., Venneri, A., 2008. Neuroanatomical correlates of neuropsychiatric symptoms in Alzheimer's disease. Brain 131 (Pt 9), 2455–2463.

Bushara, K.O., Grafman, J., Hallett, M., 2001. Neural correlates of auditory-visual stimulus onset asynchrony detection. J. Neurosci. 21 (1), 300–304.

Calvert, G.A., 2001. Crossmodal processing in the human brain: insights from functional neuroimaging studies. Cereb. Cortex 11 (12), 1110–1123.

Cascella, N.G., Gerner, G.J., Fieldstone, S.C., Sawa, A., Schretlen, D.J., 2011. The insula-claustrum region and delusions in schizophrenia. Schizophr. Res. 133 (1–3), 77–81.

Crespo-Facorro, B., Kim, J., Andreasen, N.C., O'Leary, D.S., Bockholt, H.J., Magnotta, V., 2000. Insular cortex abnormalities in schizophrenia: a structural magnetic resonance imaging study of first-episode patients. Schizophr. Res. 46 (1), 35–43.

Flaum, M., O'Leary, D.S., Swayze II, V.W., Miller, D.D., Arndt, S., Andreasen, N.C., 1995. Symptom dimensions and brain morphology in schizophrenia and related psychotic disorders. J. Psychiatr. Res. 29 (4), 261–276.

Gaser, C., Nenadic, I., Volz, H.P., Buchel, C., Sauer, H., 2004. Neuroanatomy of "hearing voices": a frontotemporal brain structural abnormality associated with auditory hallucinations in schizophrenia. Cereb. Cortex 14 (1), 91–96.

Goldstein, J.M., Goodman, J.M., Seidman, L.J., Kennedy, D.N., Makris, N., Lee, H., et al., 1999. Cortical abnormalities in schizophrenia identified by structural magnetic resonance imaging. Arch. Gen. Psychiatry 56 (6), 537–547.

Hashimoto, T., Volk, D.W., Eggan, S.M., Mirnics, K., Pierri, J.N., Sun, Z., et al., 2003. Gene expression deficits in a subclass of GABA neurons in the prefrontal cortex of subjects with schizophrenia. J. Neurosci. 23 (15), 6315–6326.

Jackson, J.H., 1958.. In: Taylor, J. (Ed.), Selected Writings of John Hughlings Jackson Basic Books, New York.

Koutsouleris, N., Gaser, C., Jager, M., Bottlender, R., Frodl, T., Holzinger, S., et al., 2008. Structural correlates of psychopathological symptom dimensions in schizophrenia: a voxel-based morphometric study. Neuroimage 39 (4), 1600–1612.

Laird, A.R., Fox, P.M., Price, C.J., Glahn, D.C., Uecker, A.M., Lancaster, J.L., et al., 2005. ALE meta-analysis: controlling the false discovery rate and performing statistical contrasts. Hum. Brain Mapp. 25 (1), 155–164.

Lancaster, J.L., Woldorff, M.G., Parsons, L.M., Liotti, M., Freitas, C.S., Rainey, L., et al., 2000. Automated Talairach atlas labels for functional brain mapping. Hum. Brain Mapp. 10 (3), 120–131.

Lancaster, J.L., Tordesillas-Gutierrez, D., Martinez, M., Salinas, F., Evans, A., Zilles, K., et al., 2007. Bias between MNI and Talairach coordinates analyzed using the ICBM-152 brain template. Hum. Brain Mapp. 28 (11), 1194–1205.

McCarley, R.W., Shenton, M.E., O'Donnell, B.F., Faux, S.F., Kikinis, R., Nestor, P.G., et al., 1993. Auditory P300 abnormalities and left posterior superior temporal gyrus volume reduction in schizophrenia. Arch. Gen. Psychiatry 50 (3), 190–197.

Nenadic, I., Smesny, S., Schlosser, R.G., Sauer, H., Gaser, C., 2010. Auditory hallucinations and brain structure in schizophrenia: voxel-based morphometric study. Br. J. Psychiatry 196 (5), 412–413.

Paillère-Martinot, M., Caclin, A., Artiges, E., Poline, J.B., Joliot, M., Mallet, L., et al., 2001. Cerebral gray and white matter reductions and clinical correlates in patients with early onset schizophrenia. Schizophr. Res. 50 (1–2), 19–26.

Penfield, W., 1974. The mind and the highest brain mechanism. Am. Scholar 43, 237–246.

Shapleske, J., Rossell, S.L., Chitnis, X.A., Suckling, J., Simmons, A., Bullmore, E.T., et al., 2002. A computational morphometric MRI study of schizophrenia: effects of hallucinations. Cereb. Cortex 12 (12), 1331–1341.

Shenton, M.E., Kikinis, R., Jolesz, F.A., Pollak, S.D., LeMay, M., Wible, C.G., et al., 1992. Abnormalities of the left temporal lobe and thought disorder in schizophrenia. A quantitative magnetic resonance imaging study. N. Engl. J. Med. 327 (9), 604–612.

Wright, I.C., McGuire, P.K., Poline, J.B., Travere, J.M., Murray, R.M., Frith, C.D., et al., 1995. A voxel-based method for the statistical analysis of gray and white matter density applied to schizophrenia. Neuroimage 2 (4), 244–252.

10

Clinical Relations: Epilepsy

Michael E. Corcoran

Department of Anatomy, and Cell Biology, University of Saskatchewan, Saskatoon, Canada

INTRODUCTION

For reasons that have as yet to be fully explicated, neurons can engage in aberrant hypersynchronous activity. In the EEG, this is seen as epileptiform spiking, which can occur for a short time and then subsist. At times, however, the spiking (seizure discharge) can be sustained and persistent, again for reasons that are not yet completely understood.

Seizure discharge can remain localized to the generator area, the epileptic focus, resulting in behavioral and functional consequences that are circumscribed and fairly unobtrusive. Thus nonconvulsive seizures may be the result, without any motor signs of seizure discharge evident in behavior as, for example, in the absence attack, formerly termed "petit mal," in which spike-wave discharge circulating at 3 per second in thalamocortical circuits results in interruption of ongoing behavior without any obvious convulsive signs. On the other hand, if motor symptoms (partial seizures) do occur they may be limited to uncontrolled focal motor manifestations such as clonic jerking in a particular region of the body. When discharge propagates to and invades other areas, however, more obvious and intense consequences can ensue, resulting, for example, in more extensive partial seizures or hemiconvulsions. In the case of seizure discharge involving structures in the temporal lobe such as the amygdala or hippocampus, seizures may result in complex seizures, which involve impairment in behavioral responsiveness – "consciousness" – and amnesia for events occurring during the ictal event. Complex seizures of temporal lobe origin can also spread to other structures, resulting in partial motor seizures or even secondary generalization to full-blown convulsive seizures involving movements of limbs on both sides of the body. Seizures

J. Smythies, L. Edelstein and V. Ramachandran (Eds):
The Claustrum.

DOI: http://dx.doi.org/10.1016/B978-0-12-404566-8.00010-6

originating in other sites can result in primary generalized manifestations, in which tonic-clonic motor responses can occur bisymmetrically and bisychronously.

For a long time epileptologists have asked whether certain structures in the brain play an important role in amplifying or, conversely, dampening the spread of seizure discharge and thereby affect generalization. That is, because of their morphological connections or other physiological characteristics, do some structures play a privileged role in facilitating or suppressing seizure generalization? Candidate structures that have been proposed at various times include the brainstem reticular formation, amygdala, hippocampus, deep prepiriform cortex or area tempestas, substantia nigra pars reticulata, locus coeruleus, piriform cortex, and perirhinal cortex (for reviews see Corcoran and Moshé, 2005; Corcoran and Teskey, 2009). The CLA is attractive as a potential facilitator of seizure generalization, on account of its widespread reciprocal connections to a variety of cortical and subcortical sites. Surprisingly, however, only a few investigators have, in recent decades, turned their attention to the claustrum (CLA) and its role in seizure generalization.

J. A. Wada's laboratory provided seminal evidence for a role of the CLA in epilepsy. Kudo and Wada (1995) exposed cats treated with D,L-allylglycine to intermittent light stimulation (ILS), which resulted in either myoclonic jerking or bisymmetrical generalized tonic-clonic seizures. A subsequent unilateral lesion of the CLA converted the ILS-induced bisymmetrical generalized convulsion into a partial onset, secondarily generalized seizure, suggesting that the CLA is critical for access of visual afferents to the substrates and mechanisms of generalized seizure.

KINDLING

In later experiments Wada's group employed the kindling paradigm, in which intermittent application of brief trains of electrical stimulation to various sites results in development of generalized seizures and a permanent state of susceptibility to seizures in a variety of species, from amphibians to mammals, including primates (Corcoran and Moshé, 2005; Corcoran and Teskey, 2009).

Kindling typically involves application of 1-second trains of high-frequency electrical stimulation via implanted electrodes once daily, although some experiments have employed longer trains, more frequent applications of stimulation, or even low-frequency stimulation. Because the intensity of stimulation is typically applied at or just above the threshold for focal afterdischarge and is not altered during the course of the experiment, kindling clearly involves a change in the brain's response to the stimulation.

Sites susceptible to kindling are found throughout the forebrain, including neocortex, as well as in the brainstem (Lam et al., 2010), although the rate of kindling and the topography of kindled seizures vary as a function of site and species under investigation. In rats, for example, development of seizures proceeds through a fairly predictable sequence, as captured in Racine's scale (1972):

stage 1: head nodding
stage 2: mouth movements
stage 3: clonus of the contralateral limbs (hemiconvulsions)
stage 4: generalized clonus
stage 5: generalized clonus with rearing and falling.

The increased susceptibility to seizures characteristic of kindling is permanent and lasts as long as the subjects are alive (Wada et al., 1974), and kindling is associated with profound and long-lasting changes in the behaviors thought to be regulated by the sites subjected to stimulation (Teskey and Corcoran, 2009).

Kindling can also be associated with the development of recurrent spontaneous seizures unprovoked by stimulation, as a function of species and the age of the individual, and in this sense can be considered a valid "animal model" of the epilepsies. For example, adult rats, guinea pigs, and rhesus monkeys are generally resistant to the development of spontaneous seizures, whereas kittens, adult gerbils, and adult baboons are quite susceptible (reviewed in Corcoran and Teskey, 2009). Adult cats are of intermediate susceptibility, and indeed the first observation of spontaneous seizures associated with kindling was in adult cats (Wada et al., 1974). In the nearly 50 years since Goddard's original description of the phenomenon (Goddard, 1967; Goddard et al., 1969), kindling has become a dominant preparation for studying the processes of and influences on epileptogenesis, as well as the factors underlying and influencing established epileptic seizures (Corcoran and Moshé, 2005).

With kindling of the amygdala, Wada and Kudo (1997) found that unilateral lesions of the anterior CLA slowed subsequent kindling in cats. When induced after kindling of generalized seizures with stimulation of the amygdala, unilateral CLA lesions destabilized generalized seizures in cats and completely eliminated generalized seizures in baboons (Wada and Kudo, 1997; Wada and Tsuchimochi, 1997). Wada's results point to the involvement of the CLA in generalization when seizures are triggered by limbic or photic stimulation, but they do not indicate whether the CLA plays a unique role or whether it is part of a circuit of structures that promote generalization of seizures. They also do not address the epileptogenicity of the CLA itself.

RELATION BETWEEN CLAUSTRUM
PHYSIOLOGY AND SEIZURES

What characteristics of the CLA's physiology might be expected if the CLA plays an important role in seizure generalization? One could make several predictions:

A. *The current required to trigger afterdischarge, or the afterdischarge threshold, might be low.* Although this prediction seems intuitively reasonable, other research suggests that its utility is questionable. That is, because afterdischarge threshold probably depends largely on properties of local circuits in the area stimulated rather than to the projections of the area and its consequent ability to recruit seizure discharge in target structures, the relation between afterdischarge threshold and seizure susceptibility is not straightforward. For example, the hippocampus displays very low afterdischarge threshold, but is relatively resistant to kindling of generalized clinical seizures (Goddard et al., 1969; Racine, 1972). Furthermore, some seizure susceptible sites that kindle quickly, such as anterior neocortex and brainstem (Lam et al., 2010; Seidel and Corcoran, 1986), have high afterdischarge thresholds, and this further muddies the waters.

B. *Repeated stimulation of the site should lead to rapid development of generalized seizures.* Put simply, if a site plays an important role in seizure generalization, it should be very susceptible to kindling.

C. *Kindling of the site in question should lead to positive transfer of susceptibility to other sites.* In other words, transfer is an important characteristic of kindling, whereby prekindling a primary site generally leads to facilitation of kindling of generalized seizures in a secondary site, even after ablation of the primary site (Racine, 1972). Because stimulation of the secondary site can lead to immediate triggering of generalized seizures, transfer presumably reflects sensitization of the circuitry underlying seizure generalization.

D. *Lesion of the site in question would disrupt generalized seizures and perhaps delay kindling from stimulation of other sites.*

In the sections below I review the evidence we have collected that addresses each of the predictions. The primary focus of my laboratory's research was on characterizing the susceptibility of the CLA to kindling, although we have attempted to consider several of the other predictions as well.

The Afterdischarge Threshold of the Claustrum

Our interest in the role of the CLA in epileptogenesis was aroused serendipitously. Paul Mohapel in my laboratory was examining the phenomenon of kindling antagonism, where kindling stimulation is applied alternately to two sites in the rat's brain (e.g., Kirkby et al., 1993). Typically, kindling proceeds normally at one site, the dominant site, whereas it is arrested at the other (suppressed) site. As shown in Figure 10.1, he found that kindling of the amygdala was arrested by alternating stimulation of the posterior perirhinal cortex only when electrodes were located in the deep layers of the that cortex (Mohapel and Corcoran, 1996). Mohapel realized that the deep perirhinal placements were in fact located in, or very near to, the posterior end or tail of the CLA, and he decided to

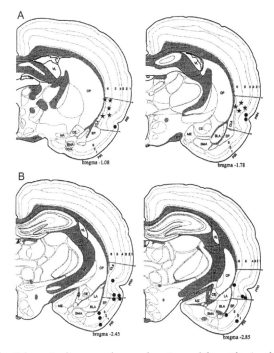

FIGURE 10.1   Schematic diagram of coronal sections of the rat brain, showing the location of cortical electrode tips in each rat that received alternating stimulation of the amygdala (AM) and cortex. A: electrode placements for the anterior limbic cortical group (AC). B: electrode placements for the posterior limbic cortical group (PC). Note that two electrode tips overlapped; hence only 13 placements are shown. *Solid circles* indicate electrode placements that did not suppress AM kindling, and the *solid stars* indicate electrode placements that did suppress AM kindling (i.e. dominant sites). Note that the suppression was always of the relative [5] form of antagonism. BLA, basolateral nucleus of amygdala; BMA, basomedial nucleus of amygdala; CE, central nucleus of amygdala; CLA, claustrum; COA, cortical nucleus of amygdala; CP, caudate-putamen; EC, external capsule; EP, endopiriform nucleus; INS, insular cortex; INT, internal capsule; LA, lateral nucleus of amygdala; MA, magnocellular preoptic nucleus; ME, medial nucleus of amygdala; OPT, optic tract; PIR, piriform cortex; PRH, perirhinal cortex; ST, stria terminalis; VL, lateral ventricle. *From Mohapel and Corcoran (1996), used with permission.*

TABLE 10.1   Characteristics of Kindling at Claustral and Other Sites

Group	Afterdischarge Threshold (μA)	Number of Afterdischarges to 1st Early Stage-5 Seizure	Afterdischarge Duration at 1st Early Stage-5 Seizure
AMYG	86[a]	12.0[a]	79.6[a]
Posterior CLA	1715[a]	2.8[a]	11.0[a]
Anterior CLA	471[b]	5.6[b]	20.5[b]
INS	5381[a]	3.9[a]	15.9[a]
PRH	3050[a]	4.3[a]	15.8[a]
NSUB	3802[c]	1.0[c]	19.2[c]

AMYG, amygdala; CLA, claustrum; INS, insula; NSUB, nucleus submedius; PRH, perirhinal cortex.
[a]*Data from Mohapel et al. (2001).*
[b]*Data from Zhang et al. (2001).*
[c]*Data from Sheerin et al. (unpublished).*

target electrodes to the posterior CLA intentionally, to examine its susceptibility to kindling.

In this and subsequent studies (except where indicated otherwise), we employed 1-second trains of balanced biphasic square-wave stimulation at 60 pps to trigger afterdischarge. As shown in Table 10.1, the initial afterdischarge threshold in the posterior CLA is over one order of magnitude higher than in the amygdala (Mohapel et al., 2001); with repeated stimulation it declined to much lower levels, but remained higher than in limbic sites (data not shown). In related research, Xia Zhang applied kindling stimulation to the anterior end or head of the CLA in rats (Zhang et al., 2001). Although lower than in the posterior CLA, the afterdischarge threshold of the anterior CLA was still considerably higher than in the amygdala (see Table 10.1). An example of the patterns of afterdischarge triggered during kindling of the anterior CLA is shown in Figure 10.2.

Zhang et al. (2001) also employed iontophoretic application of the tracers *Phaseolus vulgaris* leucoagglutinin (PHA-L) and FluoroGold (FG) to examine the efferent and afferent connections, respectively, of the anterior CLA. He detected widespread and often reciprocal connections of the CLA to a variety of cortical and subcortical structures, many of which have been implicated in seizure generalization, including frontal and motor cortex, limbic cortex, amygdala, endopiriform nucleus, olfactory areas, nucleus accumbens, and brainstem nuclei such as the substantia nigra and dorsal raphe nucleus. Unexpectedly, however, he discovered a very dense projection of the anterior CLA to the nucleus submedius of thalamus (NSUB), a projection that to our knowledge has not been described previously (Figure 10.3).

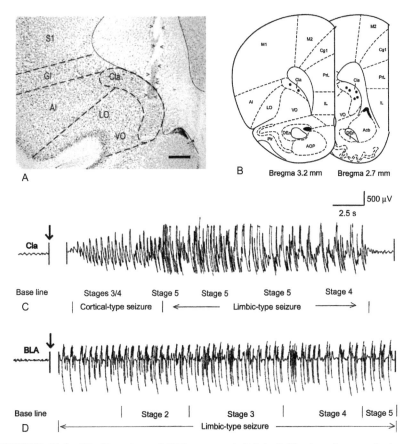

FIGURE 10.2 Kindling site and EEG pattern. A: Bright-field microphotograph showing the location of an implanted electrode. *Arrowheads* indicate the track of the electrode,
with the electrode tip located within the anterior claustrum. B: Schematic representation
of the location of electrode tips (*black dots*) in the anterior claustrum in the seven rats that
received kindling stimulation. C: A typical EEG profile associated with a seizure kindled
from the anterior claustrum. After electrical stimulation the rat almost immediately exhibited stages 3/4 seizures, several episodes of stage-5 seizures, and one episode of stage-4
seizures in sequence, which were accompanied by a relatively short period of cortical-type
EEG seizure and a relatively long-lasting limbic-type seizure. D: A typical EEG profile of a
seizure kindled from the basolateral amygdaloid nucleus. After stimulation the rat showed
a limbic-type EEG seizure lasting for 2 minutes, whereas no obvious behavioral seizure
appeared until 10 seconds after stimulation, at which time the rat began to sequentially
show stages 2, 3, 4, and 5 seizures. Acb, Accumbens nucleus; AI, agranular insular cortex;
AOP, anterior olfactory nucleus, posterior part; Cg1, cingulate cortex, area 1; Cla, claustrum; DEn, dorsal endopiriform nucleus; GI, granular insular cortex; IL, infralimbic cortex;
LO, lateral orbital cortex; M1, primary motor cortex; M2, secondary motor cortex; Pir, piriform cortex; PrL, prelimbic cortex; S1, primary somatosensory cortex; VO, ventral orbital
cortex. Scale bar, 520 mm. *From Zhang et al. (2001), used with permission.*

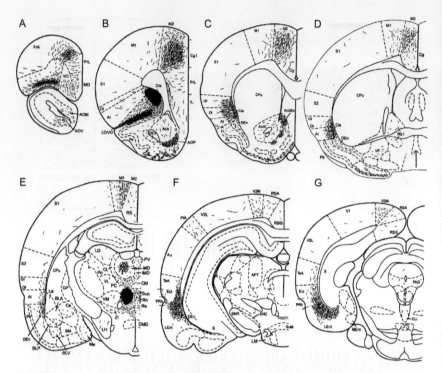

FIGURE 10.3 Projections of the anterior claustrum in rat 99217. Schematic representation of the density and localization of PHA-L-labeled fibers determined by immunohistochemistry was plotted from seven representative coronal sections corresponding to bregma 4.70, 2.70, 1.20, 20.26, 22.56, 24.80, and 26.80 mm (A–G, respectively). The *irregular dark area* in B indicates the ejection site, which is illustrated in Figure 10.3A. The full extent of labeled axons contralateral to the injection is not shown. AcbC, Accumbens nucleus, core; AcbSh, accumbens nucleus, shell; AOM, anterior olfactory nucleus, medial part; AOV, anterior olfactory nucleus, ventral part; APT, anterior pretectal nucleus; Au, secondary auditory cortex; BLA, basolateral amygdaloid nucleus, anterior part; BLP, basolateral amygdaloid nucleus, posterior part; BLV, basolateral amygdaloid nucleus, ventral part; BM, basomedial amygdaloid nucleus; BST, bed nucleus of the stria terminalis; Ce, central amygdaloid nucleus; Cg, cingulate cortex; CL, centrolateral thalamic nucleus; Cli, caudal linear nucleus of the raphe; CM, central medial thalamic nucleus; CPu, caudate putamen; DI, dysgranular insular cortex; DMD, dorsomedial hypothalamic nucleus, dorsal part; Ect, ectorhinal cortex; FrA, frontal association cortex; GP, globus pallidus; IMD, intermediodorsal thalamic nucleus; La, lateral amygdaloid nucleus; LD, laterodorsal thalamic nucleus; LH, lateral hypothalamic area; LM, lateral mammillary nucleus; MD, mediodorsal thalamic nucleus; Me, medial amygdaloid nucleus; Ment, medial entorhinal cortex; MO, medial orbital cortex; PAG, periaqueductal gray; PC, paracentral thalamic nucleus; Po, posterior thalamic nuclear group; PtA, parietal association cortex; PV, paraventricular thalamic nucleus; Re, reunions thalamic nucleus; Rh, rhomboid thalamic nucleus; RS, retrosplenial cortex; RSA, retrosplenial agranular cortex; RSG, retrosplenial granular cortex; S, subiculum; S2, secondary somatosensory cortex; SNC, substantia nigra, pars compacta; SNR, substantia nigra, pars reticulata; SuM, supramammillary nucleus; TeA, temporal association cortex; Tu, olfactory tubercle; V1, primary visual cortex; V2L, secondary visual cortex, lateral part; V2M, secondary visual cortex, medial part; VL, ventrolateral thalamic nucleus; VM, ventromedial thalamic nucleus; VP, ventral pallidum; ZI, zona incerta. *From Zhang et al. (2001), used with permission.*

The unexpected projection from the anterior CLA to the NSUB led Sheerin, Zhang, and Corcoran (unpublished) to ask whether the NSUB might itself be susceptible to kindling. Because short trains of high-frequency stimulation evoked severe motor responses, for NSUB kindling we instead employed low-frequency stimulation, consisting of a 5-second train of constant current, biphasic square wave pulses at 3 pps. Although stimulation of the NSUB triggered fully generalized seizures on the first afterdischarge (see section below), the afterdischarge threshold was exceptionally high (see Table 10.1), roughly comparable to the high afterdischarge threshold of the insular and perirhinal cortices (Mohapel et al., 2001).

Rate and Profile of Kindling from the Claustrum

Contrary to what one might expect from their very high afterdischarge thresholds, the anterior CLA and posterior CLA are highly susceptible to kindling of generalized convulsive seizures. Mohapel et al. (2001) found that stimulation of the posterior CLA led to development of generalized stage 5 convulsive seizures in fewer than 3 afterdischarges (see Table 10.1), in contrast to the 12 afterdischarges required for kindling from amygdaloid stimulation. Similarly, as shown in Table 10.1, Zhang et al. (2001) found that stimulation of the anterior CLA kindled brief generalized seizures in fewer than 6 afterdischarges. Seizures kindled from both sites in the CLA were characterized by predominantly clonic convulsive movements of the forelimbs and hindlimbs, with infrequent occurrence of tonic manifestations. The observations in freely moving rats were confirmed in rats suspended in a sling and subjected to kindling of the posterior CLA (Mohapel et al., 2001). In both groups, the generalized convulsive seizures displayed during the early phase of CLA kindling were relatively short in duration (under 20 seconds), as compared to generalized seizures kindled from the amygdala, which were closer to 80 seconds in duration (see Table 10.1).

Figure 10.2 illustrates another important feature of kindling with stimulation of either the anterior or posterior CLA. With repeated triggering of seizures, the early short or "focal cortical-type" seizures were replaced by longer and more "limbic-type" generalized convulsive seizures resembling the seizures kindled from the amygdala. The transition from early- to late-phase seizures occurred abruptly and was associated with stage 5 seizures that exceeded 30 seconds in duration of afterdischarge. The limbic-type seizures were more violent than the early-phase seizures and involved multiple episodes of rearing and falling. A similar pattern of two phases of kindling occurs when stimulation is applied to other non-limbic sites, including the anterior neocortex (Seidel and Corcoran, 1986). In the latter case, the development of limbic-type seizures is correlated with the appearance of afterdischarge in the amygdala that is independent of cortical afterdischarge. Further research is required to determine whether this relationship applies to kindling of the CLA.

As noted above, our observation of a dense projection from the anterior CLA to the NSUB of thalamus (Zhang et al., 2001) led us to examine the susceptibility of the structure to kindling (Sheerin et al., unpublished). The standard kindling stimulation of 1-second trains at 60 pps led to violent forced motor responses during the stimulation, so we explored alternative parameters of stimulation and were successful with 5-second trains of stimulation at 3 pps. The train of NSUB stimulation at these parameters evoked only moderate motor responses, which included rhythmic tonic hopping, progressing to obvious twisting along the rostral-caudal axis and culminating in forelimb clonus during the latter part of the train and subsequent afterdischarge. No kindling was required, in that fully generalized seizures appeared on the first afterdischarge. As with the early phase of CLA kindling, initial seizures were relatively brief, lasting considerably less than 20 seconds, and involved rhythmic hopping with bilateral forelimb clonus, and rolling on the side accompanied by bilateral forelimb and hindlimb tonic-clonic ictal movements. The seizures did not include the myoclonus of the head and facial musculature that is typical of early stages of limbic kindling. In a subgroup of rats subjected to 30 stimulations that triggered afterdischarges and generalized seizures, a significant increase in afterdischarge duration and subsequent generalization to late-stage limbic-type seizures occurred, which included bilateral forelimb clonus, rearing, and repeated falling. Figure 10.4 illustrates the pattern of increase in afterdischarge duration at various times. It was not until the tenth afterdischarge that we saw a significant increase in afterdischarge duration (mean = 20.9 seconds) compared to the first afterdischarge (mean = 15.6

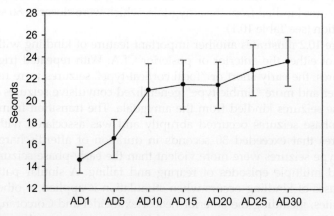

FIGURE 10.4 Durations of afterdischarge in rats stimulated in the nucleus submedius once daily for 30 days. *(Sheerin et al., unpublished).*

seconds; $P = 0.027$). After the tenth afterdischarge, duration of afterdischarge ranged from 20.5 seconds to 22.1 seconds, with no statistically significant increases in duration from 10 to 30 afterdischarges.

The immediate triggering of generalized seizures during the first NSUB afterdischarge allowed us to examine the potential anatomical substrate of the seizures. That is, the very high intensities of stimulation required to trigger afterdischarge in the NSUB would produce a large field of current, and the consequences of stimulation could be due to effects on fibers of passage or to neurons distant to the electrode tips, rather than to intrinsic neurons in the vicinity of the electrode tips. In a preliminary experiment, therefore, we employed a chemical lesioning paradigm to determine the substrate of the effects of the seizure-eliciting stimulation. To explain briefly, when employing this method, a relatively small volume of concentrated ibotenic acid selectively destroys the cell bodies it comes into contact with, but spares fibers of passage (Schwarcz et al., 1979). We reasoned that applying ibotenic acid to the NSUB before stimulation would allow us to determine whether the epileptiform effects of NSUB stimulation are due to stimulating neurons intrinsic to the region, or whether fibers of passage or distant neurons play a role.

Figure 10.5A shows a schematic illustration of the extent of the lesions. Ibotenic acid resulted in consistent damage to the intrinsic cell bodies in the NSUB. Occasionally, the lesions extended into more medial, lateral, and slightly more dorsal areas of the ventral thalamic region, but the adjacent mammillothalamic tract appeared to be undamaged in all cases. Lesions tended to be bilaterally symmetrical in size, but in several cases the contralateral lesion was produced closer to the midline, sparing much

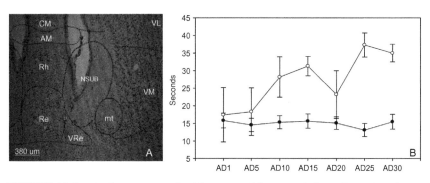

FIGURE 10.5   A: Photomicrograph of ibotenic acid lesion and electrode track with outline of the region of interest superimposed. Rh, rhomboid nucleus; Re, reuniens; mt, mammillothalamic tract; Vre, ventral reuniens; AM, anteromedial nucleus; CM; centromedial nucleus; VL, ventrolateral thalamus; VM, Ventromedial thalamus; NSUB, nucleus submedius. B: Duration of afterdischarge s 1through 30 in ibotenic acid lesioned rats (●-●) and non-lesioned control rats (○-○).

of the NSUB from damage. Overall, the entire NSUB within the region of interest was destroyed in 11 rats, and the kindling electrodes had been accurately placed in 16 rats (including 6 controls). Although ibotenic acid lesioned rats took much longer to recover immediately following surgery, no persistent gross behavioral changes were observed. Lesioned rats displayed lower afterdischarge thresholds than controls (mean values: lesioned, 6100.0 µA; controls, 7500 µA; $P = 0.039$). During the first afterdischarge, lesioned and control rats both displayed brief generalized motor seizures that did not differ between the groups, suggesting that neuronal elements outside of the area of destruction in the NSUB (i.e. fibers of passage or distant neurons) were responsible for the initial component of the ictal response. With repeated application of stimulation to the NSUB, however, control rats developed limbic-type generalized seizures by the seventh afterdischarge, whereas lesioned rats failed to develop limbic-type seizures, even after the triggering of 30 afterdischarges (Figure 10.5B), suggesting that neurons intrinsic to the NSUB critically participate in development of the delayed kindling component of the ictal response.

Transfer of Seizure Susceptibility

Mohapel et al. (2001) kindled late-stage generalized seizures from stimulation of the posterior CLA and then subjected the rats to secondary-site kindling of the contralateral amygdala. Our expectation was that CLA kindling would produce robust transfer to the amygdala, particularly given that advanced limbic-type seizures were being triggered from the posterior CLA. To our surprise, there was no evidence of transfer to the amygdala, with both groups requiring around 11 afterdischarges to display the first generalized seizure. For comparison, in separate groups we looked for transfer to amygdaloid kindling from prior kindling of the insula and perirhinal cortex, and again found no evidence of transfer (Mohapel).

In subsequent research, we determined whether prior kindling of the amygdala would lead to transfer to anterior CLA kindling and whether prior kindling of the anterior CLA would facilitate subsequent kindling of the amygdala (Sheerin et al., 2004). As shown in Figure 10.6, primary-site kindling of the amygdala produced transfer to secondary-site CLA kindling, but primary kindling of the CLA failed to transfer to the secondary-site amygdala.

In the final experiment of this series, we examined transfer between the NSUB and either the amygdala or the anterior CLA (Sheerin, Zhang, & Corcoran, unpublished). Primary-site kindling of the NSUB did not affect secondary kindling of the amygdala (i.e. no transfer was evident) and unexpectedly delayed secondary kindling of the CLA, with NSUB-kindled rats requiring nearly twice as many afterdischarges as

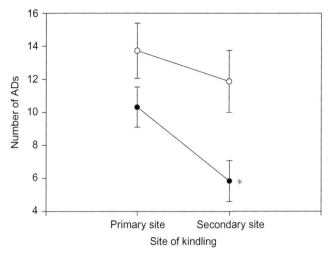

FIGURE 10.6   Transfer of kindling, assessed by comparing the rate of kindling of primary-site claustrum kindling to secondary-site claustrum kindling (●/●), and primary-site amygdaloid kindling to secondary-site amygdaloid kindling (○/○). *$P < 0.05$. From *Sheerin et al. (2004), used with permission.*

controls to develop generalized seizures in response to CLA kindling. Thus NSUB kindling produced negative transfer to the anterior CLA.

Effects of Claustrum Lesions on Kindling

As noted above, early research by Wada's group demonstrated that lesions of the CLA in cats and baboons can destabilize or block established seizures and delay kindling from the amygdala. We (Mohapel et al., 2000) decided to compare systematically the effects of prior and subsequent combined radiofrequency lesions of anterior and posterior CLA on amygdaloid kindling in rats, paying careful attention to quantification of the size of the lesions. In order to avoid nonspecific damage that extended well beyond the target structure, we ended up producing relatively small lesions of the CLA that destroyed only about 13 percent of the structure's volume. These small lesions nonetheless significantly delayed subsequent kindling from the amygdala, with controls requiring about 12 afterdischarges and lesioned rats about 19 afterdischarges to develop stage-5 generalized seizures. The volume of the lesions was found to be negatively correlated with the number of amygdaloid stimulations required for kindling ($r = -0.58$, $P < 0.03$), such that smaller lesions of CLA accounted for a larger delay in kindling. There is no obvious explanation for this counter-intuitive finding. Further analysis indicated that disruption of kindling was correlated only with damage to ventromedial CLA ipsilateral to the amygdala subjected to kindling.

We then examined the effects on kindling of lesions restricted to the anterior CLA (Sheerin et al., 2004). The lesions in this series of rats were small but variable and destroyed 10 to 50 percent of the anterior CLA. Although lesions generated before amygdaloid kindling had no effects on afterdischarge threshold or duration of the initial afterdischarge in the amygdala, they did significantly delay the development of generalized seizures, with controls requiring 10.1 afterdischarges and lesioned rats 16.6 afterdischarges. Lesions of the anterior CLA induced after kindling in other rats had no effect on the clinical signs of generalized seizures, but did shorten significantly the duration of the afterdischarge associated with the seizures, from 109.2 seconds (control) to 84.4 seconds (lesioned).

We also examined the effects of lesions of the NSUB on kindling with stimulation of either the amygdala or anterior CLA (Sheerin et al., unpublished). With amygdaloid stimulation, NSUB lesions significantly delayed kindling of stage-3 hemiconvulsions (controls: 9.9 afterdischarges; lesioned rats, 12.4 afterdischarges), but failed to affect the rate of development of stage-5 generalized seizures. However, the duration of stage-5 seizures was significantly reduced in lesioned rats (50.6 seconds) as compared to controls (77.5 seconds). With stimulation of the anterior CLA, NSUB significantly delayed the kindling of both stage-3 hemiconvulsions and stage-5 generalized seizures (11.5 and 13.0 afterdischarges, respectively) as compared to controls (8.1 and 9.5 afterdischarges, respectively), but there were no differences in the duration or intensity of generalized seizures when they emerged.

SUMMARY AND INTERPRETATION

In this chapter I have reviewed studies from my laboratory that address the role of the CLA in epilepsy. The results are at least in part consistent with the hypothesis that the CLA is a component of a network of structures that contribute to generalization of seizures. However, the results raise more questions than they answer, and additional research is clearly called for.

With regard to afterdischarge threshold in the CLA, we need to determine why afterdischarge threshold in different regions varies so significantly and seems to bear little relation to susceptibility to kindling. The CLA and NSUB exhibit high afterdischarge thresholds and yet are highly susceptible to kindling of generalized clinical seizures, whereas limbic sites exhibit much lower afterdischarge thresholds but kindle much more slowly. Our results with ibotenic acid induced lesions of the NSUB suggest that fibers of passage or distant neurons can play a role in afterdischarge threshold, and the paradigm should be extended to kindling of the CLA and indeed other structures. The high intensities of stimulation required to trigger afterdischarge in the CLA and NSUB are of general concern, on account of the issue of spread of current. If applied to other structures, the paradigm of prior lesioning of the stimulated site with ibotenic acid may help to address the concern.

Our results further show that afterdischarge threshold is clearly dissociated from susceptibility to kindling of generalized clinical seizures. The CLA is quite susceptible to kindling, and the NSUB is even more so. One could argue that both sites show less plasticity than, for example, the amygdala because of their high baseline of susceptibility. However the results are characterized, they are in my view consistent with a role of the CLA in seizure generalization, but much more needs to be learned. For example, we require evidence of cellular behavior in the CLA during kindling. With limbic kindling, we need to determine whether CLA neurons passively propagate discharge to many distant sites, or whether they become bursting neurons whose firing becomes independent of the epileptic focus, and whether they thereby amplify seizure discharge in distant sites.

Our observations of transfer after kindling of the CLA or NSUB are perhaps the most perplexing of our results, and the picture that emerges is unclear. We found an asymmetrical relation between CLA and limbic kindling, with primary amygdaloid kindling producing positive transfer to secondary kindling of the CLA, whereas primary kindling of the CLA failed to transfer to the amygdala. Even more surprising, primary kindling of the NSUB not only failed to transfer to secondary kindling of the amygdala but also actually delayed secondary kindling of the CLA. Perhaps the closer a structure is to seizure generalization, the less plastic it is, leading to less positive transfer or even negative transfer (Sheerin et al., 2004). Or perhaps, as Gale (2005) has suggested, kindling in the CLA normally does not involve the amygdala, and prior amygdaloid kindling establishes an alternate route that is abnormal and does not facilitate subsequent CLA kindling.

As Wada and colleagues reported, we found that selective CLA lesions disrupted established kindled seizures and delayed kindling from the amygdala. A limitation of these studies is that we employed radiofrequency lesions, which would destroy fibers of passage as well as intrinsic neurons. The participation of the CLA in limbic kindling would be better tested with ibotenic acid lesions that spare fibers of passage. Furthermore, in order to maintain anatomical specificity, our lesions were of necessity small, and eliminated less than 15 percent of the volume of the CLA. It remains to be determined whether more complete lesions of the CLA could be consistently achieved with infusions of ibotenic acid.

Thus the overall picture that emerges points to the CLA as a structure very susceptible to epileptogenesis, a structure that is likely involved in seizure generalization in important ways. Whatever clarification emerges from future research should bring us closer to understanding the role of this fascinating structure not only in epilepsy but also in higher brain function in general.

Acknowledgments

I dedicate this chapter to Juhn A. Wada, eminent neurologist and epileptogologist, and pioneering researcher on kindling. For over 60 years he has employed his amazing creativity, perseverance, and sheer intelligence to provide neuroscience with important insights into the mechanisms, characteristics, and treatment of epilepsy. I was privileged to work with him for 6 years, and, in this the 40th anniversary of my research in experimental epilepsy, it is a pleasure to acknowledge my indebtedness to him. I also thank the CIHR and NSERC for their support of my research over several decades, and I am particularly thankful to Justin Botterill for sharing his expertise with Photoshop. Finally, I gratefully acknowledge the dedication and intelligence of the students and colleagues who contributed to the results reported in this chapter: Paul Mohapel, Xia Zhang, Darren Hannesson, Deborah Saucier, and Aaron Sheerin.

References

Corcoran, M.E., Teskey, G.C., 2009. Characteristics and mechanisms of kindling In: Schwartzkroin, P.A. (Ed.), Encyclopedia of Basic Epilepsy Research, 2. Academic Press, Oxford, pp. 741–746.

Corcoran, M.E. Moshé, S.L. (Eds.), 2005. In: Kindling, 6. Springer, New York.

Gale, K., 2005. Designated discussion In: Corcoran, M.E. Moshé, S.L. (Eds.), Kindling, 6. Springer, New York, pp. 123.

Goddard, G.V., 1967. Development of epileptic seizures through brain stimulation at low intensity. Nature 214, 1020–1021.

Goddard, G.V., McIntyre, D.C., Leech, C.K., 1969. A permanent change in brain function resulting from daily electrical stimulation. Exp. Neurol. 25, 295–330.

Kirkby, R.D., Gilbert, T.H., Corcoran, M.E., 1993. Kindling antagonism: mapping of susceptible sites. Brain Res. 616, 17–24.

Kudo, T., Wada, J.A., 1995. Effects of unilateral claustral lesion on intermittent light stimulation-induced convulsive response in D,L-allylglycine treated cats. Electroencephalogr. Clin. Neurophysiol. 95, 63–68.

Lam, A., Whelan, N., Corcoran, M.E., 2010. Susceptibility of brainstem to kindling and transfer to the forebrain. Epilepsia 51, 1736–1744.

Mohapel, P., Corcoran, M.E., 1996. Kindling antagonism: interactions of the amygdala with the piriform, perirhinal, and insular cortices. Brain Res. 733, 211–218.

Mohapel, P., Hannesson, D.K., Armitage, L.L., Gillespie, G.W., Corcoran, M.E., 2000. Claustral lesions delay amygdaloid kindling in the rat. Epilepsia 41, 1095–1101.

Mohapel, P., Zhang, X., Gillespie, G.W., Chlan-Fourney, J., Hannesson, D.K., Cooley, S.M., et al., 2001. Kindling of claustrum and insular cortex: comparison to perirhinal cortex in the rat. Eur. J. Neurosci. 13, 1501–1519.

Racine, R.J., 1972. Modification of seizure activity by electrical stimulation. II. Motor seizure. Electroencephalogr. Clin. Neurophysiol. 32, 281–294.

Schwarcz, R., Hökfelt, T., Fuxe, K., Jonsson, G., Goldstein, M., Terenius, L., 1979. Ibotenic acid-induced neuronal degeneration: a morphological and neurochemical study. Ecperimental Brain Res. 37, 199–216.

Seidel, W.T., Corcoran, M.E., 1986. Relations between amygdaloid and anterior neocortical kindling. Brain Res. 385, 375–378.

Sheerin, A.H., Nylen, K., Zhang, X., Saucier, D.M., Corcoran, M.E., 2004. Further evidence for a role of the anterior claustrum in epileptogenesis. Neuroscience 125, 57–62.

Teskey, G.C., Corcoran., M.E., 2009. Interictal behavioral comorbidities in a model of epilepsy In: Schwartzkroin, P.A. (Ed.), Encyclopedia of Basic Epilepsy Research, 3 Academic Press, Oxford, pp. 1254–1260.

Wada, J.A., Kudo, T., 1997. Involvement of the claustrum in the convulsive evolution of temporal limbic seizure in feline amygdaloid kindling. Electroencephalogr. Clin. Neurophysiol. 103, 249–256.

Wada, J.A., Tsuchimochi, H., 1997. Role of the claustrum in convulsive evolution of visual afferent and partial nonconvulsive seizure in primates. Epilepsia 38, 897–906.

Wada, J.A., Sato, M., Corcoran, M.E., 1974. Persistent seizure susceptibility and recurrent spontaneous seizures in kindled cats. Epilepsia 15, 465–478.

Zhang, X., Hannesson, D.K., Saucier, D.M., Wallace, A.E., Howland, J., Corcoran, M.E., 2001. Susceptibility to kindling and neuronal connections of the anterior claustrum. J. Neurosci. 21, 3674–3687.

CHAPTER

11

The Claustrum and Alzheimer's Disease

Annalena Venneri[1] *and Michael Shanks*[1,2]

[1]Department of Neuroscience, University of Sheffield, UK [2]Academic Unit of Psychiatry, University of Sheffield, UK

INTRODUCTION

The claustrum offers functionality of a higher order, enabling the organism to rapidly adapt to the subtleties and nuances of its ever-changing environment. In humans, a loss of any of these multisensory and heterotopic attributes may yet be demonstrated to be involved in some aspects of dementia, attention and other perturbations or disturbances of higher-order functions.

Edelstein and Denaro, 2004

The quotation above anticipates that loss of the proposed integrative functions of the claustrum might cause distinctive cognitive dysfunctions. Disconnection and degeneration of this structure might then contribute to cognitive and neuropsychiatric symptoms including those appearing in Alzheimer's disease (AD). That the claustrum is important for at least some elements of the neuropsychological and neuropsychiatric syndromes of AD, and probably other dementias, is supported not only by its connections but also by neuropathological and in vivo neuroimaging correlative studies. Reported changes in claustral activity in neuroimaging studies are frequent, but largely little remarked upon and discussed. In this chapter the very limited literature about the claustrum in AD will be reviewed, and some speculative inferences about claustral dysfunction made in the context of the neuropathological progression of AD, and what is known about the wider relationship of regional and network brain dysfunction to the neuropsychological and neuropsychiatric manifestations of the disease.

J. Smythies, L. Edelstein and V. Ramachandran (Eds):
The Claustrum.

DOI: http://dx.doi.org/10.1016/B978-0-12-404566-8.00011-8

CLAUSTRAL CONNECTIONS AND ALZHEIMER'S DISEASE

In AD there is early degeneration of the hippocampal/entorhinal complex followed by progressive degeneration of associative neocortex, with widespread disruption to intrahemispheric cortico-cortical, commissural and subcortical connections (Braak and Braak, 1991). In a recent high-resolution diffusion tensor imaging (DTI) study of normal primate brain the complex sheet of neurones and neuropil which constitutes the claustrum had the strongest probabilistic connections with the entorhinal cortex, but was also connected with most other cortical regions and some subcortical structures (Park et al., 2012). A tractography study of the human brain with fiber dissection showed that the claustrum had topographically organized connections with superior frontal, precentral, postcentral and posterior parietal cortices (Fernández-Miranda et al., 2008). Traditional neuroanatomical tracing methods, which can more precisely determine the direction, origin, and termination of fibers in other primate species and in the cat, broadly align with these data and found topically ordered reciprocal connections (mainly ipsilateral) with all areas of the neocortex and, again important in the present context, with the entorhinal cortex (Insausti et al., 1987; Witter et al., 1988). Such techniques of course are not possible with the human brain, but it seems unlikely, given the enhanced development of both the claustrum and the cerebral cortex in man (see Chapter 7), that claustral connectivity is any less complex and neocortically pervasive in our own species. It follows that both hippocampal/entorhinal and progressive neocortical degeneration in AD can have anterograde and retrograde effects on regional claustral neuron populations in addition to any intrinsic participation in the neurodegenerative process. Studies of immunoreactive terminal species in the claustrum suggest extensive connectivity within the claustrum itself (Rahman and Baizer, 2007).

CHOLINERGIC PATHWAYS AND THE CLAUSTRUM

Cholinergic innervations extend to all parts of the human cerebral cortex. Cholinergic projections from the nucleus basalis of Meynert extend to the cortex via two pathways, one passing medially and another laterally within each hemisphere. In the human brain there is a strong cholinergic innervation into the claustrum (Mesulam M, personal communication) and the perisylvian division of the lateral cholinergic pathway originating in the nucleus basalis passes through the claustrum and supplies neurons in the frontoparietal operculum, insula and superior temporal gyrus (Selden et al., 1998). There is no direct evidence of claustral/entorhinal connectivity in man but this may be strongly inferred from primate studies (Insausti et al., 1987).

Studies in different species provide evidence of widespread connectivity between the entorhinal cortex and subcortical structures, including basal ganglia, thalamus, hypothalamus, and amygdala as well as strong reciprocal connections with the claustrum (Canto et al., 2008). Extensive links with other structures in the limbic system and indirect connectivity via other limbic structures must also be present in humans (Price, 2007; Smythies et al., 2012), and there is histological evidence of anatomical contiguity between the intermediate part of the entorhinal area and the preamygdalar part of the claustrum (Heinsen et al., 1994).

NEUROPATHOLOGY

The claustrum, in common with other cortical and subcortical structures (e.g. nucleus basalis of Meynert, entorhinal cortex) does show degenerative changes in ageing and AD. A study of age-related changes of the claustrum in older dogs showed amyloid deposition and loss of neurons (Moryś et al., 1994). In the human brain there is evidence of age-related changes in the claustrum of non-demented people with neuronal loss and decreased volume in all subfields (Moryś et al., 1996). A small number of amyloid deposits in the para-amygdalar zone of the claustrum were found in the oldest people in the sample. No evidence of neurofibrillary tangles was reported. Neurofibrillary tangles, however, are said to accumulate and develop in the claustrum as well as in the amygdala and the thalamus at a later stage of AD and are found at Braak and Braak stage IV (Serrano-Pozo et al., 2011). The limited neuropathological studies in AD found the most severe changes in the ventral part of the claustrum, inferred from primate studies to have reciprocal connections with the entorhinal cortex. Claustral areas reciprocally connected to the neocortex were reported to show less cell loss and pathological change (Moryś et al., 1996). In a study of amyloid deposition in the human brain, amyloid deposits were found in the claustrum in all 14 cases of AD studied. The deposits were numerous type 2 plaques with some type 1 and type 3 plaques (Ogomori et al., 1989). Occasional but rare plaques were found in normal controls. The clinical heterogeneity of AD syndromes and issues of staging qualify interpretation of this evidence, however. Claustral degeneration was particularly marked in familial cases of AD caused by a presenilin mutation (Gustafson et al., 1998). In a neuropathological study of other neurodegenerative syndromes, amyloid and alpha-synuclein pathology in the claustrum was strongly related to dementia diagnosis in both Parkinson's disease and dementia with Lewy bodies (Kalaitzakis et al., 2009). There is evidence of increased muscarinic M4 receptor activity in the claustrum and the cortex of patients with Lewy body dementia and AD (Piggott et al., 2003), possibly reflecting cholinergic dysfunction.

THE CLAUSTRUM AND NETWORK CONNECTIVITY IN AD

The march of neuronal degeneration in AD is paralleled by an increasing disruption of structural and functional connectivity. Intra- and interregional integrity is compromised, and in turn larger scale neurocognitive networks are affected (Sporns, 2011; Xie and He, 2011). Assessment of resting-state brain connectivity together with measures of gray matter volume and diffusion tensor imaging (DTI) has already begun to delineate these anatomically and functionally correlated networks in healthy brains (e.g. Greicius et al., 2009). For example, a network that is important for binding posterior cingulate, precuneal and medial temporal cortices, and which is activated during episodic memory tasks, is pathologically and functionally degraded early in AD and in mild cognitive impairment (e.g. Buckner et al., 2008; Greicius et al., 2004). Evidence of structural connectivity change in early AD has been obtained by DTI, and impairments in the medial prefrontal, posterior parietal, and insular cortex best differentiate between AD and healthy ageing (Shao et al., 2012). Disruption of structural connectivity in these regions has wider consequences for functional networks involved in memory processes, attention, awareness of self and the wider environment. Degradation of these primarily cholinergic connections can promote neuropsychiatric symptoms in AD as well as in dementia with Lewy bodies (Piggott et al., 2003), and abnormalities of the claustrum and the insula have been associated with the presence of positive symptoms in both neurodegenerative and psychiatric conditions (Bruen et al., 2008; Cascella et al., 2011).

In the field of connectivity and memory, functional MRI (fMRI) studies of AD have shown deactivation of the claustrum as well as of the insula, posterior cingulate, temporal and lateral parietal cortex in patients and healthy older adults during a paired associates learning paradigm, both at encoding and at retrieval (Gould et al., 2006). Some of the regions that show task-induced deactivation are not part of the default mode network and their deactivation in memory encoding and retrieval may develop a broader understanding of memory deficits in AD. The claustrum is not part of the default mode system, but it does appear to be interconnected with a network of structures directly involved in memory retrieval. These structures include a distinctive neuronal type, the Von Economo neuron (VEN; Figure 11.1), which is abundant in humans

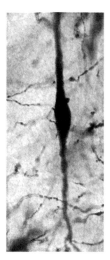

FIGURE 11.1   Photomicrographs of soma and proximal dendrites of a Von Economo neuron stained using the Golgi method. *Reproduced with permission from Watson et al. (2006) Dendritic architecture of the Von Economo neurons, Neuroscience, 141, 1107–1112.*

and richly represented in both the insula and the claustrum (Williamson, 2007). This author suggested that the activity of VEN in the claustrum would influence the function of a strongly interconnected and extensive network of heteromodal cortex. This activity might in turn facilitate the interactions between the default-mode network and task-related networks. Reduction of activity in the claustrum, therefore, could influence the functioning of task-related networks and, in particular, the efficiency of a memory-related network in AD.

Again, Park et al. (2012) suggested that the loss of spatial memory in AD (Didic et al., 2011) might be due not only to degeneration of the entorhinal cortex and the hippocampus but also to impairment of the related integrating function of the claustrum. They point to the claustrum's extensive connections with cortical areas and the likely involvement in more global aspects of perception and consciousness.

COGNITION AND THE CLAUSTRUM IN AD

The advent of fMRI and other functional imaging techniques have increasingly provided evidence about the involvement of the claustrum (either directly or indirectly via its connections with key structures) in cognition. This issue is more comprehensively reviewed in other chapters, but as suggested above, changes in claustral activity in neuroimaging studies are frequently reported but draw little or no comment. The

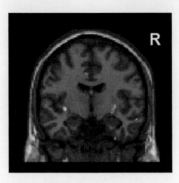

FIGURE 11.2   Significant clusters of correlation between gray matter density values in the claustrum bilaterally and confrontation naming scores in mild AD.

broad picture emerging from the available literature from animal and human studies suggests a crucial involvement in tasks in which integration of information processed by the brain via different sensory modalities (Banati et al., 2000; Hadjikhani and Roland, 1998) or across modalities (Baugh et al., 2011; Olson and Graybiel, 1980) is needed. There is also evidence against this, with other studies suggesting that, although the claustrum processes multisensory information, there are no multisensory cells in this structure, hence the multisensory integration of information cannot be mediated by the claustrum (Remedios et al., 2010). Others have suggested that its role might be better defined in the sense of ascertaining multisensory conceptual congruency (Naghavi et al., 2007). In AD, volumetric changes in both right and left claustrum were found to be significantly correlated with performance on confrontation naming and visuoconstructive tasks (Venneri et al., 2008) (Figure 11.2).

Similar findings were reported by another recent voxel-based correlation study with patients in the mild cognitive impairment (MCI) preclinical stage of AD. In this study, volumetric changes in both right and left claustrum were found to correlate, as previously found, with confrontation naming, but also with scores on the category fluency test (Gardini et al., 2013). No correlation in the claustrum was found when a similar analysis was carried out in age-matched controls. This pattern of findings might indicate that neuronal loss in the claustrum is interfering with efficient performance on this task and that the claustrum may provide a supportive role in semantic/phonological retrieval. The most probable process would be mediating the integration of visual, conceptual and phonological information to ascertain congruency between these different modes, to achieve successful confrontation naming.

A link between variance in cholinergic function in the claustrum and verbal memory learning scores in MCI patients was detected in a study of nicotinic acetylcholine receptor binding in a preclinical stage patient group, providing additional evidence that early changes in this structure might modulate cognitive performance in patients with either established or prodromal AD (Terrière et al., 2010).

The claustrum may also contribute to the neurological substrate of cognitive reserve in healthy elderly, and a negative correlation between claustrum activity during a visual memory task and an index of cognitive reserve was found by a neuroimaging study which investigated the biological underpinning of cognitive reserve in normal and abnormal ageing (Solé-Padullés et al., 2009). However, no correlation in the claustrum was found in mild AD patients in that study. The different correlation with healthy and with abnormal ageing, might be a disease effect. One explanation of the difference might be that progression of neuropathology would diminish or eliminate the claustrum's supportive contribution to this aspect of cognitive reserve, as the integrative mechanisms involved degrade. Functional neuroimaging studies of episodic memory in healthy ageing and AD have also highlighted a claustral contribution to episodic memory retrieval. A recent quantitative meta-analysis showed that, during memory retrieval, healthy controls showed greater levels of activity in several brain regions, including the claustrum, than patients with AD (Schwindt and Black, 2009). This meta-analysis suggests that defective retrieval in AD is the product of dysfunction in a wide network of cortical and subcortical structures, including the claustrum. Of interest also are the findings of a recent study of deep brain stimulation (DBS) in AD. Following one year of stimulation, increases in regional brain metabolism were observed not only in temporal and parietal cortex but also in bilateral paracentral lobule and right precentral gyri, bilateral postcentral gyri, right lingual gyrus, and left claustrum, while decreased metabolism was detected in the left anterior cingulate, left middle frontal, and right inferior frontal gyri (Laxton et al., 2010). This increase in metabolism in areas of the default mode network, and decrease in areas of the salience network, suggested that treatment with DBS was contributing to a re-balancing of these networks, and this could have cognitive benefit for treated patients. These authors, however, were not able to detect any behavioral change because of lack of statistical power and small sample size. Based on the regional metabolic increases in structures such as the anterior cingulate and the claustrum, it might be a reasonable speculation that positive effects on cognitive executive control, memory retrieval, and also assessment of cognitive congruency might be expected.

Discrete and cross-modal neuropsychological dysfunctions involving episodic or semantic memory, visuospatial function or other cognitive domains are not the only or even the earliest symptoms which draw attention to, or develop, in the course of this neurodegenerative disease (Cummings, 1985, 2005). An important group of symptoms arises from more fundamental changes in the nature of the AD patients' experience of reality. These symptoms are arguably facilitated by the progressive degradation of personal episodic memories in the course of AD, so that a plausible (to the subject) and usually less threatening version of reality can emerge in consciousness. Patients may also demonstrate, however, a more systematic impairment of world representation with loss of the ability to distinguish the animate from the inanimate or to judge the plausibility of beliefs. This group of symptoms, which includes neglect, anosognosia, misidentification, fantastic confabulation, totemism and other forms of delusional awareness, indicate dysfunction in other brain networks supporting the normal structures of conscious awareness (Feinberg et al., 2010; Venneri and Shanks, 2010). These symptoms have been subject to only rudimentary phenomenological description and analysis so far, and this is one reason why related neural networks are less well explored and understood. Lesion studies as well as limited structural and functional studies in AD suggest an important component of these networks will involve right hemisphere dorsolateral frontal, orbitofrontal and parietal cortices (Bruen et al., 2008; Feinberg et al., 2010). In a morphometric study in AD, delusional misidentifications of place or of person with related confabulations or more persistent delusional memories correlated significantly with low gray-matter density values in the right inferior frontal gyrus and inferior parietal lobule. In the left hemisphere there was significant correlation in the inferior and medial frontal gyri, but there was also a significant cluster of correlation in the left claustrum (Bruen et al., 2008) (Figure 11.3). In discussion, the probable involvement of the claustrum in higher-order, integrative functions including facilitation of rapid transfer of information along its anteroposterior and ventrodorsal axes and the instantiation

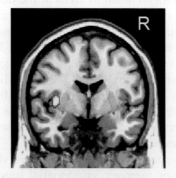

FIGURE 11.3   Significant clusters of correlation between gray matter density values in the left claustrum and misidentification delusion scores in mild AD. *Reproduced with permission from Bruen et al. (2008) Neuroanatomical correlates of neuropsychiatric symptoms in Alzheimer's disease. Brain, 131, 2455–2463.*

of multimodal (cognitive, perceptual, motor) syntheses was cited (Crick and Koch, 2005). It was further speculated that claustral dysfunction might impair (directly or indirectly) the normal synchronization of perceptual and cognitive experiences in AD patients, contributing to delusional misinterpretation of percepts, memories or mental images.

Additional evidence bearing on a possible claustral contribution to normal awareness and disorders of awareness in AD comes from studies in frontotemporal dementia (FTD) and schizophrenia, as well as the arguments in other chapters about the claustrum's special role in the emergence of self-reflective consciousness (see, for example, Chapters 9 and 14).

The importance of VEN in the insula and claustrum was discussed above in the context of task-related functions, but VEN are also phylogenetically increased in man, especially in fronto-insular cortex and in the right hemisphere. A histopathological study of VEN in FTD showed a highly significant and specific degeneration of VEN which was related to the early and nosologically distinctive impairment of self-awareness, empathy and reasoning in these patients (Seeley et al., 2006). The distinction from AD controls was emphasized, but these controls did not have frontal symptoms and had preservation of social awareness and empathy. This is true of many AD patients, but VEN do degenerate as AD progresses and the disease may present in some cases with early changes of environmental and social awareness. It seems reasonable to suggest that fronto-insular dysfunction, in which the claustrum will be implicated or will contribute, will be an important determinant of changes in self-awareness in both neurodegenerative disorders. There is convergent evidence from Lewy body dementia supporting the role of the claustrum and its connections in the correct interpretation of visual percepts. A recent study reported that Lewy body pathology in the claustrum, detected using alpha-synuclein immunohistochemistry, correlated highly with pathological changes in other regions of the visual association cortex, such as Brodmann areas (BA) 18 and 19, as well as in the insular and transentorhinal cortex (Yamamoto et al., 2007). These authors suggested that Lewy body pathology in the claustrum is more closely related to dysfunction in visual areas and that misidentification in Lewy body dementia might be the product of damage in the visuo-claustral pathway as well as the visuo-amygdaloid pathway. They also suggested that these two pathways might act as relay stations and that paralimbic structures, including insular and transentorhinal cortex, may mediate connections between visual areas and limbic areas. The relationship of claustral atrophy to misidentification delusions in AD reported by Bruen et al. (2008) might be interpreted in a similar way. Degeneration of claustral cholinergic pathways, resulting from interference with the claustrum's critical mediating role, might impair the veridical interpretation of visual inputs and contribute to delusional misidentification.

More supportive evidence for a central role of claustral structural/ functional damage in the genesis of psychotic symptoms derives from the appearance of psychoses following transient dysfunction of the claustrum in a case of severe transient encephalopathy (Sperner et al., 1996). In this case the symptoms cleared once dysfunction in gray matter in the claustrum had resolved. Again, morphometric findings similar to those observed in AD were reported in a voxel based correlational analysis of delusions in schizophrenia (Cascella et al., 2011). In this study, significant inverse correlations between the severity of delusions and gray matter volume values in the left claustrum and right insula were found in a large sample of adult patients with schizophrenia, corroborating the evidence from neurodegenerative diseases that integrity of the claustrum may be critical for the correct interpretation of visual percepts, mental images and memories.

CONCLUSION AND FUTURE DIRECTIONS

This review chapter has presented converging evidence from studies of cognitive and neuropsychiatric symptoms which flag up the emerging importance of degeneration of the claustrum in AD, and which suggests a critical role of the neuropil and connectivity of this structure in a range of functions which are essential for the maintenance of efficient cognition and a veridical interpretation of reality and the environment. The review, however, has also highlighted a fundamental and curious neglect of this structure in the neurodegenerative neuroscience literature. Claustrum involvement is often found in structural and functional neuroimaging studies of AD as well as in studies of ageing and of healthy young adults, but the reasons for claustrum deficits, dysfunction or differential activation are rarely analyzed, and even less frequently interpreted in terms of higher-order cognitive functions. This chapter, while inevitably to some degree speculative, has attempted to fill this gap and, while reviewing the available findings, has also tried to provide a preliminary theoretical rationale which might be used as a framework to make inferences about the physiological function of the claustrum in normal cognition and its possibly central role in some of the symptoms observed in neurodegenerative diseases like AD.

The evidence reviewed above helps corroborate theories that explore the importance of the disconnection between cortical and subcortical structures which follows the degeneration of cholinergic neurons (including those in the claustrum) and related cholinergic pathways, isolating mediotemporal structures from associative areas of the neocortex (Smith, 2002). In broad terms, efficient cognition would be disrupted with a breakdown of the normally aligned and synchronized computational processes in different brain circuits to which the claustrum seems to make a central contribution. This distinctive kind of multimodal disruption may make a critical contribution to both the neuropsychological and neuropsychiatric symptoms in AD. This hypothesis would of course need prospective testing, and modern structural and functional connectivity techniques will be the essential methods for clarifying the role of the claustrum in healthy and pathological states.

Acknowledgment

The authors thank Dr. William McGeown for his help with the figures in this chapter.

References

Banati, R.B., Goerres, G.W., Tjoa, C., Aggleton, J.P., Grasby, P., 2000. The functional anatomy of visual-tactile integration in man: a study using positron emission tomography. Neuropsychologia 38, 115–124.

Baugh, L.A., Lawrence, J.M., Marotta, J.J., 2011. Novel claustrum activation observed during a visuomotor adaptation task using a viewing window paradigm. Behav. Brain Res. 223, 395–402.

Braak, H., Braak, E., 1991. Neuropathological stageing of Alzheimer-related changes. Acta Neuropathol. 82, 239–259.

Bruen, P.D., McGeown, W.J., Shanks, M.F., Venneri, A., 2008. Neuroanatomical correlates of neuropsychiatric symptoms in Alzheimer's disease. Brain 131, 2455–2463.

Buckner, R.L., Andrews-Hanna, J.R., Schacter, D.L., 2008. The brain's default network: anatomy, function, and relevance to disease. Ann. N. Y. Acad. Sci. 1124, 1–38.

Canto, C.B., Wouterlood, F.G., Witter, M.P., 2008. What does the anatomical organization of the entorhinal cortex tell us? Neural Plast. 2008, 381243.

Cascella, N.G., Gerner, G.J., Fieldstone, S.C., Sawa, A., Schretlen, D.J., 2011. The insula-claustrum region and delusions in schizophrenia. Schizophr. Res. 133, 77–81.

Crick, F.C., Koch, C., 2005. What is the function of the claustrum? Philos. Trans. R. Soc. Lond., B, Biol. Sci. 360, 1271–1279.

Cummings, J.L., 1985. Organic delusions: phenomenology, anatomical correlations, and review. Br. J. Psychiatry 146, 184–197.

Cummings, J.L., 2005. Behavioral and neuropsychiatric outcomes in Alzheimer's disease. CNS Spectr. 10, 22–25.

Didic, M., Barbeau, E.J., Felician, O., Tramoni, E., Guedj, E., Poncet, M., et al., 2011. Which memory system is impaired first in Alzheimer's disease? J. Alzheimers Dis. 27, 11–22.

Edelstein, L.R., Denaro, F.J., 2004. The claustrum: a historical review of its anatomy, physiology, cytochemistry and functional significance. Cell. Mol. Biol. (Noisy-le-grand) 50, 675–702.

Feinberg, T.E., Venneri, A., Simone, A.M., Fan, Y., Northoff, G., 2010. The neuroanatomy of asomatognosia and somatoparaphrenia. J. Neurol. Neurosurg. Psychiatr. 81, 276–281.

Fernández-Miranda, J.C., Rhoton Jr., A.L., Kakizawa, Y., Choi, C., Alvarez-Linera, J., 2008. The claustrum and its projection system in the human brain: a microsurgical and tractographic anatomical study. J. Neurosurg. 108 (4), 764–774.

Gardini, S., Cuetos, F., Fasano, F., Ferrari Pellegrini, F., Marchi, M., Venneri, A., et al., 2013. Brain structural substrates of semantic memory decline in mild cognitive impairment. Curr. Alzheimer Res. 10, 373–389.

Gould, R.L., Brown, R.G., Owen, A.M., Bullmore, E.T., Howard, R.J., 2006. Task-induced deactivations during successful paired associates learning: an effect of age but not Alzheimer's disease. Neuroimage 31, 818–831.

Greicius, M.D., Srivastava, G., Reiss, A.L., Menon, V., 2004. Default-mode network activity distinguishes Alzheimer's disease from healthy aging: evidence from functional MRI. Proc. Natl. Acad. Sci. USA 101, 4637–4642.

Greicius, M.D., Supekar, K., Menon, V., Dougherty, R.F., 2009. Resting-state functional connectivity reflects structural connectivity in the default mode network. Cereb. Cortex 19, 72–78.

Gustafson, L., Brun, A., Englund, E., Hagnell, O., Nilsson, K., Stensmyr, M., et al., 1998. A 50-year perspective of a family with chromosome-14-linked Alzheimer's disease. Hum. Genet. 102, 253–257.

Hadjikhani, N., Roland, P.E., 1998. Cross-modal transfer of information between the tactile and the visual representations in the human brain: a positron emission tomographic study. J. Neurosci. 18, 1072–1084.

Heinsen, H., Henn, R., Eisenmenger, W., Gotz, M., Bohl, J., Bethke, B., et al., 1994. Quantitative investigations on the human entorhinal area: left-right asymmetry and age-related changes. Anat. Embryol. (Berl) 190, 181–194.

Insausti, R., Amaral, D.G., Cowan, W.M., 1987. The entorhinal cortex of the monkey: III. Subcortical afferents. J. Comp. Neurol. 264, 396–408.

Kalaitzakis, M.E., Pearce, R.K., Gentleman, S.M., 2009. Clinical correlates of pathology in the claustrum in Parkinson's disease and dementia with Lewy bodies. Neurosci. Lett. 461, 12–15.

Laxton, A.W., Tang-Wai, D.F., McAndrews, M.P., Zumsteg, D., Wennberg, R., Keren, R., et al., 2010. A phase I trial of deep brain stimulation of memory circuits in Alzheimer's disease. Ann. Neurol. 68, 521–534.

Morys, J., Narkiewicz, O., Maciejewska, B., Wegiel, J., Wisniewski, H.M., 1994. Amyloid deposits and loss of neurones in the claustrum of the aged dog. Neuroreport 5, 1825–1828.

Morys, J., Bobinski, M., Wegiel, J., Wisniewski, H.M., Narkiewicz, O., 1996. Alzheimer's disease severely affects areas of the claustrum connected with the entorhinal cortex. J. Hirnforsch. 37, 173–180.

Naghavi, H.R., Eriksson, J., Larsson, A., Nyberg, L., 2007. The claustrum/insula region integrates conceptually related sounds and pictures. Neurosci. Lett. 422, 77–80.

Ogomori, K., Kitamoto, T., Tateishi, J., Sato, Y., Suetsugu, M., Abe, M., 1989. Beta-protein amyloid is widely distributed in the central nervous system of patients with Alzheimer's disease. Am J Pathol 134, 243–251.

Olson, C.R., Graybiel, A.M., 1980. Sensory maps in the claustrum of the cat. Nature 288, 479–481.

Park, S., Tyszka, J.M., Allman, J.M., 2012. The claustrum and insula in Microcebus murinus: a high resolution diffusion imaging study. Front. Neuroanat. 6, 21.

Piggott, M.A., Owens, J., O'Brien, J., Colloby, S., Fenwick, J., Wyper, D., et al., 2003. Muscarinic receptors in basal ganglia in dementia with Lewy bodies, Parkinson's disease and Alzheimer's disease. J. Chem. Neuroanat. 25, 161–173.

Price, J.L., 2007. Definition of the orbital cortex in relation to specific connections with limbic and visceral structures and other cortical regions. Ann. N. Y. Acad. Sci. 1121, 54–71.

Rahman, F.E., Baizer, J.S., 2007. Neurochemically defined cell types in the claustrum of the cat. Brain Res. 1159, 94–111.

Remedios, R., Logothetis, N.K., Kayser, C., 2010. Unimodal responses prevail within the multisensory claustrum. J. Neurosci. 30, 12902–12907.

Schwindt, G.C., Black, S.E., 2009. Functional imaging studies of episodic memory in Alzheimer's disease: a quantitative meta-analysis. Neuroimage 45, 181–190.

Seeley, W.W., Carlin, D.A., Allman, J.M., Macedo, M.N., Bush, C., Miller, B.L., et al., 2006. Early frontotemporal dementia targets neurons unique to apes and humans. Ann. Neurol. 60, 660–667.

Selden, N.R., Gitelman, D.R., Salamon-Murayama, N., Parrish, T.B., Mesulam, M.M., 1998. Trajectories of cholinergic pathways within the cerebral hemispheres of the human brain. Brain 121, 2249–2257.

Serrano-Pozo, A., Frosch, M.P., Masliah, E., Hyman, B.T., 2011. Neuropathological alterations in Alzheimer disease. Cold Spring Harb. Perspect. Med. 1, a006189.

Shao, J., Myers, N., Yang, Q., Feng, J., Plant, C., Bohm, C., et al., 2012. Prediction of Alzheimer's disease using individual structural connectivity networks. Neurobiol. Aging 33, 2756–2765.

Smith, A.D., 2002. Imaging the progression of Alzheimer pathology through the brain. Proc. Natl. Acad. Sci. USA 99, 4135–4137.

Smythies, J., Edelstein, L., Ramachandran, V., 2012. Hypotheses relating to the function of the claustrum. Front. Integr. Neurosci. 6, 53.

Solé-Padullés, C., Bartrés-Faz, D., Junqué, C., Vendrell, P., Rami, L., Clemente, I.C., et al., 2009. Brain structure and function related to cognitive reserve variables in normal aging, mild cognitive impairment and Alzheimer's disease. Neurobiol. Aging 30, 1114–1124.

Sperner, J., Sander, B., Lau, S., Krude, H., Scheffner, D., 1996. Severe transitory encephalopathy with reversible lesions of the claustrum. Pediatr. Radiol. 26, 769–771.

Sporns, O., 2011. Networks of the Brain. MIT Press.

Terrière, E., Dempsey, M.F., Herrmann, L.L., Tierney, K.M., Lonie, J.A., O'Carroll, R.E., et al., 2010. 5-(123)I-A-85380 binding to the alpha4beta2-nicotinic receptor in mild cognitive impairment. Neurobiol. Aging 31, 1885–1893.

Venneri, A., Shanks, M.F., 2010. A neurology of awareness and belief: so near so far? Neuropsychoanalysis 12, 185–189.

Venneri, A., McGeown, W.J., Hietanen, H.M., Guerrini, C., Ellis, A.W., Shanks, M.F., 2008. The anatomical bases of semantic retrieval deficits in early Alzheimer's disease. Neuropsychologia 46, 497–510.

Williamson, P., 2007. Are anticorrelated networks in the brain relevant to schizophrenia? Schizophr. Bull. 33, 994–1003.

Witter, M.P., Room, P., Groenewegen, H.J., Lohman, A.H., 1988. Reciprocal connections of the insular and piriform claustrum with limbic cortex: an anatomical study in the cat. Neuroscience 24, 519–539.

Xie, T., He, Y., 2011. Mapping the Alzheimer's brain with connectomics. Front. Psychiatry 2, 77.

Yamamoto, R., Iseki, E., Murayama, N., Minegishi, M., Marui, W., Togo, T., et al., 2007. Correlation in Lewy pathology between the claustrum and visual areas in brains of dementia with Lewy bodies. Neurosci. Lett. 415, 219–224.

Parkinson's Disease and the Claustrum

Michail E. Kalaitzakis

Neuropathology Unit, Division of Brain Sciences,
Department of Medicine, Imperial College London, UK

INTRODUCTION

In 1817 at the age of 62, James Parkinson, a British physician, published *An Essay on the Shaking Palsy*, a monograph about a condition he called "paralysis agitans" (Parkinson, 2002). This work is now considered a medical classic because of the eloquent and detailed description of the main clinical features of a disease that is now known as Parkinson's disease (PD). Today, PD is viewed as a progressive neurodegenerative disorder (the second most common after Alzheimer's disease (AD) (Lees, Hardy, & Revesz, 2009) that is characterized clinically by tremor, rigidity, and bradykinesia as well as a variety of non-motor complications including neuropsychiatric and autonomic/vegetative deficits (Chaudhuri, Odin, et al., 2011; Poewe, 2008; Simuni & Sethi, 2008). The prevalence and incidence of PD varies widely across epidemiological studies, with door-to-door surveys demonstrating a higher prevalence and incidence than studies using medical records (Campenhausen et al., 2005; Muangpaisan et al., 2011). However, most studies estimate that the prevalence and incidence rates of PD in Europe are approximately 108 to 257/100,000 and 11 to 19/100,000 per year, respectively. In the older age groups (i.e. >60 years) the rates of prevalence and incidence are much higher: 1280 to 1500/100,000 and 346/100,000, respectively

J. Smythies, L. Edelstein and V. Ramachandran (Eds):
The Claustrum.

DOI: http://dx.doi.org/10.1016/B978-0-12-404566-8.00012-X

(Campenhausen et al., 2005). As the disease leads to profound physical and mental disability and affects the quality of life of patients and caregivers (Chen, 2010), it merits intensive investigation to elucidate the underlying etiology and pathogenesis.

THE VARIABLE CLINICAL PICTURE OF PARKINSON'S DISEASE

PD is characterized as a hypokinetic movement disorder with four cardinal motor features, namely tremor at rest, rigidity, akinesia (or bradykinesia) and postural instability (Jankovic, 2008). A precise definition of three levels of diagnostic confidence for the diagnosis of PD has been proposed (Gelb, Oliver, & Gilman, 1999; Hughes et al., 1992). A patient diagnosis is classified as "probable" when the patient has clinical features that are absolutely typical of PD, at least 3 years of parkinsonian symptoms without the development of atypical features, and a clear clinical response to dopaminergic treatment. The criteria for "possible" PD are less strict: patients are not required to have as many typical features of PD, symptom duration can be less than 3 years, and patients need not have received an adequate trial of dopaminergic therapy. Patients in either category who have autopsy findings consistent with PD are designated as "definite."

The non-motor clinical manifestations of PD are summarized in Box 12.1. These features frequently go unnoticed or untreated in as many as 50 percent of patients (Chaudhuri, Odin, et al., 2011). Given that non-motor complications predict an increased risk for nursing home placement, increased mortality, greater caregiver burden, and reduced patient and caregiver quality of life (Aarsland et al., 1999; Aarsland et al., 2000; Levy, et al., 2002), identifying and treating these problems is extremely relevant to optimal case management. A prospective 15-year-follow-up multicentre study indicated that although 95% of PD patients experienced motor complications, it was the non-motor symptoms of the disease that were the predominating cause of disability (Hely et al., 2005).

The prevalence of these complications in PD is clearly illustrated in recent studies using newly developed questionnaires and scales for the assessment of non-motor features of the disease, indicating that between 50 and 100% of PD patients exhibit non-motor complications during the course of their illness (Shulman et al., 2001). These studies also demonstrate that non-motor features of the disease are more frequently seen in PD than in age-matched controls, occur at early stages of the disease and increase in number and severity during the course of the illness (Chaudhuri, et al., 2006, 2007; Martinez-Martin, et al., 2007). Collectively, these studies highlight the ubiquity of non-motor complications and their impact in the natural history of PD.

BOX 12.1

SUMMARY OF NON-MOTOR CLINICAL MANIFESTATIONS IN PD

Autonomic Deficits

- Cardiovascular dysfunction (e.g. orthostatic hypotension)
- Urologic dysfunction (e.g. urinary frequency and urgency)
- Thermoregulatory dysfunction (e.g. excessive sweating)
- Pupillary changes
- Sexual dysfunction
- Gastrointestinal dysfunction (e.g. constipation)

Sleep Disturbances

- Excess daytime somnolence
- Insomnia
- Rapid eye movement behavior disorder
- Restless legs syndrome and periodic limb movements
- Sleep apnea
- Vivid dreams

Mood and Neuropsychiatric Disturbances

- Depression
- Anxiety
- Apathy
- Hallucinations (e.g. visual and auditory)
- Delusions
- Illusions

Cognitive Impairment and Dementia

Other

- Fatigue
- Weight loss
- Blurred vision

Data compiled from the following sources: Bayulkem & Lopez, 2010; Bonnet et al., 2012; Chaudhuri, Prieto-Jurcynska, et al., 2011; Rektorova et al., 2012).

Non-motor complications have also been identified as prodromal features of PD. The disease is considered to have a premotor phase in which a variety of non-motor features predate the development of the classical motor symptoms of the illness. Premotor features that have been strongly linked to PD include olfactory deficits (Ross et al., 2008), excessive daytime sleepiness (Abbott et al., 2005), rapid eye movement behavior disorder (Postuma et al., 2009), constipation (Abbott et al., 2001) and depression (Leentjens et al., 2003). These prodromal features are not universal and are present to a variable degree among patients before a clinical diagnosis of PD is made.

As the disease is characterized by motor and non-motor complications, PD is considered a relatively heterogeneous clinical syndrome. In recent years, this clinical variability has become the focus of intense research efforts, and studies using cluster analysis have tried to identify more homogeneous PD subtypes. Accordingly, four main PD subtypes have been identified:

- early onset
- tremor-dominant
- postural instability and gait-dominant, and
- old-age onset.

These are based on age at disease onset, duration of disease/rate of progression, severity/type of motor symptoms and presence or absence of cognitive impairment (Van Rooden et al., 2010).

Even though PD is defined by the presence of motor complications, we have entered an era where non-motor features of the disease are being increasingly identified as important clinical determinants of disability in the course of PD. Dopaminergic treatment strategies can often control motor aspects of the illness but treating non-motor features of the disease has been problematic. Therefore, a better understanding of the pathophysiological mechanisms that underlie the development of non-motor complications in PD is warranted.

NEUROPATHOLOGY OF PARKINSON'S DISEASE

The main classical pathological hallmark of PD is degeneration of the substantia nigra pars compacta (SNc). It is necessary to deplete 80–85% of striatal dopamine and to lose 50–70% of dopaminergic neurons before clinical symptoms appear. Although loss of neurons in the SNc remains a *sine qua non* for the pathological confirmation of PD, extranigral pathological lesions involve a variety of mainly subcortical neuronal systems causing multiple neurotransmitter dysfunctions that might explain the complex clinical picture of the disease (Kalaitzakis & Pearce, 2009).

In PD, apart from depletion of dopamine in the striatum, the neurotransmitters noradrenaline (norepinephrine), serotonin, and acetylcholine, which originate from the locus coeruleus, raphe nuclei, and nucleus basalis of Meynert (NBM), respectively, are markedly reduced due to degeneration of these nuclei (Jellinger, 1991, 1997, 2000). Although the pathology of PD affects a variety of neuronal systems, substantial brain tissue loss and atrophy is not normally seen in PD, due to the fact that neuronal loss is restricted to specific neuronal cell populations (Fearnley & Lees, 1991; Pedersen et al., 2005).

Another defining pathological feature of PD is the presence of neuronal intracytoplasmic inclusions known as Lewy bodies (LB), and inclusions confined to neuronal processes known as Lewy neurites (LN) in brainstem neurons (Figure 12.1). The discovery that α-synuclein (αSyn) gene mutations can cause PD (Kruger et al., 1998; Polymeropoulos et al., 1997), and that the encoded protein, which is natively unfolded and still of unknown physiological function, constitutes a major component of LB and LN (Spillantini et al., 1997) has provided the basis for a molecular definition of the disease and a better understanding of its pathogenesis. In fact, the term "synucleinopathy" has been used to designate a spectrum of degenerative diseases that share the presence of abnormal αSyn immunoreactive inclusion bodies in neurons and/or glial cells (e.g. multiple system atrophy (MSA), dementia with Lewy bodies (DLB), the Lewy body variant of Alzheimer's disease (LBVAD), neurodegeneration with brain iron accumulation type 1 (NBIA I)), and with PD as the most frequently occurring synucleinopathy (Arawaka et al., 1998; Bayer et al., 1999; Dickson et al., 1999; Gai et al., 1998; Papp & Lantos, 1994).

Although neuronal loss and the presence of LB and LN in SNc and other brainstem regions such as the locus coeruleus are diagnostic for PD, the disease is a "multisystem" illness with pathology occurring in a variety of brain and extra-CNS regions including but not limited to: the olfactory bulb and related areas, the spinal cord, the dorsal motor nucleus of the vagus nerve, the pedunculopontine nucleus, thalamic nuclear nuclei with diffuse projections to cortical and subcortical regions, intralaminar and midline thalamic nuclei, the amygdala, NBM, transentorhinal cortex, claustrum, striatum, hippocampal formation, and the temporal and frontal cortices as well as autonomic nerves (e.g. cardiac and abdominopelvic autonomic plexuses) (Bertrand et al., 2004; Braak & Braak, 2000; Braak, Sastre, Bohl, et al., 2007; Jellinger, 1991; Kalaitzakis, Christian, et al., 2009; Kalaitzakis et al., 2008b, 2008c; Kalaitzakis, Pearce, & Gentleman, 2009; Kalaitzakis et al., 2011; Minguez-Castellanos et al., 2007; Rub et al., 2002). This widespread pathology in the PD brain most likely accounts for the non-motor complications.

The widespread presence of LB and LN in PD posed the question of how αSyn pathology progresses within the PD brain. In 2003 Braak

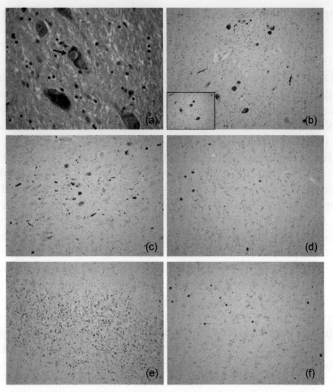

FIGURE 12.1   Photomicrographs of Lewy pathology in cortical and subcortical regions in the Parkinson's disease brain. (a) Hematoxylin and eosin (H&E) stain in the substantia nigra pars compacta. A "classical" Lewy body (arrow) is seen in the neuronal perikaryon of a pigmented neuron. (b) αSyn immunohistochemistry in the substantia nigra pars compacta. Many αSyn positive lesions, i.e. Lewy bodies (LB) and Lewy neurites (LN), can be seen. Note that some pigmented neurons contain multiple LB in their perikaryon. Inset higher magnification showing αSyn lesions in the substantia nigra pars compacta. (c) αSyn pathology in the nucleus basalis of Meynert. (d) αSyn pathology in the amygdala. (e) Selective vulnerability of the cornu ammonis 2 sector of the hippocampus to LN-type pathology. (f) αSyn pathology in the cingulate gyrus. Magnification: a: 40×, b: 10×, inset: 20×, c-f: 10×.

and his colleagues, using αSyn immunohistochemistry in neurologically unimpaired subjects and in subjects with PD, suggested that αSyn pathology in PD does not occur at random but follows a predictable and stereotyped pattern and they proposed six distinct stages of PD (Braak et al., 2003). According to this design, the earliest pathology is to be observed in the dorsal motor nucleus of the vagus nerve and olfactory bulb (stage 1), from where αSyn pathology is thought to proceed in a rostral direction via the pons (stage 2) to the midbrain (stage 3), and then to the basal forebrain and mesocortex (stage 4) before finally spreading to and involving the neocortex (stages 5–6). Although the Braak staging system has attracted much attention, its validity has been questioned both on methodological grounds as well as regarding clinical relevance (Attems & Jellinger, 2008; Jellinger, 2008; Kalaitzakis et al., 2008a, 2008c; Kalaitzakis, Graeber, Gentleman, & Pearce, 2009; Parkkinen, Pirttila, & Alafuzoff, 2008).

αSyn pathology is commonly found in sporadic and familial AD as well as Down syndrome (patients with a triplicate of amyloid precursor protein (APP)), with estimates ranging from 50 percent to 60 percent (Hamilton, 2000; Lippa et al., 1998). Similarly, AD-related pathology (amyloid plaques and neurofibrillary tangles) is often found in PD brains to a varying degree (Kalaitzakis, Christian, et al., 2009) and this pathology has also been suggested as underlying cognitive impairment and dementia in PD patients (Kalaitzakis & Pearce, 2009; Kalaitzakis et al., 2011). Therefore, in the PD brain widespread αSyn positive LB and LN are frequently accompanied by concomitant AD-type changes. A molecular interaction between these abnormally deposited proteins (αSyn (i.e. LB and LN), tau (i.e. neurofibrillary tangles) and Aβ (i.e. amyloid plaques) in the brain has also been demonstrated both in vitro and in vivo (Giasson et al., 2003; Jensen et al., 1999; Kalaitzakis, Christian, et al., 2009; Masliah et al., 2001; Pletnikova et al., 2005).

CLAUSTRAL PATHOLOGY IN PARKINSON'S DISEASE

Few studies have been reported in the literature treating claustral pathology in PD. The majority of these publications have been primarily concerned with other brain regions and not the claustrum itself, which tends to confuse rather than clarify the subject. Nevertheless, in recent years, with the use of modern techniques, it has become apparent that the claustrum can indeed become pathological in PD (Braak et al., 2001; Braak, Sastre, & Del Tredici, 2007; Kalaitzakis, Pearce, & Gentleman, 2009).

Lewy pathology (LB and LN) represents the principal pathological change observed in the claustrum of PD brains. The morphology of LB in the claustrum is of the so-called cortical type (Van Duinen et al., 1999). Generally, LB are divided into two types: classical and cortical. The shape and size of LB depends on the neuronal perikaryon they occupy but usually LB have a spherical configuration of around 5–25 μm in diameter. In hematoxylin and eosin staining, LB in the brainstem are considered to have a classical morphology consisting of a dark eosinophilic centre (body) surrounded by a paler halo (classical LB; see Figure 12.1a). In the cortex, amygdala and claustrum on the other hand, LB

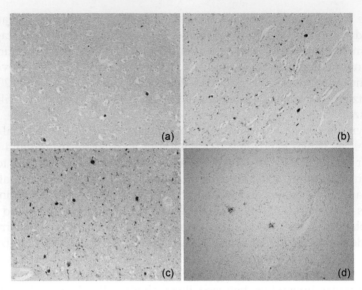

FIGURE 12.2   Photomicrographs of Lewy and Aβ pathology in the claustrum in the Parkinson's disease (PD) and dementia with Lewy bodies (DLB) brain. αSyn positive lesions in the claustrum of a non-demented PD case (a), of a demented PD case (b), and from a DLB case (c). Note the more extensive severity of αSyn deposits in the demented PD and DLB brains. (d) Diffuse Aβ deposits in the claustrum of a non-demented PD case. Magnification: a, b, c: 20× and d: 10×. *Reproduced from Kalaitzakis, Pearce, & Gentleman, 2009 with permission from Elsevier.*

have a less conspicuous morphology than their brainstem counterparts (cortical LB; Figures 12.1 and 12.2). The introduction of αSyn immunohistochemistry as the gold standard for the detection and visualization of LB and LN has revealed a variety of αSyn inclusions previously unrecognized (i.e. during the pre-αSyn era). αSyn inclusions in the claustrum appear principally as αSyn immunoreactive deposits with irregular shape, sometimes as small particulate deposits and, less frequently, as fully compact spherical inclusions (i.e. corresponding to classical LB, see Figure 12.2). αSyn inclusions in the claustrum can also be seen extracellularly with no obvious association to a neuronal cell. These probably correspond to neuronal intracytoplasmic inclusions that remain after neuronal death. Pathological involvement in neuronal processes (i.e. LN) are predominantly seen in the CA2 sector of the hippocampus, NBM, amygdala and lower brainstem structures but also in the claustrum (see Figure 12.2c). LN in the claustrum display a range of morphologies varying from segmental serpentine to continuous serpentine and pearl-like arrangements (Kalaitzakis et al., 2008c). In a recent clinico-pathological study, we investigated αSyn pathology in the claustrum in PD, PD with dementia (PDD) and DLB patients. We observed claustral αSyn lesions in 75 percent of PD cases without dementia, and in 100 percent of PDD and DLB cases (Kalaitzakis, Pearce, & Gentleman, 2009).

Astrocytic αSyn immunoreactive inclusions have also been described in the claustrum of PD patients (Braak, Sastre, & Del Tredici, 2007). Braak and colleagues studied 14 autopsy cases with a clinical and neuropathological diagnosis of PD and 6 control cases without a history of psychiatric or neurological disorder. All PD cases demonstrated αSyn immunoreactive astrocytes in the claustrum, whereas these were absent in the control group. Braak and colleagues also observed αB-crystallin-immunoreactive neurons in the ventral claustrum of PD patients (Braak et al., 2001). αB-crystallin is a heat-shock protein and as such its upregulation indicates a cellular response to stress (Iwaki, et al., 1992). In addition, Mochizuki and colleagues reported a case with abundant αSyn-positive glial cytoplasmic inclusions (GCIs) in the claustrum of a patient who was clinically diagnosed with PD and neuropathologically with PD and early stage MSA (striatonigral degeneration) (Mochizuki, Komatsuzaki, & Shoji, 2002). GCIs are the pathological hallmark of multiple system atrophy (MSA), a sporadic, adult-onset, progressive neurodegenerative disorder that is characterized clinically by parkinsonism, ataxia, pyramidal signs and autonomic dysfunction (Gilman et al., 2008).

Neurofibrillary tangles (NFTs) and amyloid plaques represent the two major neuropathological footprints of AD (Braak et al., 2006). The former are aggregates of hyperphosphorylated tau protein and, using tau immunohistochemistry, they appear as flame-shaped or globoid masses of fibrous intraneuronal material, often accompanied by abnormal neuronal processes (neuropil threads (NTs) and dystrophic neurites) that are also tau immunoreactive. It appears that tau pathology in the claustrum (i.e. NFTs, NTs and dystrophic neurites) is negligible in PD patients with or without dementia (Kalaitzakis, Pearce, et al., 2009). Amyloid diffuse plaques, on the other hand, represent extracellular deposits of amyloid-β-peptide (Aβ) and these are observed in the claustrum (see Figure 12.2d). In a recent study we found that 25 percent of PD, 58 percent of PDD and 100 percent of DLB cases studied exhibited claustral Aβ deposits (Kalaitzakis, Pearce, & Gentleman, 2009).

In summary, the available literature on claustral pathology in the PD brain is rather limited. Nevertheless, PD-related pathology as well as abnormal deposition of other proteins that are known to co-occur with PD-associated lesions (i.e. Aβ protein of the AD-type), have been described in the claustrum of PD brains. In addition, astrocytic αSyn immunoreactive inclusions and αB-crystallin have also been observed in claustral neurons. Collectively, these studies indicate that the claustrum becomes pathologically involved in PD and possibly contributes to the development of some of the clinical features seen in these patients.

Clinical Correlates of Claustral Pathology in Parkinson's Disease

As described above, PD is a disabling movement disorder. Traditionally, attention has focused on the motor symptoms of the disease, but it is now appreciated that significant non-motor symptoms affecting neuropsychiatric, sleep, autonomic and sensory domains occur in the majority of PD patients (Poewe, 2008). Two of the most common of these complications are dementia, with a cumulative prevalence ranging between 48 percent and 78 percent (Hely et al., 2005), and visual hallucinations, which occur in up to 60 percent of sufferers (Diederich, Goetz, & Stebbins, 2005). These features are also found in DLB, a neurodegenerative disorder accounting for 15 percent to 25 percent of all dementia cases (McKeith, et al., 2004) and now recognized as the second most common cause of dementia after AD (McKeith et al., 2005). It is characterized by progressive cognitive decline and parkinsonism. The core features of DLB include fluctuating cognition, attention and alertness, visual hallucinations and motor features of parkinsonism (McKeith et al., 2005).

In a recent clinico-pathological study, our research group looked at claustral pathology in PD, PDD and DLB cases. A total of 39 cases fulfilling the clinical and neuropathological diagnostic criteria for these conditions were studied (20 PD, 12 PPD and 7 DLB) and the extent of claustral pathology was examined using αSyn, tau and Aβ immunohistochemistry with the aim of identifying a possible relationship between pathology in the claustrum and the presence of dementia or visual hallucinations, or both (Kalaitzakis, Pearce, & Gentleman, 2009). The claustrum was sampled at the level of the nucleus accumbens, as shown in Figure 12.3. Assessment of pathology was carried out in a semi-quantitative fashion based on a four-point scale from 0 to 3 (absent to frequent). Our findings demonstrated that all PDD and DLB cases and 75 percent of non-demented PD cases exhibited αSyn lesions in the claustrum. Interestingly, compared to PD cases without dementia, PDD cases showed significantly greater αSyn burden in the claustrum, and there was even greater αSyn deposition in DLB cases. A similar hierarchy, PD < PDD < DLB was seen in terms of Aβ burden in the claustrum, while tau deposition was negligible in all cases. Comparison of αSyn and Aβ burden in those cases with and without visual hallucinations did not reveal any significant associations.

Yamamoto and colleagues (2007), in another clinico-pathological study, investigated the relationship between claustral Lewy pathology and the presence of visual hallucinations in DLB patients. A total of 20 pathologically verified DLB cases were studied and the extent of claustral pathology was examined using αSyn immunohistochemistry. The claustrum showed many LB- and LN-related pathologies in both

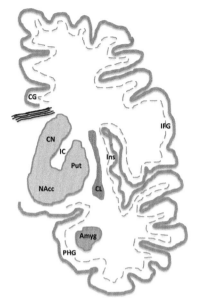

Abbreviations:
Amyg: Amygdala
CG: Cingulate gyrus
CL: Claustrum
CN: Caudate nucleus
IC: Internal capsule
IFG: Inferior frontal gyrus
Ins: Insula
Nacc: Nucleus accumbens
PHG: Parahippocampal gyrus
Put: Putamen

FIGURE 12.3   Simplified diagram showing the anatomy of the claustrum at the level of the nucleus accumbens. *Reproduced from Kalaitzakis, Pearce, & Gentleman, 2009 with permission from Elsevier.*

the ventral and dorsal parts. The insular and inferior temporal cortices, amygdala, Brodmann areas 18 and 19, transentorhinal and cingulate cortices demonstrated stronger or similar Lewy pathology compared to that observed in the claustrum. On the other hand, Brodmann area 17, precentral, postcentral and transverse temporal cortices showed weaker Lewy pathology than that seen in the claustrum. Statistical analysis demonstrated a higher correlation coefficient between the claustrum and Brodmann areas 18 and 19 as well as insular and transentorhinal cortices. These findings indicate that Lewy pathology in the claustrum relates more closely to visual areas than motor, auditory or somatosensory cortical regions and that dysfunction of the visuo-claustral pathway may participate in the formation of visual hallucinations in DLB. However, Yamamoto and colleagues did not find significant differences between cases with and without hallucinations with respect to Lewy pathology in the claustrum.

A recent case report by Ishii and colleagues further supports a role of the claustrum in the formation of visual hallucinations (Ishii, Tsuji, & Tamaoka, 2011). The authors reported a 21-year-old patient with mumps virus encephalitis who experienced visual hallucinations (the patient reported seeing people and cats who were not present). Brain magnetic resonance indicated symmetric high-intensity lesions in the bilateral claustrum. As the authors argued, although it is difficult to pinpoint the specific mechanism involved for the development of visual hallucinations, this case is important as it demonstrates that bilateral symmetric claustral lesions are associated with visual hallucinations.

Although dementia in PD and DLB is a well-documented non-motor complication, its underlying pathoanatomical substrate(s) remains as yet undefined. Lewy pathology (LB and LN) in cortical and limbic regions has been suggested as a possible cause of cognitive dysfunction and dementia (Kalaitzakis, Christian, et al., 2009; Kovari, et al., 2003) as well as Aβ deposition in the cortex and in the striatum (Jellinger, 2012; Kalaitzakis, Christian, et al., 2009; Kalaitzakis, et al., 2008b, 2011). On the other hand, others have found that severe LB pathology in cortical or limbic structures does not necessarily predict dementia in PD (Colosimo et al., 2003; Parkkinen et al., 2005). While acknowledging this ongoing controversy, the findings from our study and the work from others indicate that pathology in the claustrum relates to the presence of dementia in PD and DLB.

The development of visual hallucinations is well recognized in PD and DLB, but only a few studies have looked at the association with specific neuropathological changes. Lewy pathology in the parahippocampal and inferior temporal cortices has been associated with early visual hallucinations in DLB (Harding, Broe, & Halliday, 2002) while, on the other hand, well-formed visual hallucinations in DLB have been related to Lewy pathology in the amygdala and parahippocampal cortex (Harding, Broe, & Halliday, 2002). In PD, Lewy pathology in the amygdala has been shown to relate to the presence of visual hallucinations with or without dementia (Harding, Stimson, et al., 2002; Kalaitzakis, Christian, et al., 2009). Yamamoto et al. (2007) demonstrated that claustral pathology in DLB disturbs the visuo-claustral pathway, thus contributing to the development of visual hallucinations in DLB.

Given our lack of knowledge regarding either the physiological function of the claustrum or its relevance in neurodegeneration, attempting to explain the clinical consequence of claustral involvement in PD and DLB must of necessity also focus on the interrelations between this structure and other brain regions. Neuroanatomical studies have demonstrated connections between the claustrum and neocortical areas such as the frontal cortex (Kievit & Kuypers, 1975), visual cortical fields including the striate cortex (Baizer, Lock, & Youakim, 1997), temporal cortex (Webster, Bachevalier & Ungerleider, 1993), entorhinal cortex (Insausti, Amaral, & Cowan, 1987), parieto-occipital cortex (Shipp, Blanton, & Zeki, 1998), and the parietal cortex (Pearson et al., 1982). The claustrum is also reciprocally connected with limbic structures such as the hippocampus (Amaral & Cowan, 1980) and the amygdala (Amaral & Insausti, 1992) as well as the caudate nucleus and putamen (Andersen, 1968; Arikuni & Kubota, 1985), regions that have been shown to relate to dementia as well as to visual hallucinations in PD (Churchyard & Lees, 1997; Harding, Stimson, et al., 2002; Kalaitzakis, Christian, et al., 2009; Kalaitzakis, et al., 2008b, 2011). It has recently been suggested that the claustrum is critical in "binding" information from cortical and subcortical neural circuits (Crick & Koch, 2005). This supports a connectionist view of the claustrum. Therefore, in PD, the claustrum may not necessarily be an essential epicentre for the development of dementia but, as part of a large-scale neural network in which the claustrum "binds" information, its pathological involvement could contribute to functional disturbances of these complex cortical and subcortical regulatory loops. In addition, the claustrum has been shown to be involved in higher-order cognitive function including memory storage, learning, addiction (repetitive behavior), fear recognition and emotional and behavioral responses (Bennett & Baird, 2006; Chachich & Powell, 2004; Morys et al., 1996; Naqvi et al., 2007) and this evidence further supports a clinical relevance of the claustrum in the development of cognitive dysfunction and dementia in PD.

Intriguingly, the existence of specific connections between the claustrum and not only visual areas (LeVay & Sherk, 1981a, 1981b; Sherk & LeVay, 1981a, 1981b) but also subcortical nuclei modulating visual processing according to cognition and emotion (i.e. the amygdala) (Amaral & Insausti, 1992) place the claustrum within circuitry that implies a relevance of claustral pathology to the development of visual hallucinations. This is further supported by imaging studies indicating that the claustrum, through its extensive connections, plays an important role in multimodal representation of objects (Naghavi et al., 2007). Therefore, at the very least, the claustrum is situated within a network of connections subserving cognitive function and visual processing and thus claustral pathology may likely contribute to the presence of some of the most common non-motor complications in PD.

Possible Clinical Correlates of Claustral Pathology in Parkinson's Disease

Impulse-control disorders are characterized by an inability to resist an impulse, drive or temptation that is harmful to the individual or to others. The behaviors are often motivated by the prospect of pleasure or gratification and, in people with Parkinson's, most typically involve sex and gambling. Both pathological gambling and hypersexuality can occur in patients with PD (Ferrer et al., 2012). A positron emission tomography (PET) study investigated regional blood flow in healthy males presented with visual sexual stimuli (Qurrat-ul-Ain & Abidi, 2005). The results of the study indicated that such stimuli triggered high levels of activity in the claustrum of healthy males. Therefore, claustral pathology may well contribute to disinhibited behaviour in PD patients, manifested as gambling or hypersexuality.

One very common non-motor disturbance in PD is that of disordered sleep. Sleep-related problems occur frequently in PD, representing a major cause of disability and adversely affecting the quality of life of patients and partners (Dhawan et al., 2006). Périco et al. (2005) examined the relationship between resting regional cerebral blood flow (rCBF) patterns in patients with major depressive disorder and specific symptom clusters derived from ratings on the Hamilton Rating Scale for Depression and the Mini Mental State Examination (Périco et al., 2005). The results of the study indicated that insomnia severity inversely correlated with rCBF in the right rostral and subgenual anterior cingulate cortices, insula and claustrum. The claustrum is reciprocally connected with sensory cortical brain regions and receives afferents from the hypothalamus, thalamus and locus coeruleus. This suggests that the claustrum may be involved in arousal modulation of sensory processing. Consequently, claustral pathology in PD may contribute to disturbed sleep and therefore clinico-pathological studies are warranted to investigate this association.

Among the most common affective and behavioral disorders in patients with PD is depression (Tandberg et al., 1996). A PET imaging study by Dunn et al. (2002) examined 31 unipolar and 27 bipolar, all medication-free, mildly to severely depressed patients (Dunn et al., 2002). The results of the study indicated that in both unipolar and bipolar depression psychomotor-anhedonia symptoms correlated with lower absolute metabolism in the right insula, claustrum, anteroventral caudate/putamen, and temporal cortex, and with higher normalized metabolism in the anterior cingulate. It is therefore possible that claustral pathology may underlie some depressive symptoms that also occur in PD patients (i.e. psychomotor retardation and anhedonia).

CONCLUSION

PD is a common and disabling disorder, characterized by clinical and pathological complexity. While degeneration of the substantia nigra pars compacta results in the classical motor clinical signs and symptoms of PD, the pathology is widespread, affecting many cortical and subcortical nuclei. This widespread pathology is thought to underlie the variable non-motor manifestations that are commonly seen in PD patients. The claustrum has only recently attracted attention in PD. Studies using modern histological techniques indicate that the claustrum can be affected in the PD and DLB brain. The evidence obtained from clinico-pathological studies indicates that both Lewy and AD-type pathologies are observed in the claustrum and the severity of pathological lesions relates to the development of dementia in both PD and DLB. Destruction of the visuo-claustral pathway may well contribute to the development of visual hallucinations in PD, PDD and DLB. In addition, imaging studies provide important evidence of involvement of the claustrum in the development of impulse-control disorders, disturbed sleep, and depression, features that are also seen in PD. However, the claustrum's physiological function remains largely unknown in man and its involvement in neurodegeneration is therefore somewhat enigmatic. Undoubtedly, the presence of extensive connections between the claustrum and cortical and subcortical nuclei has important implications for normal cognitive function as well as in the setting of neurodegenerative disorders such as PD. In order to further our understanding of the claustrum in PD, more detailed neuroanatomical studies are needed to clarify the mechanisms and brain circuits in which the claustrum is involved. In addition, retrospective as well as prospective clinico-pathological studies in PD and DLB brains are warranted to investigate further the correlation between claustral pathology in PD and clinical symptomatology. Developing a greater understanding of the claustrum in PD will perhaps hold promise for the development of more rational therapeutic interventions to combat the variable clinical picture of this disabling disease.

Acknowledgments

The author would like to thank Parkinson's UK, registered charity 948776, for their continual support. The valuable comments of Dr. Ronald K.B. Pearce and Professor Stephen M. Gentleman are also greatly appreciated.

References

Aarsland, D., Larsen, J.P., Karlsen, K., Lim, N.G., Tandberg, E., 1999. Mental symptoms in Parkinson's disease are important contributors to caregiver distress. Int. J. Geriatr. Psychiatry. 14 (10), 866–874.

Aarsland, D., Larsen, J.P., Tandberg, E., Laake, K., 2000. Predictors of nursing home placement in Parkinson's disease: a population-based, prospective study. J. Am. Geriatr. Soc. 48 (8), 938–942.

Abbott, R.D., Petrovitch, H., White, L.R., Masaki, K.H., Tanner, C.M., Curb, J.D., et al., 2001. Frequency of bowel movements and the future risk of Parkinson's disease. Neurology 57 (3), 456–462.

Abbott, R.D., Ross, G.W., White, L.R., Tanner, C.M., Masaki, K.H., Nelson, J.S., et al., 2005. Excessive daytime sleepiness and subsequent development of Parkinson disease. Neurology 65 (9), 1442–1446.

Amaral, D.G., Cowan, W.M., 1980. Subcortical afferents to the hippocampal formation in the monkey. J. Comp. Neurol. 189 (4), 573–591.

Amaral, D.G., Insausti, R., 1992. Retrograde transport of D-[3H]-aspartate injected into the monkey amygdaloid complex. Exp. Brain Res. 88 (2), 375–388.

Andersen, D.L., 1968. Some striatal connections to the claustrum. Exp. Neurol. 20 (2), 261–267.

Arawaka, S., Saito, Y., Murayama, S., Mori, H., 1998. Lewy body in neurodegeneration with brain iron accumulation type 1 is immunoreactive for alpha-synuclein. Neurology 51 (3), 887–889.

Arikuni, T., Kubota, K., 1985. Claustral and amygdaloid afferents to the head of the caudate nucleus in macaque monkeys. Neurosci. Res. 2 (4), 239–254.

Attems, J., Jellinger, K.A., 2008. The dorsal motor nucleus of the vagus is not an obligatory trigger site of Parkinson's disease. Neuropathol. Appl. Neurobiol. 34 (4), 466–467.

Baizer, J.S., Lock, T.M., Youakim, M., 1997. Projections from the claustrum to the prelunate gyrus in the monkey. Exp. Brain Res. 113 (3), 564–568.

Bayer, T.A., Jakala, P., Hartmann, T., Havas, L., McLean, C., Culvenor, J.G., et al., 1999. Alpha-synuclein accumulates in Lewy bodies in Parkinson's disease and dementia with Lewy bodies but not in Alzheimer's disease beta-amyloid plaque cores. Neurosci. Lett. 266 (3), 213–216.

Bayulkem, K., Lopez, G., 2010. Nonmotor fluctuations in Parkinson's disease: clinical spectrum and classification. J. Neurol. Sci. 289 (1-2), 89–92.

Bennett, C.M., Baird, A.A., 2006. Anatomical changes in the emerging adult brain: a voxel-based morphometry study. Hum. Brain Mapp. 27 (9), 766–777.

Bertrand, E., Lechowicz, W., Szpak, G.M., Lewandowska, E., Dymecki, J., Wierzba-Bobrowicz, T., 2004. Limbic neuropathology in idiopathic Parkinson's disease with concomitant dementia. Folia Neuropathol. 42 (3), 141–150.

Bonnet, A.M., Jutras, M.F., Czernecki, V., Corvol, J.C., Vidailhet, M., 2012. Nonmotor symptoms in Parkinson's disease in 2012: relevant clinical aspects. Parkinson's Dis. 2012, 198–316.

Braak, H., Alafuzoff, I., Arzberger, T., Kretzschmar, H., Del Tredici, K., 2006. Staging of Alzheimer disease-associated neurofibrillary pathology using paraffin sections and immunocytochemistry. Acta Neuropathol. 112 (4), 389–404.

Braak, H., Braak, E., 2000. Pathoanatomy of Parkinson's disease. J. Neurol. 247 (Suppl. 2), II3–10.

Braak, H., Del Tredici, K., Rub, U., de Vos, R.A., Jansen Steur, E.N., Braak, E., 2003. Staging of brain pathology related to sporadic Parkinson's disease. Neurobiol. Aging. 24 (2), 197–211.

Braak, H., Del Tredici, K., Sandmann-Kiel, D., Rub, U., Schultz, C., 2001. Nerve cells expressing heat-shock proteins in Parkinson's disease. Acta Neuropathol. 102 (5), 449–454.

Braak, H., Sastre, M., Bohl, J.R., de Vos, R.A., Del Tredici, K., 2007. Parkinson's disease: lesions in dorsal horn layer I, involvement of parasympathetic and sympathetic pre- and postganglionic neurons. Acta Neuropathol. 113 (4), 421–429.

Braak, H., Sastre, M., Del Tredici, K., 2007. Development of alpha-synuclein immunoreactive astrocytes in the forebrain parallels stages of intraneuronal pathology in sporadic Parkinson's disease. Acta Neuropathol. 114 (3), 231–241.

Campenhausen, S., von, Bornschein, B., Wick, R., Botzel, K., Sampaio, C., Poewe, W., et al., 2005. Prevalence and incidence of Parkinson's disease in Europe. Eur. Neuropsychopharmacol. 15 (4), 473–490.

Chachich, M.E., Powell, D.A., 2004. The role of claustrum in Pavlovian heart rate conditioning in the rabbit (Oryctolagus cuniculus): anatomical, electrophysiological, and lesion studies. Behav. Neurosci. 118 (3), 514–525.

Chaudhuri, K.R., Martinez-Martin, P., Brown, R.G., Sethi, K., Stocchi, F., Odin, P., et al., 2007. The metric properties of a novel non-motor symptoms scale for Parkinson's disease: results from an international pilot study. Mov. Disord. 22 (13), 1901–1911.

Chaudhuri, K.R., Martinez-Martin, P., Schapira, A.H., Stocchi, F., Sethi, K., Odin, P., et al., 2006. International multicenter pilot study of the first comprehensive self-completed nonmotor symptoms questionnaire for Parkinson's disease: the NMSQuest study. Mov. Disord. 21 (7), 916–923.

Chaudhuri, K.R., Odin, P., Antonini, A., Martinez-Martin, P., 2011. Parkinson's disease: the non-motor issues. Parkinsonism Relat. Disord. 17 (10), 717–723.

Chaudhuri, K.R., Prieto-Jurcynska, C., Naidu, Y., Mitra, T., Frades-Payo, B., Tluk, S., et al., 2011. The nondeclaration of nonmotor symptoms of Parkinson's disease to health care professionals: an international study using the nonmotor symptoms questionnaire. Mov. Disord. 25 (6), 704–709.

Chen, J.J., 2010. Parkinson's disease: health-related quality of life, economic cost, and implications of early treatment. Am. J. Manag. Care, 16 Suppl. Implications, S87–93.

Churchyard, A., Lees, A.J., 1997. The relationship between dementia and direct involvement of the hippocampus and amygdala in Parkinson's disease. Neurology 49 (6), 1570–1576.

Colosimo, C., Hughes, A.J., Kilford, L., Lees, A.J., 2003. Lewy body cortical involvement may not always predict dementia in Parkinson's disease. J. Neurol. Neurosurg. Psychiatry. 74 (7), 852–856.

Crick, F.C., Koch, C., 2005. What is the function of the claustrum? Philos. Trans. R. Soc. Lond. B. Biol. Sci. 360 (1458), 1271–1279.

Dhawan, V., Healy, D.G., Pal, S., Chaudhuri, K.R., 2006. Sleep-related problems of Parkinson's disease. Age Ageing 35 (3), 220–228.

Dickson, D.W., Liu, W., Hardy, J., Farrer, M., Mehta, N., Uitti, R., et al., 1999. Widespread alterations of alpha-synuclein in multiple system atrophy. Am. J. Pathol. 155 (4), 1241–1251.

Diederich, N.J., Goetz, C.G., Stebbins, G.T., 2005. Repeated visual hallucinations in Parkinson's disease as disturbed external/internal perceptions: focused review and a new integrative model. Mov. Disord. 20 (2), 130–140.

Dunn, R.T., Kimbrell, T.A., Ketter, T.A., Frye, M.A., Willis, M.W., Luckenbaugh, D.A., et al., 2002. Principal components of the Beck Depression Inventory and regional cerebral metabolism in unipolar and bipolar depression. Biol. Psychiatry. 51 (5), 387–399.

Fearnley, J.M., Lees, A.J., 1991. Ageing and Parkinson's disease: substantia nigra regional selectivity. Brain 114, 2283–2301.

Ferrer, I., Lopez-Gonzalez, I., Carmona, M., Dalfo, E., Pujol, A., Martinez, A., 2012. Neurochemistry and the non-motor aspects of PD. Neurobiol. Dis. 46 (3), 508–526.

Gai, W.P., Power, J.H., Blumbergs, P.C., Blessing, W.W., 1998. Multiple-system atrophy: a new alpha-synuclein disease? Lancet 352 (9127), 547–548.

Gelb, D.J., Oliver, E., Gilman, S., 1999. Diagnostic criteria for Parkinson disease. Arch. Neurol. 56 (1), 33–39.

Giasson, B.I., Forman, M.S., Higuchi, M., Golbe, L.I., Graves, C.L., Kotzbauer, P.T., et al., 2003. Initiation and synergistic fibrillization of tau and alpha-synuclein. Science 300 (5619), 636–640.

Gilman, S., Wenning, G.K., Low, P.A., Brooks, D.J., Mathias, C.J., Trojanowski, J.Q., et al., 2008. Second consensus statement on the diagnosis of multiple system atrophy. Neurology 71 (9), 670–676.

Hamilton, R.L., 2000. Lewy bodies in Alzheimer's disease: a neuropathological review of 145 cases using alpha-synuclein immunohistochemistry. Brain Pathol. 10 (3), 378–384.

Harding, A.J., Broe, G.A., Halliday, G.M., 2002. Visual hallucinations in Lewy body disease relate to Lewy bodies in the temporal lobe. Brain 125 (Pt 2), 391–403.

Harding, A.J., Stimson, E., Henderson, J.M., Halliday, G.M., 2002. Clinical correlates of selective pathology in the amygdala of patients with Parkinson's disease. Brain 125 (Pt 11), 2431–2445.

Hely, M.A., Morris, J.G., Reid, W.G., Trafficante, R., 2005. Sydney multicenter study of Parkinson's disease: non-L-dopa-responsive problems dominate at 15 years. Mov. Disord. 20 (2), 190–199.

Hughes, A.J., Daniel, S.E., Kilford, L., Lees, A.J., 1992. Accuracy of clinical diagnosis of idiopathic Parkinson's disease: a clinico-pathological study of 100 cases. J. Neurol. Neurosurg. Psychiatry. 55 (3), 181–184.

Insausti, R., Amaral, D.G., Cowan, W.M., 1987. The entorhinal cortex of the monkey: III. Subcortical afferents. J. Comp. Neurol. 264 (3), 396–408.

Ishii, K., Tsuji, H., Tamaoka, A., 2011. Mumps virus encephalitis with symmetric claustrum lesions. AJNR. Am. J. Neuroradiol. 32 (7), E139.

Iwaki, T., Wisniewski, T., Iwaki, A., Corbin, E., Tomokane, N., Tateishi, J., et al., 1992. Accumulation of alpha B-crystallin in central nervous system glia and neurons in pathologic conditions. Am. J. Pathol. 140 (2), 345–356.

Jankovic, J., 2008. Parkinson's disease: clinical features and diagnosis. J. Neurol. Neurosurg. Psychiatry 79 (4), 368–376.

Jellinger, K.A., 1991. Pathology of Parkinson's disease. Changes other than the nigrostriatal pathway. Mol. Chem. Neuropathol. 14 (3), 153–197.

Jellinger, K.A., 1997. Morphological substrates of dementia in parkinsonism. A critical update. J. Neural. Transm. Suppl. 51, 57–82.

Jellinger, K.A., 2000. Morphological substrates of mental dysfunction in Lewy body disease: an update. J. Neural. Transm. Suppl. 59, 185–212.

Jellinger, K.A., 2008. A critical reappraisal of current staging of Lewy-related pathology in human brain. Acta Neuropathol. 116 (1), 1–16.

Jellinger, K.A., 2012. Neuropathology of sporadic Parkinson's disease: evaluation and changes of concepts. Mov. Disord. 27 (1), 8–30.

Jensen, P.H., Hager, H., Nielsen, M.S., Hojrup, P., Gliemann, J., Jakes, R., 1999. alpha-synuclein binds to Tau and stimulates the protein kinase A-catalyzed tau phosphorylation of serine residues 262 and 356. J. Biol. Chem. 274 (36), 25481–25489.

Kalaitzakis, M.E., Christian, L.M., Moran, L.B., Graeber, M.B., Pearce, R.K., Gentleman, S.M., 2009. Dementia and visual hallucinations associated with limbic pathology in Parkinson's disease. Parkinsonism Relat. Disord. 15 (3), 196–204.

Kalaitzakis, M.E., Graeber, M.B., Gentleman, S.M., Pearce, R.K., 2008a. Controversies over the staging of alpha-synuclein pathology in Parkinson's disease. Acta Neuropathol. 116 (1), 125–128. author reply 129–131.

Kalaitzakis, M.E., Graeber, M.B., Gentleman, S.M., Pearce, R.K., 2008b. Striatal beta-amyloid deposition in Parkinson disease with dementia. J. Neuropathol. Exp. Neurol. 67 (2), 155–161.

Kalaitzakis, M.E., Graeber, M.B., Gentleman, S.M., Pearce, R.K., 2008c. The dorsal motor nucleus of the vagus is not an obligatory trigger site of Parkinson's disease: a critical analysis of alpha-synuclein staging. Neuropathol. Appl. Neurobiol. 34 (3), 284–295.

Kalaitzakis, M.E., Graeber, M.B., Gentleman, S.M., Pearce, R.K., 2009. Evidence against a reliable staging system of alpha-synuclein pathology in Parkinson's disease. Neuropathol. Appl. Neurobiol. 35 (1), 125–126.

Kalaitzakis, M.E., Pearce, R.K., 2009. The morbid anatomy of dementia in Parkinson's disease. Acta Neuropathol. 118 (5), 587–598.

Kalaitzakis, M.E., Pearce, R.K., Gentleman, S.M., 2009. Clinical correlates of pathology in the claustrum in Parkinson's disease and dementia with Lewy bodies. Neurosci. Lett. 461 (1), 12–15.

Kalaitzakis, M.E., Walls, A.J., Pearce, R.K., Gentleman, S.M., 2011. Striatal Abeta peptide deposition mirrors dementia and differentiates DLB and PDD from other parkinsonian syndromes. Neurobiol. Dis. 41 (2), 377–384.

Kievit, J., Kuypers, H.G., 1975. Subcortical afferents to the frontal lobe in the rhesus monkey studied by means of retrograde horseradish peroxidase transport. Brain Res. 85 (2), 261–266.

Kovari, E., Gold, G., Herrmann, F.R., Canuto, A., Hof, P.R., Bouras, C., et al., 2003. Lewy body densities in the entorhinal and anterior cingulate cortex predict cognitive deficits in Parkinson's disease. Acta Neuropathol. 106 (1), 83–88.

Kruger, R., Kuhn, W., Muller, T., Woitalla, D., Graeber, M., Kosel, S., et al., 1998. Ala30Pro mutation in the gene encoding alpha-synuclein in Parkinson's disease. Nat. Genet. 18 (2), 106–108.

Leentjens, A.F., Van den Akker, M., Metsemakers, J.F., Lousberg, R., Verhey, F.R., 2003. Higher incidence of depression preceding the onset of Parkinson's disease: a register study. Mov. Disord. 18 (4), 414–418.

Lees, A.J., Hardy, J., Revesz, T., 2009. Parkinson's disease. Lancet 373 (9680), 2055–2066.

LeVay, S., Sherk, H., 1981a. The visual claustrum of the cat. I. Structure and connections. J. Neurosci. 1 (9), 956–980.

LeVay, S., Sherk, H., 1981b. The visual claustrum of the cat. II. The visual field map. J. Neurosci. 1 (9), 981–992.

Levy, G., Tang, M.X., Louis, E.D., Cote, L.J., Alfaro, B., Mejia, H., et al., 2002. The association of incident dementia with mortality in PD. Neurology 59 (11), 1708–1713.

Lippa, C.F., Fujiwara, H., Mann, D.M., Giasson, B., Baba, M., Schmidt, M.L., et al., 1998. Lewy bodies contain altered alpha-synuclein in brains of many familial Alzheimer's disease patients with mutations in presenilin and amyloid precursor protein genes. Am. J. Pathol. 153 (5), 1365–1370.

Martinez-Martin, P., Schapira, A.H., Stocchi, F., Sethi, K., Odin, P., MacPhee, G., et al., 2007. Prevalence of nonmotor symptoms in Parkinson's disease in an international setting; study using nonmotor symptoms questionnaire in 545 patients. Mov. Disord. 22 (11), 1623–1629.

Masliah, E., Rockenstein, E., Veinbergs, I., Sagara, Y., Mallory, M., Hashimoto, M., et al., 2001. Beta-amyloid peptides enhance alpha-synuclein accumulation and neuronal deficits in a transgenic mouse model linking Alzheimer's disease and Parkinson's disease. Proc. Natl. Acad. Sci. USA 98 (21), 12245–12250.

McKeith, I.G., Dickson, D.W., Lowe, J., Emre, M., O'Brien, J.T., Feldman, H., et al., 2005. Diagnosis and management of dementia with Lewy bodies: third report of the DLB Consortium. Neurology 65 (12), 1863–1872.

McKeith, I., Mintzer, J., Aarsland, D., Burn, D., Chiu, H., Cohen-Mansfield, J., et al., 2004. Dementia with Lewy bodies. Lancet Neurol. 3 (1), 19–28.

Minguez-Castellanos, A., Chamorro, C.E., Escamilla-Sevilla, F., Ortega-Moreno, A., Rebollo, A.C., Gomez-Rio, M., et al., 2007. Do alpha-synuclein aggregates in autonomic plexuses predate Lewy body disorders? a cohort study. Neurology 68 (23), 2012–2018.

Mochizuki, A., Komatsuzaki, Y., Shoji, S., 2002. Association of Lewy bodies and glial cytoplasmic inclusions in the brain of Parkinson's disease. Acta Neuropathol. 104 (5), 534–537.

Morys, J., Bobinski, M., Wegiel, J., Wisniewski, H.M., Narkiewicz, O., 1996. Alzheimer's disease severely affects areas of the claustrum connected with the entorhinal cortex. J. Hirnforsch. 37 (2), 173–180.

Muangpaisan, W., Mathews, A., Hori, H., Seidel, D., 2011. A systematic review of the worldwide prevalence and incidence of Parkinson's disease. J. Med. Assoc. Thai. 94 (6), 749–755.

Naghavi, H.R., Eriksson, J., Larsson, A., Nyberg, L., 2007. The claustrum/insula region integrates conceptually related sounds and pictures. Neurosci. Lett. 422 (1), 77–80.

Naqvi, N.H., Rudrauf, D., Damasio, H., Bechara, A., 2007. Damage to the insula disrupts addiction to cigarette smoking. Science 315 (5811), 531–534.

Papp, M.I., Lantos, P.L., 1994. The distribution of oligodendroglial inclusions in multiple system atrophy and its relevance to clinical symptomatology. Brain 117, 235–243.

Parkinson, J., 2002. An essay on the shaking palsy. 1817. J. Neuropsychiatry Clin. Neurosci. 14 (2), 223–236; discussion 222.

Parkkinen, L., Kauppinen, T., Pirttila, T., Autere, J.M., Alafuzoff, I., 2005. Alpha-synuclein pathology does not predict extrapyramidal symptoms or dementia. Ann. Neurol. 57 (1), 82–91.

Parkkinen, L., Pirttila, T., Alafuzoff, I., 2008. Applicability of current staging/categorization of alpha-synuclein pathology and their clinical relevance. Acta Neuropathol. 115 (4), 399–407.

Pearson, R.C., Brodal, P., Gatter, K.C., Powell, T.P., 1982. The organization of the connections between the cortex and the claustrum in the monkey. Brain Res. 234 (2), 435–441.

Pedersen, K.M., Marner, L., Pakkenberg, H., Pakkenberg, B., 2005. No global loss of neocortical neurons in Parkinson's disease: a quantitative stereological study. Mov. Disord. 20 (2), 164–171.

Perico, C.A., Skaf, C.R., Yamada, A., Duran, F., Buchpiguel, C.A., Castro, C.C., et al., 2005. Relationship between regional cerebral blood flow and separate symptom clusters of major depression: a single photon emission computed tomography study using statistical parametric mapping. Neurosci. Lett. 384 (3), 265–270.

Pletnikova, O., West, N., Lee, M.K., Rudow, G.L., Skolasky, R.L., Dawson, T.M., et al., 2005. A beta deposition is associated with enhanced cortical alpha-synuclein lesions in Lewy body diseases. Neurobiol. Aging 26 (8), 1183–1192.

Poewe, W., 2008. Non-motor symptoms in Parkinson's disease. Eur. J. Neurol. 15 (Suppl. 1), 14–20.

Polymeropoulos, M.H., Lavedan, C., Leroy, E., Ide, S.E., Dehejia, A., Dutra, A., et al., 1997. Mutation in the alpha-synuclein gene identified in families with Parkinson's disease. Science 276 (5321), 2045–2047.

Postuma, R.B., Gagnon, J.F., Vendette, M., Fantini, M.L., Massicotte-Marquez, J., Montplaisir, J., 2009. Quantifying the risk of neurodegenerative disease in idiopathic REM sleep behavior disorder. Neurology 72 (15), 1296–1300.

Qurrat ul, A., Abidi, T.S., 2005. Unraveling the function of claustrum. J. Pak. Med. Assoc. 55 (3), 123–125.

Rektorova, I., Aarsland, D., Chaudhuri, K.R., Strafella, A.P., 2012. Nonmotor symptoms of Parkinson's disease. Parkinsons Dis. 2011, 351–461.

Ross, G.W., Petrovitch, H., Abbott, R.D., Tanner, C.M., Popper, J., Masaki, K., et al., 2008. Association of olfactory dysfunction with risk for future Parkinson's disease. Ann. Neurol. 63 (2), 167–173.

Rub, U., Del Tredici, K., Schultz, C., Ghebremedhin, E., de Vos, R.A., Jansen Steur, E., et al., 2002. Parkinson's disease: the thalamic components of the limbic loop are severely impaired by alpha-synuclein immunopositive inclusion body pathology. Neurobiol. Aging 23 (2), 245–254.

Sherk, H., LeVay, S., 1981a. The visual claustrum of the cat. III. Receptive field properties. J. Neurosci. 1 (9), 993–1002.

Sherk, H., LeVay, S., 1981b. Visual claustrum: topography and receptive field properties in the cat. Science 212 (4490), 87–89.

Shipp, S., Blanton, M., Zeki, S., 1998. A visuo-somatomotor pathway through superior parietal cortex in the macaque monkey: cortical connections of areas V6 and V6A. Eur. J. Neurosci. 10 (10), 3171–3193.

Shulman, L.M., Taback, R.L., Bean, J., Weiner, W.J., 2001. Comorbidity of the nonmotor symptoms of Parkinson's disease. Mov. Disord. 16 (3), 507–510.

Simuni, T., Sethi, K., 2008. Nonmotor manifestations of Parkinson's disease. Ann. Neurol. 64 (Suppl. 2), S65–80.

Spillantini, M.G., Schmidt, M.L., Lee, V.M., Trojanowski, J.Q., Jakes, R., Goedert, M., 1997. Alpha-synuclein in Lewy bodies. Nature 388 (6645), 839–840.

Tandberg, E., Larsen, J.P., Aarsland, D., Cummings, J.L., 1996. The occurrence of depression in Parkinson's disease. A community-based study. Arch. Neurol. 53 (2), 175–179.

Van Duinen, S.G., Lammers, G.J., Maat-Schieman, M.L., Roos, R.A., 1999. Numerous and widespread alpha-synuclein-negative Lewy bodies in an asymptomatic patient. Acta Neuropathol. 97 (5), 533–539.

Van Rooden, S.M., Heiser, W.J., Kok, J.N., Verbaan, D., van Hilten, J.J., Marinus, J., 2010. The identification of Parkinson's disease subtypes using cluster analysis: a systematic review. Mov. Disord. 25 (8), 969–978.

Webster, M.J., Bachevalier, J., Ungerleider, L.G., 1993. Subcortical connections of inferior temporal areas TE and TEO in macaque monkeys. J. Comp. Neurol. 335 (1), 73–91.

Yamamoto, R., Iseki, E., Murayama, N., Minegishi, M., Marui, W., Togo, T., et al., 2007. Correlation in Lewy pathology between the claustrum and visual areas in brains of dementia with Lewy bodies. Neurosci. Lett. 415 (3), 219–224.

CHAPTER
13

Hypotheses Relating to the Function of the Claustrum

John R. Smythies[1], Lawrence R. Edelstein[2], and Vilayanur S. Ramachandran[1]

[1]Center for Brain and Cognition, Department of Psychology, UC San Diego, La Jolla, CA, USA [2]Medimark Corporation, Del Mar, CA, USA

INTRODUCTION

A central feature of consciousness is one's sense of being a unified person despite being confronted with a diversity of sense-impressions from different sense organs. Except in disease states we also experience ourselves as a single person. Where and how this unification occurs has been a subject of considerable debate. There are philosophers who have argued that this subjective experience of unity should not be taken as evidence for anatomical convergence in some single brain location because it would imply the fallacy of a Cartesian theater with a homunculus inspecting it. In this chapter, following the lead of Crick and Koch (2005), we take the opposite view and argue that there may indeed be an anatomical structure (or structures) involved somehow in the unity of conscious experience. Reasonable candidates would include the angular gyrus – sitting strategically at the crossroads between occipital (vision), temporal (hearing), and somatosensory cortices (touch and propriocep-tion) – and a curious, and until recently little known, structure called the claustrum. Crick and Koch (2005) shone a sudden spotlight on this hith-erto much neglected organ by suggesting that the claustrum might play a key role in information processing in the brain by correlating the sep-arate activity in the different sensory cortices into one coherent activity

J. Smythies, L. Edelstein and V. Ramachandran (Eds):
The Claustrum.

DOI: http://dx.doi.org/10.1016/B978-0-12-404566-8.00013-1

that "binds" separate sensations into the unitary objects that we experience in consciousness.

We should state here that by "consciousness" we refer in this chapter to what Längsjo et al. (2012) refer to as the "contents of consciousness," and not to what they refer to as the "conscious state." This group has shown, by PET scanning experiments on subjects awakening from anesthesia, that the conscious state is related to a function of a net that includes the brainstem, thalamus, hypothalamus, anterior cingulate, basal forebrain and the inferior parietal lobule – but not, at the level of resolution used, the claustrum. Only lesions of brain areas (such as the brainstem) that relate to the conscious state induce coma (Smythies, 1997). Lesions of brain areas that are related to the contents of consciousness do not induce coma. Viral infections of the claustrum induce hallucinations and seizures, but not coma.

The claustrum is broadly divided into three compartments: an anterior dorsal area connected with the somatosensory and motor cortices; a posterior dorsal area (visual cortex connections) and a ventral area (auditory cortex connections) (see Crick and Koch, 2005; Edelstein and Denaro, 2004; LeVay and Sherk, 1981; Sherk, 1986). The claustrum has reciprocal, widely distributed anatomical projections to almost all regions of the cortex, as well as to many subcortical structures. A recent high-definition diffusion imaging study (Park et al., 2012) reports that the claustrum has connections with the frontal, premotor, ventral anterior cingulate, ventral temporal, visual, motor, somatosensory, and olfactory cortices, and most strongly with the entorhinal cortex. It also has connections with some subcortical structures including the putamen, globus pallidus, and lateral amygdala. Until recently, however, the connections between the claustrum and the primary visual cortex (V1) were the subject of some disagreement. Day-Brown et al. (2009) have now published evidence, based on experiments that employed injections of cholera toxin B (CTB), and biotinylated dextran amine tracers, application of antibodies to γ-aminobutyric acid (GABA) and glutamic acid decarboxylase, and confocal and electron microscopy, that the two have reciprocal connections. They also noted that four times as many cells were labeled in the claustrum than in the dorsal lateral geniculate nucleus following V1 CTB injections.

The auditory cortex may have somewhat different claustral connections than the visual and somatosensory cortices. Clarey and Irvine (1986) report that, in the cat, the input from the primary auditory area to the claustrum is sparse and comes rather from higher auditory areas. Beneyto and Prieto (2001) studied the connections between the claustrum and the auditory cortex by tracer injection methods. They report that all auditory areas have reciprocal connections with the ipsi- and contralateral claustrum. They state, "These findings suggest that the intermediate region of the claustrum integrates inputs from all auditory cortical areas, and then sends the result of such processing back to every auditory cortical field." By means of differential bidirectional tracer injection experiments in rat somatosensory cortex, Smith et al. (2012) failed to find any significant anatomical projections from the primary somatosensory cortex (S1) to the claustrum, whereas they detected dense projections from the same part of the claustrum to both S1 and the motor cortex. They concluded that this shows that the claustrum, in the rat, does not function as an integrator of somesthetic and motor function. A study conducted by Gomes et al. (2012), using the fluorescent retrograde markers Diamidino Yellow and Fast Blue, showed that territories in the claustrum that project to widely separate parts of the cortex have overlapping sources in the claustrum. The authors suggest that this provides evidence of a multimodal operating system.

The experiments of Remedios et al. (2010), however, showed that there is a rapid functional connection between the auditory system and the claustrum. Moreover, Zhang et al. (2001), using a similar technique, were able to demonstrate bidirectional connections between the claustrum, on the one hand, and both primary and secondary visual, somatosensory and auditory cortices on the other.

The claustrum has a well-marked, retinotopically organized map of the visual field, as well as an equivalent map of the somatosensory field (LeVay and Sherk, 1981; Olson and Graybiel, 1980; Sherk, 1986). It has been claimed that areas of the claustrum in the cat send "precisely reciprocal" projections back to those area(s) of the cortex whence its inputs derived (LeVay and Sherk, 1981; Sherk, 1986). It has also been claimed that the projections to and from the claustrum are diffuse, but quite specific, in both directions (one point to specific points) (Divac, 1979; Divac et al., 1978; Sloniewski et al., 1986). Single cells in the claustrum have been reported to send branched axons to several cortical areas. The same cells receive input from these areas (Rahman and Baizer, 2007). However, there are marked species differences in the anatomy of the claustrum, and others claim that the cortico-claustral projection is more diffuse. Crick and Koch summarize their view of the situation thus: "Most regions of the cortex send a projection to the claustrum, usually to many parts of it. Thus their mappings are far from being a precise local mapping and tend to be somewhat global (that is, all to all), though not completely so."

The ventral claustrum is also connected to limbic structures, such as the amygdala, subiculum and cingulate cortex. In addition, the claustrum has a relatively uniform microanatomical structure that would allow what Crick and Koch describe as "widespread intra-claustral interactions." These, they suggest, may be in the form of waves of information involving dendrodendritic synapses and networks of gap-junction linked neurons. They also suggest that claustral neurons "could be especially sensitive to the timing of the inputs." Lastly, and most significantly, they suggest that the function of the claustrum may be to synchronize cortical oscillations. Rahman and Baizer (2007) also suggest that the claustrum mediates "integration across compartments mediated by inhibitory interneurons." On the afferent side, previous reports stated that the majority (75 percent) of claustral neurons are multisensory in that they respond to stimuli in more than one sensory modality, whereas 25 percent are unimodal (Spector et al., 1969, 1974). However, in the same species more recent experiments failed to find multisensory responses (Olson and Graybiel, 1980; Sherk and LeVay, 1981). The injection of three different fluorescent tracers into the occipital, frontal and cingulate cortices results in double- and triple-labeled cells in the claustrum (Li et al. (1986)). By contrast, in a recent investigation in awake primates, Remedios et al. (2010) found that the great majority of claustral neurons are unimodal. On the efferent side, using similar techniques, Minciacchi et al. (1985) reported that single claustral neurons project to both anterior and posterior cortex, some ipsilaterally and some contralaterally. This suggests that single claustral cells project to more than one major cortical area. We will return to this point later.

This chapter considers what these "waves of information" might be, and what type of timing-related, or other, computations and integrations they might perform. Neurocomputations may take various forms, for example those based on fixed neural networks (Churchland and Sejnowski, 1999), or non-linear dynamics (Freeman, 2001), or synchronized oscillations (Uhlhaas et al., 2009). In this chapter we will focus on the last in order to supply some detail to Crick and Koch's specific focus on the role of the claustrum in the synchronization of cortical oscillations. In addition, the claustrum may not be involved solely in sensory "binding," but it may also be concerned with synchrony detection and modulation in connection with salience processing and a wide range of motor and cognitive processes.

OUR HYPOTHESIS

Our basic hypothesis suggests that the claustrum functions as a synchrony detector, and modulator and integrator of synchronized oscillations (Smythies et al., 2012b). Remedios et al. (2010) have suggested that the function of the claustrum is related to salience and salience detection. This was based on their own experiments conducted on awake primates. The stimuli were naturalistic video recordings, naturalistic auditory recordings, and both presented together. Recordings were taken from individual neurons in various parts of the claustrum. Their results confirmed the fact that visual neurons and auditory neurons are located in different loci of the claustrum (ventral and dorso-central, respectively). However, they reported that the neurons reacted to visual or auditory stimuli, but not to both. They added some other interesting details. In all cases the response latencies were short (circa 50 ms) and the responses were brief and stereotyped – a short strong transient followed by a diminished sustained response. This was in contrast to what the same group, using the same stimuli, had found in higher sensory cortex, where considerable fractions of integrating or bimodal neurons were found in the auditory cortex (Kayser et al., 2008) and the superior temporal cortex (Dahl et al., 2010). Furthermore, Remedios et al. (2010) tested for multisensory integration in the claustrum by looking for instances where the response to one stimulus is modulated by a stimulus in a different modality. No such modulations were found. These workers suggested that previous reports of multimodal neurons in the claustrum were probably contaminated by influences from the insula. They concluded that the claustrum is most likely to act as a salience detector, and that their results corroborated previous reports that discrete sensory zones within the claustrum are connected to the corresponding primary sensory cortical area. They noted that inputs from the early cortices and/or thalamic nuclei are rapidly fed to the claustrum, and are rapidly directed from there to most parts of the higher cortex.

These experiments indicate that during naturalistic perception, not requiring multisensory integration of separate groups of neurons, the claustrum responds to visual and auditory stimuli with a simple peak and lower plateau response. Activation of the claustrum in this way might be expected to initiate low level reverberatory activity in specific claustro-cortical circuits that would be responsible for the plateaux. Remedios et al. (2010) suggested that this represents a salience function. The mechanism for salience detection entails, early in the afferent inflow, the incoming down-up messages being matched against the up-down messages that the brain expects. If a mismatch is detected between expectation and the input, the mechanism sends an alerting message to the rest of the brain. Liang et al. (2013) have shown, using a Bayesian model selection and functional MRI (fMRI) data, that salient sensory information is channeled directly from thalamic nuclei to multisensory cortex, bypassing primary and secondary sensory cortex.

The brain contains several systems involved in salience detection – for example the mesolimbic dopamine system (Enomoto et al., 2011; Friston et al., 2012). Another salience network includes the dorsal anterior cingulate cortex, middle and inferior temporal cortex and the fronto-insular cortex (Cauda et al., 2011; Yuan et al., 2012). Day-Brown et al. (2010) report the existence of a direct projection from the superior colliculus to the amygdala in the tree shrew that they suggest alerts the animal to the presence of danger; that is, it carries negative salience. There is at least one system involved in salience and executive control, of which the claustrum is an integral part, and that is the rubral network based on the red nucleus. This comprises (in addition to the red nucleus) the cerebellum, mesencephalon, substantia nigra, hypothalamus, pallidum, thalamus, insula, claustrum, posterior hippocampus, precuneus, and occipital, prefrontal, and fronto-opercular cortices (Nioche et al., 2009). Little is known about the functions of this circuit. It is therefore necessary to distinguish between several types of "salience." Salience can involve the process by which new and possibly important stimuli are identified and evaluated. In this form of salience detection, the direct sensory input from the sense organ is compared with feedback from the higher cortex that predicts what the input should be. It can also mean the process by which information about the reinforcement history of stimuli is processed. Is the claustrum then just one of a number of saliency detectors? Or does it exert some special function? We suggest that there is evidence to indicate that it does, and that the key function of the claustrum is to integrate activity between separate brain areas during multimodal tasks.

For example, the claustrum is involved in esthetic but not brightness judgments (Ishizu and Zeki, 2013). The authors concluded that, as the claustrum has been described as a critical station for cross-modal processing and integration of information from different perceptual modalities and sources (e.g. color and vision in vision), "Judging the beauty of two similar stimuli may be a complicated integrative process that may involve many different features, such as colours, forms, proportion, or facial expression." Or consider the case of the brain having to deal with a multimodal operation like playing the piano, in which there has to be constant integration between the visual, auditory and somatosensory inputs. The input from the senses will set up centers of activity in the three sensory cortices that will in turn generate activity in three sections of the cortico-claustral circuits. On the principle that cortical areas that work together oscillate together, these three areas will tend to synchronize their gamma oscillations.

With an ever-changing retinal input, such small groups of synchronized cells in the cortex may continually form and reform in competition. Their fate depends on the degree of subsequent attention, recruitment, and reinforcement dictated by the nature and details of the stimuli and of the task being performed. In human EEG experiments, Kaiser (2003) reported that auditory spatial "mismatch" detection that is driven bottom-up elicits gamma-band activity over the posterior parietal cortex, whereas an auditory pattern "mismatch" report leads to gamma-band enhancements over the anterior temporal and inferior frontal regions. An unexpected stimulus (e.g. "roar of tiger") also activates the reticular activating system with its widespread projection to the cortex from the nucleus basalis of Meynert (Smythies, 1997, 1999).

In this context, the observations of Pearson et al. (1982) are of crucial importance. Crick and Koch (2005) describe them as follows:

> It has been suggested by Pearson et al. (1982) for monkeys that (i) if cortical area A, which is located in either the occipital, parietal or temporal lobe, projects to cortical area B in the frontal lobe, then the target zones of cortical areas A and B in the claustrum will overlap along a dorsal ventral axis; (ii) if cortical area C in the frontal lobe is connected to cortical area D, also in the frontal lobe, their respective claustral target zones will overlap along a dorsoventral axis, and (iii) if area A is not connected to cortical area B, no matter where in the cortex they are located, their target zones in the claustral will not overlap.

This means that, if synchronized gamma oscillations develop in cortical areas A and B, then this synchronized activity is likely to be transmitted to claustral areas A and B, but such activity between cortical areas C and D will not be transmitted to claustral areas C and D. Now the cortical areas A and B, whose gamma oscillations are so synchronized, are far apart, whereas the claustral areas A and B are closer together in the claustrum, where neurons are packed closely together. It is therefore possible that, since the A B sets of neurons are connected by cortico-claustral loops, such a system would result in magnification of weak intercortical oscillations by stronger intraclaustral oscillations.

Thus intracortical synchronization could then be transmitted, via cortico-claustral loops, to the densely packed, connected cell bodies (pyramidal (P) cells and interneurons (INs)) in the claustrum. In particular GABAergic INs can form dense syncytia, being interconnected by gap junctions and electrical synapses that allow the widespread dispersion of activity. We may suppose that these are in a continual state of interacting synchronized oscillations at various frequencies in dynamic competition. In a natural environment salient stimuli rarely affect only one sensory channel. So, when the signal from visual X (e.g. "sight of tiger") reaches the interior of the claustrum, it may be accompanied by an auditory signal ("roar of tiger") and an olfactory one ("smell of tiger"). These oscillatory activities may then interact in the internal P cell/IN syncytium, and a derived signal is then sent to the motor cortex ("run for it!"). Surface and intracerebral recordings have revealed bursts of synchronized gamma activity that peak just before or around movement onset (Van Wijk et al., 2012). These appear only in the hemisphere contralateral to the movement and have a focal somatotropic distribution. We suggest that some elements of these bursts may originate in the claustrum. A note may be added that, in the Remedios et al. (2010) experiment, the stimulus was a "familiar jungle scene." Therefore, the animal was expecting it, and so no salience response system was triggered.

We therefore suggest the following scenario. When an animal attends to a non-salient stimulus, as in the Remedios experiments, a signal is sent to the appropriate part of the claustrum carrying the information "Event out there." The claustrum responds by a brief local activation. If none of the salience detecting mechanisms listed above is triggered, nothing more happens. The same sequence is followed if the animal is responding to two simultaneous non-salient stimuli. However, if the input activates one of these saliency detecting mechanisms, this reinforces the claustral effect and synchronized oscillations are locally promoted. If there are two incoming stimuli then synchronized activity between them will be promoted and magnified by the Pearson mechanism we have proposed (Figure 13.1).

We suggest in fact that the brief signal to the claustrum picked up by Remedios et al (2010) is a priming signal that alerts selected areas of cortex with the message, "Saliency signal possible: stay alert." Clearly, to test this hypothesis experiments need to be carried out using the technique employed by Remedios et al. (2010) with salient stimuli.

However, the following reports on the role of attention and other factors in synchronization suggest that the actual mechanism may be more complicated.

1. Attention to a stimulus enhances frequency synchrony in visual areas (Buffalo et al., 2011). In the gamma range (40–60 Hz) this is largely confined to the superficial layers, whereas the deep layers showed maximum coherence at low frequencies (6–16 Hz). In the superficial layers of V2 and V4, gamma synchrony was increased by attention, whereas in the deep layers, alpha synchrony was reduced by attention. This indicates different roles for synchrony in different locations.

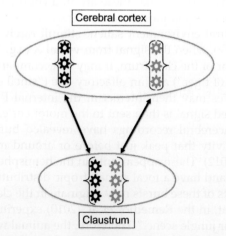

FIGURE 13.1 Diagram of the proposed magnification (and possibly modulation) by the claustrum of intracortical synchronized oscillations.

2. Cardin et al. (2009) report that light-driven activation of fast-spiking INs at varied frequencies (8–200 Hz) selectively amplifies gamma oscillations. In contrast, such stimulation of pyramidal neurons amplifies only lower frequency oscillations. They reported that the timing of a sensory input relative to a gamma cycle determines both the amplitude and the precision of evoked responses. Thus, specificity for gamma oscillations may be dependent on the type of neuron involved.

3. Chalk et al. (2010) measured the local field potential (LFP) and V1 spiking activity in monkeys during an attention-demanding detection task. They reported that, if attention was directed at a visual stimulus in the receptive field of the recorded neurons, decreased LFP gamma power and gamma synchrony resulted. The authors suggested that this decrease could be the result of an attention-mediated reduction of surround inhibition. They further suggested that modulation of synchrony in V1 could be a byproduct of reduced inhibitory drive, rather than a mechanism that directly involved perceptual processing. This report suggests a different mechanism altogether for gamma oscillations.

The next ten reports add yet more complexity.

4. Dockstader et al. (2010) used time-frequency analyses of human somatosensory oscillations with neuromagnetic recordings during stimulation of the median nerve in healthy adults. They reported that selective attention-modulated somatosensory oscillations in the alpha, beta, and gamma bands. These were either phase-locked or non-phase-locked to the stimulus. In the S1 cortex, if the subject's attention was toward the somatosensory stimulus, this resulted in increased gamma band power (30–55 Hz) that was phase-locked to stimulus onset. Such attention also produced an initial desynchronization, followed by increased synchronization, in the beta range, that was not phase-locked to the stimulus.

5. Tallon-Baudry (2009) suggested that "… gamma oscillations are not related to a single cognitive function, and are probably better understood in terms of a population mechanism taking advantage of the neuron's fine temporal tuning: the 10–30 ms time precision imposed by gamma-band rhythms could favor the selective transmission of synchronized information (attention) and foster synaptic plasticity (memory)."

6. Palva and Palva (2007) propose that simultaneous alpha-, beta- (14–30 Hz) and gamma- (30–70 Hz) frequency band oscillations are necessary for unified cognitive operations that mediate the selection and maintenance of neuronal object representations during working memory, perception and consciousness.

7. In human EEG studies Haenschel et al. (2000) reported that evoked (stimulus-locked) gamma oscillations preceded beta 1 oscillations in response to novel stimuli.

8. In 10 patients who underwent intracranial electrocorticography, somatosensory-related gamma augmentation involving the post- and precentral gyri evolved into beta and alpha augmentation. This was later followed by beta and alpha attenuation that involved the post- and precentral gyri (Fukuda et al., 2010).

9. The prediction of a forthcoming stimulus by a warning cue stimulus was associated with an increase in gamma oscillations overlying occipital areas and a decrease in beta oscillations overlying sensorimotor cortex before the stimulus was presented (Kilner et al. (2005)).

10. Gregoriou et al. (2009) suggest that long-range excitatory connections onto INs determine whether different pyramidal cell "assemblies" can synchronize at gamma frequencies, whereas excitatory connections onto pyramidal cells determine whether such assemblies can synchronize at beta frequencies.

11. Colgin et al. (2009) reported that fast and slow gamma synchronizations in the hippocampus CA1 area have differential effects. The former synchronize with fast gamma oscillations in the medial entorhinal cortex (an area that provides information about the animal's current position) whereas the latter synchronize with slow gamma oscillations in the CA3 area of the hippocampus (an area necessary for memory storage of such information).

12. Fujioka et al. (2009) studied beta and gamma band activities in the auditory cortex of humans during musical beat processing. The beta reaction peaked after each tone but showed no response after an omission. The gamma response, in contrast, peaked after both tones and omissions. This suggests that the beta response in auditory cortex under these circumstances does not signal "mismatch," and that the gamma response is related to anticipation rather than a "match" signal.

13. Burns et al. (2011) conducted a time-frequency analysis of LFPs recorded in macaque V1 cortex. They concluded that gamma oscillations cannot be used as 'clocks' for precise temporal encoding and 'binding' of signals across brain regions in the way that most theorists in this field have proposed. Instead they suggest that chaotic (noisy) gamma oscillations can still function in network computations. They concluded that: "Short high-amplitude bursts in the gamma-band, if present in multiple networks, could operate as a transient sync pulse. This pulse would not serve as a clock but may synchronize different gamma-activated networks to fire simultaneously. Furthermore, the relative timing of this sync pulse in different networks could also function in the attentional selection of different features of a stimulus."

In view of the many unanswered questions these reports raise, we will confine the formal definition of our hypothesis to limited terms with respect to the functions and frequencies of the synchronizations involved. The claustrum may be involved in detecting, integrating, promoting, and directing synchronized oscillations in a wide range of frequencies and in a number of functional categories. The details await further experimental investigations.

However, we feel that there may be certain principles relevant in all the possible variations. Let us look again in more detail at the situation where the subject is engaged in a complex task that involves coordination, say between sight and touch. Frequent activations of selected visual and tactual cortical neurons will result. On the principle that cells that work together oscillate together, the activated cells will start to oscillate in synchrony. Both sets of cells will spontaneously start synchronized oscillations and will generate axon spikes that will have a timing determined by these oscillations. Each will initially oscillate at its own preferred frequency. This concomitant activity might tend to bind their frequencies of oscillation, and spike timing patterns, into line. These cortical neurons project to the claustrum via specific layer 6 pyramidal cells. The axons of these layer 6 cells synapse on the cell bodies and dendrites of claustral pyramidal and IN cells and may induce in them oscillations at the same frequency.

Evidence that this could occur is based on the following reports. Axons can carry multiple codes in the spatiotemporal patterns of their spike trains (Kayser et al., 2009; Uhlhaas et al., 2009). Codes based on temporal spike-train patterns and spatial populations can nest additional information (codes) based on the relative phase of slow, ongoing rhythms at which these (temporal or population) responses occur. Information carried by spike trains can cross synapses and transmit to the postsynaptic neuron (Sejnowski and Paulsen, 2006). Kumar et al. (2010) stated: "The brain is a highly modular structure. To exploit modularity, it is necessary that spiking activity can propagate from one module to another while preserving the information it carries." Masuda and Aihara (2002) put it thus: "If a real brain uses spike timing as a means of information processing, other neurons receiving spatiotemporal spikes from such sensory neurons must also be capable of treating information included in deterministic interspike intervals." Asai et al. (2008) found that cortical regular-spiking neurons can propagate filtered temporal information in a reliable way through the thalamo-cortical network, and with high temporal accuracy. This evidence supports the concept that a synchronized afferent input to the claustral cells can cause them to fire at the same frequency and allow them to form reverberating cortico-claustro-cortical loops.

The layer 6 cells in the cortex in the two groups (hearing and vision) are far apart, and are directly connected only by cortico-cortical axons.

The claustral P cell bodies connected to these cortical areas are close together and are connected to each other by short axons and axon collaterals. These may effect some of the postulated interactions. The P cells are surrounded by GABAergic INs and these may form a GABAergic syncytium linked to each other by these GABAergic chemical synapses and electrical gap junctions. This would allow very fast communication. It may also allow rapid movements of electrolytes and current between them. Electrical synapses between INs have been reported to contribute to synchronized firing and network oscillations in the brain (Vervaeke et al., 2010). In the bushy cells of the cochlear nucleus Chanda and Xu-Friedman (2010) have shown that the activation of GABAergic receptors adjusts the function of these cells by suppressing the relaying of individual inputs and requiring the coincident activity of multiple inputs. This function may promote "binding." Activation of such GABAergic systems strongly promotes gamma synchronization. So co-temporal activity in the auditory and visual cortex, in this case, may result in a common frequency of oscillation throughout this simple net, under the influence of possibly weak, and (importantly) slow, cortico-cortical interactions but strong, and (importantly fast) intraclaustral interactions. The cortico-cortical interactions may be slower than the intraclaustral interactions because of the greater distances involved in the former. This particular link up may persist while this particular audiovisual task is being conducted. When the task is completed, the conjoined activity in the cortico-claustro-cortical system will die down and be replaced when a new task is undertaken, when the whole process will be repeated again elsewhere. However, the brain can compensate for these long distances to some extent by dynamical self-organization by the distant target neurons into lag-free oscillations (Vicente et al., 2008).

This process may involve local field potentials (LFPs), which mainly originate from synchronous synaptic input into the local neural neighborhood (Samengo and Montemurro, 2010). Different LFP bands represent dynamic natural stimuli on different timescales, and may provide a "potentially rich source for sensory processing and decoding of brain activity" (Belitzi et al, 2010). However, some complications are introduced by the following reports. Kraskov et al. (2007) presented evidence from microelectrode recordings in the medial temporal lobe during neurosurgery in response to the presentation of semantic categories of visual stimuli. They reported that there was little overlap between LFP and spike-selective electrodes, implying that two different coding systems were in operation. In another study using recordings of spikes and LFPs in human motor cortex during reaching movements, Flint et al. (2012) produced evidence that LFPs retain information about movement in the absence of spikes on the same electrodes. In multiple electrode recording experiments Gaucher et al. (2012) reported that LFPs reflect global functional connectivity in the thalamo-cortical auditory system, whereas spiking activities reflect more independent local processing.

Thus this model allows the claustrum to promote synchronized inter-modal gamma oscillations in widely separated parts of the cortex. In a previous model (Smythies et al., 2012a) it was thought that to do this the claustrum needed direct inputs to individual claustral cells from axons derived from the two disparate cortices. The data obtained by Remedios et al. (2010) suggested that these direct connections do not exist. However, in our present model, such direct connections are not necessary because the majority of the "binding" is done inside the claustrum. Furthermore, no complex spike codes are necessarily involved as in the previous model (Smythies et al., 2012a). In non-sensory "higher" cognitive processing operations, which require interaction between any higher cortical areas, the same mechanism may be involved. The key to our hypothesis is that weak cortico-cortical promotion of a common oscillation frequency is magnified by a strong intraclaustral promotion.

In our hypothesis, synchronized oscillations in the claustrum potentiate synchronized oscillations in the cortex, if these two form part of a cortico-claustro-cortical loop. This explains the anatomical arrangement reported by Pearson et al. (1982).

The interlinked P cells in the claustrum, embedded in a GABAergic net, may generate a complex, fluctuating, competitive and dynamic domain of shared and disparate oscillations. This might compensate for the weak axonal interconnections that exist between areas of the claustrum (Sherk, personal communication). This might be modulated by the ever-changing pattern of afferent spikes from the cortex, as well as by chemical neuromodulators such as dopamine and input from the reticular activating system and other subcortical structures. For example, positive reward leads to a widespread release of dopamine in the brain. Dopamine promotes gamma synchrony (Li et al. (2012)). Remedios et al. (2010) suggested that the short, sharp initial claustral signal from the sense organ to the cortex involves salience and does not involve "binding." We suggest the hypothesis that this short, sharp initial signal only alerts the cortico-claustral mechanism, which does, if circumstances are favorable, then effect "binding."

There is considerable debate regarding whether the projections of the cortex onto claustral neurons are point-to-point or diffuse. There are also considerable interspecies differences in the anatomy of this system. In the cat this projection is described as "precisely reciprocal" (Sherk, 1986). Little is known about this anatomy in the primate: there are indications that this projection may be more diffuse (Sherk, 1986). However, in this context, it is important to note that the only requirement of our hypothesis is that the activated claustral P cells should receive their input from the cortex via local groups of neurons. This ensures that there is no ambiguity when they fire. Their efferent targets could be back to these same neurons by direct return axons, or to other cortical neurons by branched axons. The projections of groups of single cortical neurons to the claustrum may follow a bell-shaped curve. That is, there are more connections with the closest claustral neurons, and fewer with more distal ones: the more distant the fewer the connections. Then the projection as a whole will consist of a series of partly reciprocal return projections with a degree of overlap. Each small group of claustral neurons may project back to the small group of cortical neurons that gave rise to the axons connecting the two groups in the statistical manner just described. More widespread distribution of claustral impulses may also be mediated by the intraclaustral integrative system of multiple interacting synchronized oscillations that we have described.

The question may be asked whether the connectivity between the cortex and the claustrum could be random. Corey and Scholl (2012) have shown how specific selective computations can be carried out via random connectivity in the cortex. In this context it may be relevant to note that the modality of a cortical sensory neuron is determined by the modality of its afferent axons (we can call this DAM – determination by afferent modality) (Gao and Pallas, 1999; Lomber et al., 2010; Ptito et al., 2008; Sharma et al., 2000). Therefore, the modality of a claustral neuron might also be determined by the modality of its afferent neurons. In which case 'randomness' takes on a new meaning. Without DAM, afferent axons to the claustrum cannot be random as they have to find their unique correct targets. However, with DAM this no longer holds because the afferent neurons create their own targets. So they can be "random" in the sense that they take the shortest route from the cortex to the claustrum, and so naturally reproduce there in the order of the cortical cells they arise from. Take a cortical arrangement, SAV, (somatosensory, auditory, visual) in an orderly array, each of whose neurons sends axons that take the shortest route to the claustrum – DAM ensures that the claustrum will automatically reproduce the cortical maps for SAV.

Our hypothesis seems to fit in with the general pattern of how the brain operates – by a very large number of different, small conjoined units that all carry out the one same simple operation. In our case these units are not permanent nerve nets; rather, they consist of dynamically organized collections of neurons temporarily oscillating in synchrony at the same frequency. Theodore Bullock (1993) is skeptical about some computer network models (see pp. 171 & 195–198). Instead he supports what he calls the "small set recognition mechanism" (pp. 195–196 & 198–200). He says, "The small set of equivalent cells can act as a quasi-redundant set of parallel decision units to trigger specific motor systems, action patterns or affective states" (p. 196). Sherk and LeVay (1981) suggest that, "… the claustrum performs some rather simple service that does not require a complex transformation of information it receives."

Our concept of the nature of the neural correlates of consciousness is similar to the Dynamic Core hypothesis of Tononi and Edelman (1998). They put their hypothesis as follows:

1. A group of neurons can contribute directly to conscious experience only if it is part of a distributed functional cluster that achieves high integration in hundreds of milliseconds.
2. To sustain conscious experience, it is essential that this functional cluster be highly differentiated, as indicated by high values of complexity.

They suggested that neuronal correlates of consciousness (NCCs) should be "… a large but distinct set of distributed and functionally integrated neuronal groups that interact over fractions of a second much more strongly among themselves than with the rest of the brain." These assemblies are to be found in, but not limited to, the cortico-thalamic system. They further suggest that these assemblies consist of synchronously firing neurons (intercortical and cortico-thalamic). They do not suggest what causes these neurons to enter into these highly integrated dynamic assemblies, nor do they mention either Walter Freeman's hypothesis (2001) or use the term 'non-linear dynamics' – but presumably something of that kind must be involved.

Our hypothesis suggests that we need to add the cortico-claustral system to the cortico-thalamic system to complete the dynamic core. Tononi and Edelman (1998) do not mention the claustrum. We suggest that the basic integration between the different functional groups is carried out by the mechanism we propose for the claustrum, although local dynamic factors could integrate oscillations within each functional group. They go on to say that the membership of such assemblies is not fixed, but is dynamic in that particular neurons can partake sometimes in consciousness-related activities and at other times in unconscious related activities: thus, the intrinsic properties of neurons do not have, in some mysterious way, a privileged correlation with consciousness.

FURTHER DATA ON THE ROLE OF THE CLAUSTRUM IN SYNCHRONIZED OSCILLATIONS AND MODIFICATION OF OUR HYPOTHESIS

In a magnetoencephalographic study, Emrich et al. (2006) examined the effect on synchronous oscillations in the brain associated with object perception (shape-from-motion) in six normal volunteers. The stimulus was a computer generated "object" seen against background noise. Each lasted for 4 seconds. During the MOVE epoch, the object and background moved in counter phase, followed by one of two stationary epochs. In the first – STOP – the motion stopped, and the object remained in the display along with the stationary background noise. In the second – VANISH – the motion stopped and the object was removed from the display, leaving only the stationary background noise. During the MOVE epoch synchronized gamma oscillations (35–45 Hz) appeared in the right insula, right claustrum, right superior temporal gyrus (RSTG), and right parahippocampal gyrus (RHG). During the STOP epoch the gamma synchronization remained but changed to a higher frequency (55–72 Hz). The authors proposed that claustral activity during the MOVE epoch may be related to conscious awareness of the object, and/or motion cues, and/or memory processes: and that claustral activity during the STOP epoch might be related to awareness of the object only (since the object is not moving).

However, the absence of movement by a previously moving object carries information just as much as the presence of movement does, and both are recorded in memory processes. Therefore, this experiment may actually be relevant to the binding of form and movement rather than, or in addition to, the "conscious awareness" of an object. Furthermore, in their analysis, the action of the claustrum during the MOVE epoch involves binding of form and movement, but during the STOP epoch there is no binding, since the only input to the claustrum involves form only.

Using an fMRI technique (which provides evidence of activation but not synchronization) Kavounoudias et al. (2008) demonstrated that the integration of intramodal proprioceptive and tactile information (causing kinesthetic illusions of a clockwise rotation of the right hand) resulted in activation of the superior temporal gyrus, the inferior parietal lobe, and the claustrum. They suggest that this involved detection of spatial coherence by the inferior parietal lobule and detection of temporal coincidence by the insula structure "usually linked to the relative synchrony of different stimuli." In an fMRI investigation of color synesthesia it was shown that the result was activation of the color area V4 (Nunn et al., 2002).

How does our hypothesis explain these findings? First, we can note that Emrich et al. (2006) found that the effect they reported does not involve the whole cortex, but only the RSTG and the RHG. This suggests that these two loci have a special importance for object recognition of the type examined in this experiment. The STG contains both auditory and polysensory areas (Cappe and Barone, 2005; Smiley and Falchier, 2009). In particular, it contains visual neurons sensitive to both form and the direction of movement (Oram and Perrett, 1996) and projections to the frontal eye fields (Scalaidhe et al., 1997). The RHG contains a higher visual area that processes topographic scenes. The insula has multiple involvements in interoceptive awareness, emotions and salience. We suggest that the claustrum is involved in tying this inter-center activity together by integrating the oscillations in their respective neuronal populations.

However, Emrich and colleagues measured overall spectral power, not the frequency of the oscillations. This is compatible with either (a) that there are increases across all synchronized frequencies in that range, or (b) that the determination of the exact frequency concerned (e.g. 37 or 42 Hz) cannot be made (Emrich, 2012, personal communication). Moreover, it also means that we cannot assume that the synchronies reported in the literature are all intermodal and not intramodal (local), unless this is specifically stated. That is to say that the oscillations in the four brain areas may be synchronized in the sense that they all oscillate together at the same frequency (intermodal); or they may all oscillate, but the synchrony is between the individual neurons making up that area, and the areas themselves may oscillate at different frequencies (local). For example, in a study of electrophysiological responses related to the processing of visible and invisible words in a delayed matching to sample task, Melloni et al. (2007) found that both stimuli caused a similar increase of local (gamma) oscillations in the EEG, but only perceived words induced a transient long-distance synchronization of gamma oscillations across widely separated regions of the brain. Lachaux et al. (1999) applied phase-locking statistics (PLS) methodology in a study of intracortical recordings in an epileptic patient performing a visual discrimination task. They found large scale, frequency-specific intermodal synchrony at 45 Hz between the hippocampus and the frontal cortex. A phase-locking values analysis study provided evidence for early-latency, gamma-band neuronal synchronization between the S1 and S2 areas during early stages of somatosensory processing in normal controls (Hagiwara et al., 2010).

It seems possible that the claustrum may be involved, at times, in integrating local intramodal oscillations. In this case, a rider needs to be made to our hypothesis: In some instances, the claustrum may not effect intermodal synchronization of its various inputs, but merely amplify these inputs (level 1 of our hypothesis). This model suggests that the claustrum does not integrate the synchronized oscillations it processes, it only amplifies them. In addition, stimulated by a salient inflow, gamma oscillations are set up at a frequency specific for each cortical area. These are projected to the claustrum, where they set up oscillations synchronized at that frequency. Owing to the compact nature of the claustrum, such oscillations might interact more vigorously than any promulgated by direct cortico-cortical connections. These would in turn activate the claustro-cortical projections and elevate the activity in each cortical area. This might be effective because it would give each of the two areas competitive advantage over all other areas. If the claustrum receives only one strong volley of spikes down one channel, the magnification mechanism might not be switched on, or not switched on fully. But, if it received two, or more, such volleys down different channels, the magnification mechanism might be switched on (or switched on fully) and the same effect achieved as in our main hypothesis – competitive advantage for its clients. Again to settle this point, further experiments may be needed in which specific measures of oscillation frequencies are made under various conditions.

In the EEG, fast oscillations such as gamma are in general less able to traverse large distances and are therefore more likely to be restricted to local circuits (Von Stein and Sarnthein, 2000). These workers showed that: (1) local interactions during visual processing involve gamma frequency dynamics, (2) medium range semantical interactions between temporal and parietal cortex involve beta frequency dynamics, and (3) very long range interactions during internal mental top-down processes, such as working-memory retention or visual imagery, involve interactions in a low theta or alpha frequency range. They also distinguished, in all sensory cortices, an early gamma oscillation that occurs 20–100 ms after stimulus onset, from a later gamma oscillation. They attributed the former to early sensory processes, and the latter to later cognitive processes. In our model, the priming signal from the claustrum might contribute to the former, and later cortico-claustral circuits might contribute to the latter. Our model might also explain why the cortex needs the claustrum to mediate long range gamma synchronizations, such as those reported by Emrich et al. (2006). If the LFPs of gamma oscillations in general can only traverse short distances, their long distance synchronization may have to be boosted by the axonally transmitted cortico-claustral system.

Thalamo-Claustral Connections

What we have been describing as a single cortico-cortical projection in fact consists of two parts (Sherman and Guillery, 2011). The first is the direct part we described. The second is indirect with a relay in a higher order thalamic relay (HOTR). These HOTRs include the pulvinar (mainly vision), the dorsal division of the MGN (audition), and the posterior medial nucleus (sensorimotor). In a simplified account, a pyramidal cell A in layer 5 of one cortical area projects to the relevant HOTR, and activates neuron C therein. Let us call this synapse X. The HOTR neuron sends a projection to a neuron B in another cortical area. A and B are also connected by direct cortico-cortical connections. The cortico-thalamic axon on the indirect route (A–C) also sends a branch to lower motor areas, but not to any other cortical area. The axons on the direct route (A–B), in contrast, send no branches to lower motor areas. Thus there are two information channels from A to B. The direct route goes straight from A to B. The indirect route goes via synapse X on neuron C in a HOTR. Only the indirect route (A–B–C) can be modulated by other thalamic systems. The axons on this indirect route deliver motor instructions to lower motor centers and exact copies of these to the HOTR. Furthermore, the thalamo-cortical neurons, which project from the HOTR to area of cortex B, will synapse on some of the same neurons that are the targets of the direct A–B projection. This may form the basis for a salience detection mechanism. Further complications in this account are introduced by Shipp's "Replication Principle," which we will describe below.

Synapse X forms a "transthalamic" gate with two modes of operation. The first (closed) is the linear tonic mode consisting of single spikes that faithfully carry information. The second (open) is the burst mode consisting of a short burst of very fast spikes. The advent of an unexpected motor instruction opens the switch and sends a short, sharp "wake up" call to the cortex. This, the authors say,

> ... activates cortex more strongly than does tonic firing... For this reason burst firing may represent a "wake up call" to the target cortical area that something new or unexpected is being relayed through the ... This may provide a basis for a form of coincidence detection. That is, coactivation of both pathways (i.e. the transthalamic gate is open) could lead to strong activation of the target area, activation that leads ultimately to the linking of the two target areas: conversely, if the thalamic gate is closed, the response in the target area is too weak to support such a linking."
> *Sherman and Guillery (2011)*

This is eerily reminiscent of our hypothesis of what the claustrum does, linking two cortical areas by possibly a different basic mechanism (promoting synchronized oscillations rather than simple addition). Previously we had concentrated upon sensory and limbic salience mechanisms to the neglect of the motor salience system that Sherman and Guillery (2011) brought to our notice.

We must next deal with the question of how the motor part of the system recognizes that the motor instruction is unexpected. This requires a mechanism, template, or failure-to-coincide detector that compares the pattern of neuronal excitation present in the system with what the system computes should be present. The authors suggest that their system does it when a strong stimulus is received after a period of relative quiescence. However, this may not do, as the strong stimulus has to be unexpected – and just following a period of relative quiescence does not supply that information.

In our account of the claustral system we did not specify where the saliency detecting mechanism is located except that it needs to be early. However, the sensory and motor systems need different types of salience detection. In the sensory system, because "reality" is provided by the sense organ, the detector needs to be as close to the sense organ as possible in order to provide rapid service. In contrast, in the motor system, because "reality" is provided by the motor cortex (in the form of the actual motor instructions), the detector has to be located as near downstream to the motor cortex as possible – which would seem to be in the HOTR. This location may be also where the transthalamic switch is located. If this is so, then a mismatch detected would send a signal to the rest of the cortex by the transthalamic route "unexpected motor instruction received."

In our account of the salience detection used on the sensory side by the claustrum, it seems possible that it is located in a thalamic relay nucleus or in S1 cortex. Detection of salience by this mechanism could potentiate ongoing claustro-cortical oscillations, as we described earlier. So what could be the relation between the transthalamic and claustral systems? Are they quite separate, or are they linked in some way? We suggest they are linked because:

- the claustrum and pulvinar have extensive interconnections (Druga, 1972; Flindt-Egebak and Olsen, 1978). (No information is available as to the status of the claustral connections of the other two HOTRs.)
- all cortex, including sensory areas, sends layer 6 projections to subcortical motor centers. Sherman and Guillery (2011) stated that all cortex is sensorimotor (the differences between sensory and motor are only quantitative), and that interactions between the two systems occurs at all levels.

We suggest therefore the following scenario: When an unexpected sensory input is received, claustro-cortical oscillations are potentiated by the saliency mechanisms, and recruit sensory mechanisms to deal with it, as described earlier in this chapter. The claustrum may also send impulses directly to the lateral posterior pulvinar with the message: "Unexpected (salient) stimulus has been received – prepare for unusual motor commands." Likewise, when an unexpected motor instruction is detected by the motor saliency system, it may activate parallel tracts in reverse, opening up the transthalamic gate. This relays the message to the claustrum: "An unexpected motor instruction has been detected – prepare for unusual sensations." Likewise, in the emotional and sensory sides, similar messages could be sent by the claustro-cortical system to the rest of the pulvinar (see next section).

In line with this, Cappe et al. (2012) have suggested a mechanism whereby the thalamus, rather than the claustrum, can act an integrator of synchronized oscillations. They said:

> Some restricted thalamic territories send divergent projections to cortical areas and thus could afford different sensory and/or motor inputs which can be mixed simultaneously. This pattern could support a temporal coincidence mechanism as a synchronizer between remote cortical areas, allowing a higher perceptual saliency of multimodal stimuli.

A Comparison Between the Mode of Operations of the Pulvinar and the Claustrum

There seem to be some similarities in the way synchronized oscillations are modulated in the claustrum and the pulvinar. In an extended review of the latter Shipp (2003) makes the following observations:

> The pulvinar has two visual maps. Both reproduce the topology of the contralateral visual field and both are mirror images of each other. Both V1 and V2 are connected to both maps. Layer 5 pyramidal cells are driving neurons and layer 6 cells are modulatory neurons. There is a good deal of overlap between fields.

He then presents the "Replication Principle," stating: "If two cortical areas communicate directly they are likely to have overlapping thalamic fields [in the pulvinar], if not their thalamic fields avoid each other." This is the very much the same as Pearson's rule for the claustrum. It also suggests that the C neurons in the pulvinar, which carry the X synapses (as described earlier), may consist of two subpopulations, C_1 and C_2. The C_1 neurons connect with As and Bs in the cortex that are themselves connected. In contrast, C_2s connect with As and Bs in the cortex that are not so connected. This brings the thalamic and claustral systems even closer into line. However, it is not clear if the difference between the two systems may lie in the fact that, whereas cortico-claustro-cortical connections are bidirectional, cortico-pulvinar connections may be unidirectional.

Shipp then suggests that the pulvinar may facilitate synchronized neuronal firing, and thus play a coordinating role in cortico-cortical connections. In detail, he proposes that input from layer 6 pyramidal cells in one visual area (e.g. V3) will synchronize those pulvinar cells that are activated by layer 5 pyramidal cells in another visual area (e.g. V1). This results in "coherent oscillations" between V3 and its pulvinar connection zone. In addition, feedback connections from V1 to layer 1 cells in V3 "ultimately recruit elements of V1 to the coherently oscillating neural assembly." He continues:

> ... the innate rhythmicity of thalamic neurons enables the c-p-c link to facilitate long range synchronization of cortical activity, but it is less important in specifying the sensory tuning of that activity ... Any pulvinar element can act to facilitate large-scale synchronization across several distant transcortical networks.

These networks compete with each other for P cells. This, he concludes is the basis for attention.

What are the Similarities and Differences in Function between the Claustrum and the Pulvinar?

The following facts may be relevant. To begin with, the pulvinar nucleus can be considered to consist of three functional parts:

1. The *medial pulvinar* has extensive connections with many paralimbic cortical structures and higher multisensory cortex (Mufson and Mesulam, 1984) and is involved in emotional salience estimation

(Padmala et al., 2010). It has strong connections with the amygdala (Rosenberg et al., 2009).

2. The lateral posterior area is a higher order thalamic relay nucleus involved in motor salience, as described in detail above (Sherman and Guillery, 2011). The "motor saliency" part of the pulvinar, which we have discussed in relation to Sherman and Guillery's (2011) account, would seem to be limited to only one section – the *lateral posterior pulvinar*.

3. All other parts of the pulvinar constitute a *visual area*. This is involved in promoting competitive, large-scale synchronizations across several distant transcortical networks (Shipp, 2003). These may be involved in information processing. Shumikhina and Molotchnikoff (1999) present evidence that the lateral posterior pulvinar acts to synchronize gamma oscillations in cortico-pulvinar circuits. In a recent study Saalmann et al. (2012) present evidence that the pulvinar promotes cortical alpha-band synchronization that subserves communication of attended information. During a behavioral task involving spatial discrimination of visual stimuli, two different cortico-thalamic systems were synchronized by beta oscillations at different frequencies. One was composed of cortical area 17, the ventromedial region of the caudal part of the lateral zone of the pulvinar, and the medial suprasylvian sulcus. The other was composed of area 18 and the dorsolateral part of the same region of the pulvinar (Wróbel et al., 2007). This functional mechanism seems to be eerily similar to that which we propose for the claustrum.

The pulvinar is also involved in "distractor filtering" (Fischer and Whitney, 2012; Strumpf et al., 2012), contrast processing (Cortes and van Vreeswijk, 2012), and interhemispheric integration (Shinomoto et al, 2009). Items of note are:

- The input to parts 2 and 3 of the pulvinar are mainly visual, whereas all cortical areas, and many subcortical areas, project to the claustrum.
- The cortical input to the pulvinar comes from both layer 5 and layer 6 cortical P cells, whereas the cortical input to the claustrum comes only from layer 6 pyramidal cells.

Note that Sherman and Guillery (2011) state that layer 6 pyramid cells are type 2 glutamatergic neurons, are purely modulatory, and cannot activate neurons by themselves – only type 1 glutaminergic neurons can do that. This seems to pose a problem for the claustrum:

- either the layer 6 P cells that project to the claustrum are exceptions to this general rule, which seems unsatisfactory, or
- the cortical input to the claustrum is only modulatory. So the excitatory input must come from the subcortical input to the claustrum, which seems even more unsatisfactory.

This question can be settled because type 1 and type 2 boutons are different: the former are large and scarce, the latter are small and plentiful. On this matter Sherk (1986) says only that the cortico-claustral synapses are asymmetrical, which fits both categories. More information is clearly needed.

Some other facts may also be relevant:

- The pulvinar nucleus projects to the striatum and the lateral amygdala, potentially relaying: (1) topographic visual information from the superior colliculus to the striatum to aid in guiding precise movements, and (2) non-topographic visual information from the superior colliculus to the amygdala that would alert the animal to potential danger (Day-Brown et al., 2010).
- The dorsal aspect of the pulvinar nucleus is also a critical hub for spatial attention and selection of visually guided actions (Wilke et al., 2010).
- (Kaas and Lyon, 2007) present evidence that a subset of medial nuclei in the inferior pulvinar function predominantly as a subcortical component of the dorsal stream, whereas the most lateral nucleus of the inferior pulvinar, and the adjoining ventrolateral nucleus of the lateral pulvinar are components of the ventral stream of cortical processing. The authors conclude: "These nuclei provide cortico-pulvinar-cortical interactions that spread information across areas within streams, as well as information relayed from the superior colliculus via inferior pulvinar nuclei to largely dorsal stream areas."

In summary, it does not seem clear why the brain has two such apparently similar mechanisms. One hypothesis to explain these facts might be that part 3 of the pulvinar integrates oscillations mainly for the intra-modal visual system, whereas the claustrum does so mainly on the long-range intermodal scale. Furthermore, the claustrum may be more concerned with decision making, the initiation of voluntary actions (see further below) and higher cognitive functions.

The Thalamo-cortical System Suggests a Variation of our Hypothesis Relating to the Claustrum

Our current hypothesis relating to the function of the claustrum follows the Pearson rule. There are two cortical areas A and B and two claustral areas C and D. These have bidirectional connections as follows: A connects to B and to C. B connects to A and to D. C connects to D (Figure 13.2A).

Weak oscillation synchrony between A and B is potentiated by stronger oscillation synchrony between C and D. In this model there are no necessary main connections between C and B, or between D and A. However, many neurons in the claustrum have axons that bifurcate and

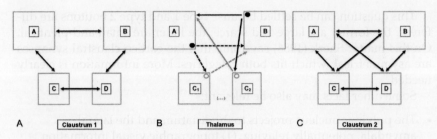

FIGURE 13.2   A: Diagram of the first level of proposed connections between two corti-cal areas A and B and corresponding claustral areas C and D according to Pearson's rule. B: Diagram of thalamo-cortical connections according to Sherman and Guillery (2011). The small ring in C_1 and C_2 represents the thalamic gate they propose. C: Diagram of second level of proposed connections between cortical areas A and B and corresponding claustral areas C and D.

supply two different areas of the cortex (Divac, 1979; Divac et al., 1978; Rahman and Baizer, 2007; Sloniewski et al., 1986).

Sherman and Guillery (2011) describe a different system in the thala-mus. Here, there are two cortical areas, A and B, and two subcortical (HOTR) areas, C_1 and C_2. These have unidirectional connections as fol-lows. A connects to B both directly and via a synapse on C_1. Likewise, B connects to A, both directly and via a synapse on C_2. In contrast, C_1 and C_2 have only modulatory connections (Figure 13.2B).

If we transpose this model to the claustrum, synchronized oscillations between A and B (driven from A over the direct route) may be potenti-ated by a second input from A routed via an indirect route involving a synapse on C in the claustrum (Figure 13.2C). Likewise, such synchro-nized oscillations between B and A, driven from B over the direct route, may be potentiated by impulses over an indirect route involving a syn-apse on D in the claustrum. This is in accord with the reported bifurca-tion of claustro-cortical efferents, and with the weak intraclaustral axonal interactions referred to by Sherk in Chapter 5 of this book. However, long-range intraclaustral connections have been reported by Smith and Alloway (2010), Behan and Haberley (1999) and Zhang et al. (2001). Furthermore, the intraclaustral connections between C and D, postulated by our main hypothesis, must be important, otherwise the Pearson rule would be invalid. Therefore, our ancillary hypothesis suggests that both mechanisms may operate at the same time, involving both direct C–B and D–A connections as well as significant modulations between C and D carried by the GABAergic IN syncytium.

However there is one major difference between the thalamo-cortical and the claustral systems that we discussed earlier. Cortico-thalamic axons derive from P cells in cortical layer 5, whereas cortico-claustral axons derive from cortical layer 6 P cells (Sherman and Guillery, 2011).

Spike-Train Coding

We said earlier that the claustrum does not necessarily have to process the spike codes it receives along its afferent axons. However, we are not ruling out the possibility that, on occasion, it may do so. In the hippocampus the spike code carried by the afferent CA3 Schaeffer pathway to CA1 is partially transmitted into the output spike code of the CA1 cells in a process that involves gamma oscillations (Fernández-Ruiz et al., 2012). If this same operation takes place in the claustrum, output spike codes of an individual claustral P cell could "carry over" the spike codes of its afferent axons (or some of these at least). It would deliver these output spikes, via its axon collaterals, to a "hierarchical assembly" of neighboring P and IN claustral cells of different cell types. The authors suggest that, even if mixed messages pass to different target cells (in this case different types of claustral P and IN cells), specialized postsynaptic machinery may be tuned to decode specific parts of them by the kinetic properties of different postsynaptic receptors and intrinsic channels. This resultant activity might then modulate the spike coding of efferent axons that the P cells return to the cortex. In this way, the claustrum may process the information contained in the spike trains it receives, in addition to, but not instead of, processing the synchronization of these spike trains themselves. This might explain the plethora of different types of INs in the claustrum better than our principal hypothesis does.

OUR HYPOTHESIS AT THREE LEVELS

Thus our hypothesis suggests that the claustrum may operate at three exclusive levels. At the first local level it just magnifies the oscillations in cortico-claustral circuits. At level two it integrates these oscillations. At level three it processes the contained spike codes. Recent evidence appears at first sight to support the level one version. Foxworthy et al. (2013) have published the first wiring diagram of a multisensory cortex, in this case the rostral posterior parietal area (PPr). The highest proportion of multisensory neurons, and the highest incidences of multisensory response enhancement, and extrinsic inputs, are located in layers 2 and 3. In contrast, cortical layer 6 shows the highest proportion of unisensory neurons and the fewest external inputs. Layer 4 is minimal.

This pattern might seem to support the simplest form of our hypothesis, in which the claustrum is involved only in the magnification and selection of competitive synchronized oscillations. In the two more complex versions, we supposed that the claustrum is also involved in integrating synchronized oscillations received in its unimodal input to form new "bound" oscillations in its output back to the cortex. In area

17 these axons go to layers 4 and 6 (Sherk, 1986). This data is not available for multisensory cortex. In the PPr, layer 4 is "mimimal," and the bulk of the input to layer 6 is described by Foxworthy et al. (2013) as unimodal. However, this discussion depends on how "unimodal" and "multimodal" are defined. One distinction that is appropriate for sensory cortex depends on how the neurons respond, either by firing or by response enhancement, to different sensory stimuli. A monosensory neuron only fires in response to activation of one sensory input. A multisensory neuron fires in response to activation of any one of more than one direct sensory inputs. However, when a cortical neuron in layer 6 of the PPr cortex is activated by a claustral axon, it is no longer acting as any kind of sensory neuron, any more than a neuron in the motor cortex activated from the claustrum is. Thus, if an experimenter records from such a neuron in layer 6 of the PPr cortex, it might respond as a monosensory neuron to any activation of the direct sensory pathway leading to it. But, if the claustro-cortical pathway leading to it were activated, it might also respond – only this time as a non-sensory neuron. Furthermore the distinction between unisensory and multisensory is not clear cut. The data from deafferentation experiments (Budinger et al., 2006; Clavagnier et al., 2004; Collignon et al., 2009; Driver and Noesselt, 2008; Falchier et al., 2002; Ptito et al., 2008; Rockland and Ojima, 2003) show that all sensory neurons are multisensory, only in some the extra ones are latent and only become functional when the principal pathway is removed.

Our hypothesis at level 1 would explain the data presented by Foxworthy et al. (2013) as follows. In multisensory cortex (e.g. the PPr), a unimodal input from V1 to layer 6 activates the layer 6 P cells. These transmit the oscillating signal on to the claustrum. Here, the different oscillations, set up by such messages from all over the cortex, interact and compete. Their efferents, carrying the magnified results, return to cortical layers 5 and 6 in the areas from which the initial impulses came, to repeat the cycle. Then, somehow in the winning cycle, this increased oscillatory activity in the winning locale modulates the activity of the overlying layer 2/3 neurons so as to increase the oscillatory activity there, and so the bound signal is transmitted to higher centers. This fits the new data and removes the need to have two binding systems – the one described by Foxworthy et al. (2013) in the cortex and another in the claustrum. This new data emphasizes the need to obtain new information as to the mode of termination of the claustro-cortical projection in multisensory cortex. However, this "fit" does not establish *per se* the level 1 hypothesis. This is because, in the second level hypothesis, for reasons just described, the claustro-cortical axons are carrying a code that is neither unisensory nor, in the ordinary sense, multisensory. Likewise, in the level 3 hypothesis, these axons are carrying spike trains whose temporal code is neither unisensory nor, in the ordinary sense, multisensory. Many more experiments are needed to determine which, if any, of these three hypotheses is correct.

CLINICAL EVIDENCE FOR BINDING

We know from the work of Schilder (1942) and others on the manner in which sight returns following injury to the occipital lobe that the visual field in consciousness is constructed by a tripartite mechanism. After such an injury, the first aspect of sight to recover is movement perception. In this the subject sees pure motion, usually rotary, without any shape or color. Then "space" or "film" colors appear floating about in visual space unattached to any objects. Next, the subject starts seeing parts of objects, say the handle of a teacup. Lastly these parts join up to form complete objects into which the space colors enter. So activity in the visual field in consciousness can be driven solely by one of these parameters. Their binding into one complete colored, moving object comes later as the cortical areas concerned recover from the injury.

Two Binding Systems?

We now know that not only higher cortex but also probably most of the S1 cortex is polysensory (Budinger et al., 2006; Clavagnier et al., 2004; Collignon et al., 2009; Driver and Noesselt, 2008; Falchier et al., 2002; Ptito et al., 2008;Rockland and Ojima, 2003). Ghazanfar and Schroeder (2006) concluded that, "The nature of most, possibly all, of the neocortex forces us to abandon the notion that the senses ever operate independently during real-world cognition." Thus the computations necessary for binding might all be done "in house" within the cortex, in which case we need to explain the need for a second claustral binding mechanism. To approach this question we will first consider the visual system.

There is evidence that the claustrum may have two different operative visual systems. The "lower" visual cortex (i.e. those areas that receive direct projections from the lateral geniculate body; LVC) connects with the dorsolateral claustrum, whereas those "higher" visual areas (HVC) that lack a direct input from the lateral geniculate body connect with the ventral claustrum (Sherk, 1986). These higher visual areas include areas 20a, 21a, 21b, and the posteromedial and posterolateral areas of the lateral suprasylvian gyrus (PMLS and PLLS, respectively), as well as the precuneus (area 7) – a multisensory area with visual and somatosensory inputs. Moreover, the precuneus itself has a topographically organized visual field map, as does its projection to the claustrum. Therefore the claustrum appears to have two visual field maps, one connected to the LVC and the other to some parts of the HVC (Sherk, 1986). It is possible that circuits between "unbound" unisensory LVC and the claustrum may have a different function than circuits connecting "bound" polysensory HVC and the claustrum. The former may have more to do with "binding," and the latter with providing the information regarding where particular objects are located. In polysensory cortex the "binding" would be performed by multimodal feed-forward nerve nets. In the LVC (neurons with small receptive fields) the sensory data is unbound, whereas in the multisensory HVC (neurons with large receptive fields) the sensory data is already bound intracortically by multiple crossing feed-forward connections. So the cortico-claustral circuits in the LVC may cooperate in the binding, whereas, in the HVC case, they may be more concerned with providing information about the spatial location of the stimuli, as described earlier. These same considerations may apply to the difference between primary unbound lower sensory cortex and higher bound sensory cortex in general.

The Role of GABAergic Neurons

Full coverage of the GABAergic system in the claustrum is the subject of Chapter 3 in this book. We will cover only some functional aspects relevant to our hypothesis in the present chapter.

Our hypothesis allots a key role to GABAergic INs in the claustrum. In the cortex and the lateral geniculate nucleus GABAergic INs have been found to form extensive polysynaptic bidirectional networks linked by electrical junctions (Fukuda et al., 2006). The authors suggested that these networks support "… the precise synchronization of neuronal populations with differing feature preferences thereby providing a temporal frame for the generation of distributed representations." Such gap-junction linked networks can either promote network synchronization or trigger rapid network desynchronization, depending on the synaptic input (Vervaeke et al., 2010).

In mouse frontal cortex Fino and Yuste (2011) report dense connectivity of GABAergic somatostatin-positive INs. They found that every IN was connected to every pyramidal cell within the range of its axonal tree. They reported:

> In fact, the complete connectivity that we observe appears in some cases deterministic, as if the circuit has been built to ensure that every IN is connected to every single local PC cell … in this way neighboring neurons would have overlapping but not identical connectivity patterns. *Fino and Yuste (2011).*

It has yet to be determined if this applies to the claustrum.

The cortico-claustro-cortical system may be an example of a strong feed-forward inhibitory circuit (FFI) (Bruno, 2011). An FFI is composed of a group of presynaptic neurons that directly excite both glutamatergic excitatory neurons (P cells) and GABAergic inhibitory INs, and provide greater synaptic input to the latter. The postsynaptic neurons are interconnected. Circuits that lack inhibition simply relay presynaptic activity to postsynaptic neurons. In contrast, postsynaptic neurons in an FFI are highly sensitive to the relative timing of action potentials, and the synchrony, transmitted by the presynaptic neurons (Bruno, 2011). Neuromodulators and feedback connections may modulate the temporal sensitivity of such circuits and gate the propagation of synchrony into other layers and cortical areas. The prevalence of strong feed-forward inhibitory circuits throughout the central nervous system suggests that synchrony codes and timing-sensitive circuits may be widespread, occurring well beyond sensory thalamus and cortex (Bruno, 2011). However, the evidence is currently lacking as to whether cortico-claustral axons provide greater synaptic input (needed for a strong FFI system) to the inhibitory neurons than to the excitatory neurons. There are considerably more P cells than INs in the claustrum. However, the relative importance of the number of INs and their degree of connectivity needs to be determined. A report by Koós and Tepper (1999) may be relevant to this point. They found that inhibitory synaptic potentials generated from single INs are sufficiently powerful to delay or entirely block the generation of action potentials in a large number of projection neurons simultaneously.

The possible role of Class 1 and Class 2 glutaminergic neurons in the claustrum have not as yet been studied. Sherman and Guillery (2011) noted that some type 2 glutaminergic neurons are inhibitory. They wrote: "Such inhibition has always been considered strictly in terms of GABAergic circuitry. The action of class 2 inputs that activate group II metabotropic glutamate receptors offers the possibility that these glutamatergic pathways can also contribute to such functions". This may be relevant to the claustrum.

Kappa Opioid Receptors

An interesting clue regarding the inhibition discussed in the previous section may be provided by recent findings concerning the psychoactive drug salvinorin A. This is a specific agonist at kappa opioid receptors and, psychologically, it induces an intense sensory synesthesia in which subjects claim that they see sounds and hear sights (Babu et al., 2008; also see Baron-Cohen, 2008, and Hubbard and Ramachandran, 2005). This may be interpreted as an inhibition of sensory binding. In our present context it is interesting that activation of the kappa opioid receptor has been shown to inhibit the release of GABA in the bed nucleus of the stria terminalis by a presynaptic mechanism (Li et al. (2012)). If the same holds in the claustrum, this would provide direct evidence that the GABAergic system in the claustrum may be related to sensory binding.

The claustrum contains particularly high levels of mRNA for the kappa receptor (Mansour et al., 1994; Meng et al., 1993). The binding of kappa1 opioid-stimulated [35S]GTPgammaS (a marker for the kappa opioid receptor) is also particularly high in the ventral claustrum (Sim-Selley et al., 1999).

The Relevance of Illusory Conjunctions and Attention

Attention plays a prominent role in binding. Vohn et al. (2007) examined the fMRI response in normal volunteers to three modes of divided-attention tasks. Two were within-modal (vision/vision and auditory/auditory) and one was cross-modal (auditory/visual). The results were that the three divided attention tasks, irrespective of sensory modality, revealed significant activation in a predominantly right hemisphere network involving the prefrontal cortex, the inferior parietal cortex, and the claustrum. Under the cross-modal condition, however, the frontal and parietal activation was more extended and more bilateral, and there also was stronger right hemisphere activation of the anterior cingulate cortex and the thalamus. This might suggest a different mode of action of the integrating function of the claustrum in some cases. In focused attention tasks, the claustrum is associated with activation of higher visual and the auditory cortices (Emrich et al., 2006). The divided attention tasks (both unimodal, and more so in cross-modal tasks) would seem to activate higher multisensory cortices.

Also relevant to the binding problem is the question of "illusory conjunctions" (Crick and Koch, 2003; Treisman and Schmidt, 1982). If separate features of an object (e.g. "red square" or "moving left") are processed in different brain areas (i.e. the color area V4 and the motion area MT), which results in the loss of their topographic location label, and if two objects are simultaneously presented – for example, a red one moving left and a green one moving right, then how does the brain compute which color goes with which motion? One possibility is that Crick's "searchlight" of attention is directed toward different portions of the visual scene at an early stage of processing when topographical information is still present (e.g. area 17 of the V1 cortex, or the claustrum). If the spotlight permits "red" and "left" to go through, then they will be bound. This explains why, if a red triangle and green square are briefly presented and masked, subjects see illusory conjunctions in 50 percent of trials. This is because there has been enough time to process the features separately, but not enough time for (in our scheme) claustro-striate iterations to ensure correct binding through attention. Crick and Treisman suggested that the nucleus reticularis thalami is involved. We would argue, from the current evidence, that ascending brainstem efferents to the claustrum interacting iteratively with topographically organized area 17 may constitute the searchlight (see also Smythies, 1997).

Using diffusion tensor imaging and local recordings in monkeys performing a visuospatial attention task, Saalmann et al. (2012) report that the pulvinar synchronizes activity between interconnected visual cortical areas according to attentional allocation. This suggests that the claustrum may specifically be involved in longer distance intermodal synchronizations, predominantly at the gamma range. This may avoid duplication between what the pulvinar and the claustrum are doing.

Some difference in the motor and sensory functions of the claustrum may be suggested by the observations that the claustral neurons projecting to the contralateral motor cortex are predominantly pyramidal in shape, whereas the predominant claustral neurons projecting to the somatosensory, visual, and auditory cortices are mainly oval in shape (Sadowski et al., 1997).

Naghavi et al. (2007) have reported that, although the claustrum/ insula area (CIA) was activated by attentionally focused congruent stimuli (e.g. the sight of a cat's face and hearing a meow) it failed to be activated by attentionally focused, non-congruent stimuli from the same location in external space (e.g. the visual stimulus of a cat's face with the contemporaneous auditory stimulus of a dog's bark). They suggest that this indicates that the CIA must be involved in the "analysis of the content of stimuli." However, it must be noted that the CIA did not react at all to the cat's face or dog's bark. So it can hardly be said to have extracted the information "cat" from its visual input, and "dog" from its auditory input, compared the two, computed that they were incongruent and rejected them from its binding operation. Furthermore, we suggest that the CIA simply does not have the computational capacity to decipher all the immense flow of information that passes through it. The decision that "cat" and "bark" do not go together must have been made in the higher cortex before it activated the CIA.

So we would suggest that the CIA does not react in this experiment to "cat/bark" and "dog/meow" because these improbable signals have been suppressed upstream. This would be an example of the same sort of mechanism that operates in the experiment reported by Kovács et al., 1996). In this experiment they took two photographs – one of a monkey's face and the second of a leafy tropical jungle – and converted these into two pastiches, each composed of portions of each photo, so that in the location where one photo showed part of the monkey's face the other showed leafy jungle. Then each pastiche was shown separately to each retina, so that retinal rivalry occurred. Under these circumstances, the subject did not see what was actually there (i.e. the two pastiches alternating) but rather a complete monkey face alternating with a complete leafy jungle. Clearly the brain had suppressed the improbable mixed pastiche in favor of what it was familiar with (and thus computed to be more probable). Many other experiments, based on stimuli such as moving plaid patterns, have shown this phenomenon, where the perception of an improbable input is suppressed by the brain, and replaced with the perception of what it computes to be a more probable one (see Ramachandran and Anstis, 1983).

THE ROLE OF THE CLAUSTRUM IN COGNITIVE PROCESSING

So far we have focused upon the original Crick-Koch hypothesis that is based on the idea that the claustrum is involved in binding features of the sensory stimulus. We have suggested that a spike-burst synchrony detecting, modulating and transmitting mechanism may be involved. However, it is clear that the claustrum is involved in many other processes besides sensory binding. The claustrum has bidirectional connections with non-sensory limbic, temporal and frontal cortices as much as with the sensorimotor cortex. The claustrum also has a massive input from all the major neuromodulator circuits (Baizer, 2001; Das, 2010; Edelstein and Denaro, 2004; Eggan and Lewis, 2007; Gill et al., 2006; Lacković et al., 1990; Meador-Woodruff et al., 1991; Sutoo et al., 1994). The ventral claustrum has extensive connections with limbic areas, such as the anterior cingulate gyrus, amygdala, hippocampus, and others. Why would this information, relating largely to cognition and emotions, reinforcement and motivation, be required by a system concerned with sensory binding? One answer might be that the claustrum is composed of synchrony modulators involved in binding between limbic operations (emotions, etc.) and the sensory and motor systems. The dorsal claustrum has connections mainly with sensory and motor cortices, the ventral claustrum mainly with limbic cortex and subcortical structures, and the rostral claustrum projects mainly with frontal cortex (LeVay and Sherk, 1981; Sherk, 1986). The anterodorsal, posterodorsal, and part of the ventral subdivisions could be concerned with sensorimotor binding in one of the two manners we described earlier (involving intermodal or intramodal synchronization, respectively; p. 315). The anterior and another part of the ventral subdivisions could be concerned with reinforcement and other limbic functions. Thus the claustrum could cooperate with the sensory cortex for sensory binding, with the limbic system for emotional coordination to allow modulation of behavior by complex patterns of reinforcement, and with the decision-making mechanisms in the prefrontal cortex to coordinate "higher brain" functions (in response in each case to specific inputs from these different loci).

Our hypothesis suggests that, in these cognitive and limbic processes, the role of the claustrum may be put as follows. The claustrum may become activated whenever a computational process involves more than one brain area. If two such areas are involved, for example, the synchronized activities in each are coordinated by the claustrum in the manner we have described – scattered weak intercortical synchronizations are potentiated and processed by strong intraclaustral synchronizations. We can call this "cognitive binding."

Recent research suggests what some of these "higher brain functions" may be. Volz et al. (2010) carried out fMRI studies of retrieval fluency in normal subjects. This is defined as how long it takes to retrieve a trace from long-term memory. If one of two objects is retrieved more quickly this indicates that that one has a higher value than the other. This measure was accompanied by another that assessed the emotional feeling in the subject of the "rightness" of that memory (a "feeling-of-knowing" judgment). The authors suggested that a number of brain processes could contribute to these measures including "… the ease with which such memories are bound together." Their results were that these procedures were associated with activation of the dorsal claustrum, but not the ventromedial prefrontal cortex. The authors suggested that their findings indicate that the claustrum may also bind semantic and emotional information in addition to sensory information. Tian et al. (2011) investigated another possible function in which the claustrum may play a role – the mental preparation of successful insight problem solving (that is the "Aha" experience as exemplified by Archimedes). Using fMRI they showed that successful preparation coding was associated with activation of a circuit that included parts of the frontal and temporal cortices, the cerebellum and the bilateral claustrum. The claustrum is also engaged in the rapid interhemispheric transmission of information needed for the bilateral coordination of movement regulation (Smith and Alloway, 2010). The mechanism we propose might affect this by "binding" information from two inputs, one from each ipsilateral motor system.

Our postulated claustral synchrony detection and augmenting mechanism, therefore, may be plugged into a number of different neural circuits engaged in different distributed computations. This might be done at two levels – local and global. At the local level the claustrum merely supplies a synchrony modulating mechanism, as described above, for each individual circuit with no integration in the claustrum between the circuits. For example, one such circuit could be the RSTG, RHG, insula, and claustrum circuit, as described by Emrich et al. (2006) quoted above. At the global level the claustrum also operates at the local level, but, in addition, internally integrates the activities generated by other circuits. For example, the circuit described by Emrich et al. (2012) would recruit a circuit involving expectancy and saliency (e.g. the anterior cingulate gyrus) and higher mental functions including memory (e.g. the prefrontal cortex), as well as the motor cortex. Sejnowski and Paulsen (2006) proposed that, in order for the brain to route sensory inputs to motor output correctly, the brain needs to have a fast and flexible system that links different cortical areas, and controls attention, expectation, and memory. We suggest that the claustrum may play a key role in providing this system.

THE SALIENCY DETECTION HYPOTHESIS

We agree with the hypothesis that one function of the claustrum may be saliency detection. However, as we have seen, it is not necessary for Crick and Koch-type binding that each claustral cell should receive axons from both cortical areas whose functions are to be bound. This binding may be effected inside the claustrum in the manner we have suggested. Also, even if the input to claustral cells is unimodal, the output axons from claustral P cells commonly bifurcate and project to more than one modal type of cortex (Divac, 1979; Divac et al., 1978; Rahman and Baizer, 2007; Sloniewski et al., 1986). The experiments reported by Remedios et al. (2010) only produce strong evidence that the input to individual claustral P cells is not intermodal. However, this evidence is not relevant to the possibility of intraclaustral binding operations, nor to the question of whether the output of single claustral P cells projects to more than one cortical modality.

Moreover, in support of the saliency hypothesis, spike-timing (synchrony) codes may carry information in addition to that relating to the properties of the stimulus. Doucette et al. (2011) suggest another account of what information synchronized spikes may be carrying. This leads to a wider concept of what the function of the postulated spike-burst synchrony detection mechanism of the claustrum might be. Doucette et al. (2011) studied spike-timing codes in the olfactory system and came up with a challenging result. They reported that the number of synchronous spikes (SS) fired by pairs of olfactory bulbar neurons signals, not stimulus properties, but the salient information of whether the odor is associated with reward or not. The SS fall below the spontaneous activity level for unrewarded odors and rise above it in the case of rewarded odors. The authors suggested that this is an easily understood and implemented population temporal code, the decoding of which simply requires downstream coincidence detectors connected to decision-making networks, which take input from both members of the neuron pair. Doucette et al. (2011) then found that the SS rate is modulated by noradrenergic input to the bulb. In this way the SS code can signal the reinforcement significance of the stimulus. Katz and Maier (2011) commented in the same issue of *Neuron*: "It is also unclear whether, when coherently firing neurons are studied in larger ensembles, the observable patterns will become more complicated." They concluded: "Thus, these are important, novel data added to a growing corpus suggesting that '"sensory"' coding is as much about the stimulus in context as what the stimulus physically is." That is to say that the reinforcement significance of most stimuli will be affected by other events happening in the environment. For example, the last meal chosen and eaten by a condemned prisoner will not carry the same reinforcement value that it should.

The role of synchronized oscillations in coding for salience may not be in the olfactory bulb. In an ingenious experiment Bosman et al. (2012) employed two almost identical visual stimuli (using colored gratings) that activated two different V1 sites and one common V4 site. If either V1 site was activated by itself, it activated the V4 site: and the V1 and V4 sites synchronized at 60–80 Hz. If both V1 sites were activated simultaneously – with the stimulus to one signaling reinforcement (A) and the other unconnected with reinforcement (B) – oscillations in V4 synchronized with oscillations in A but not with oscillations in B. The gamma peak frequency in A was 2–3 Hz higher than the peak frequency in B, and 4–6 Hz higher than in V4. These interactions were predominantly directed from V1 to V4.

So how does V4 know which of the two rival gamma synchronizations sent up from V1 in this situation is the behaviorly significant one? Either this signal must be carried by the spike trains and/or gamma oscillations of the packet from A (and not by the packet from B) or else it must come from another parallel neuromodulatory channel. The experiment was designed so that the same neurons can carry A signals or B signals depending on the stimulus. So it seems unlikely that any hard wiring evoked by epigenetic changes is involved in the differential response of the M4 site when both V1 sites are activated together. So what could be the signal? As the relevant V1 site had its gamma peak frequency 2–3 Hz higher than the irrelevant V1 site, the signal may be a higher gamma peak frequency. But how was this higher frequency imparted into the V1 population concerned? How do the molecular mechanisms that mediate reinforcement signaling lead to a higher gamma peak frequency in V1 on a dynamic, not fixed wiring basis? Also how does the reinforcement memory imprinted by the training sessions impact on this system? Dopamine is a prominent player in reinforcement mechanisms. Dopamine promotes gamma synchronization levels – but does it raise gamma peak frequencies?

A clue may be provided by the experiments reported by Douchette et al (2011), described above. These showed that the information as to the salience value of a stimulus can be carried by the degree of synchronization of the spike trains of the efferent axons from the olfactory bulb. By extrapolation to the V1 to V4 connection, we can suggest that the high-salience spikes from A are more synchronized than the low salience spikes from B. This high-salience signal may promote the observed synchronization between V4 neurons and V1 A neurons. In the olfactory bulb this response is modulated by norepinephrine (noradrenaline). Perhaps in the visual cortex it is modulated by dopamine, or indeed norepinephrine itself. Hajós et al. (2003) used extracellular single-unit recordings from the medial septum/diagonal band of Broca (MS/DBv), with simultaneous hippocampal EEG recordings of anesthetized rats. They found that systemic administration of reboxetine (a specific norepinephrine reuptake inhibitor) synchronized hippocampal EEG, resulting in a significant increase in power at theta frequency, and an increase *in frequency* and power of gamma-wave activity [our emphasis].

Perhaps, then, the claustrum could be functioning in some locations as such a reinforcement-related "downstream coincidence detector" on a global scale in larger assemblies? In some areas (such as the anterior and part of the ventral claustrum) the spikes impinging on the synchrony detectors in the claustrum may carry salient information relating to the reinforcement value of the stimulus rather than to the sensory properties of the stimulus. The function of this might be to provide a metastable network that enables the brain to compute the significance of a stimulus in its complex global context. That is, a stimulus can be signaling reward, or the opposite, depending on its contextual environment – on what surrounds it. In other words, the brain must have a global mechanism that needs input from other sensory systems to evaluate the reinforcement significance of a single stimulus presented in a complex environment.

It is possible, therefore, that the claustrum might perform both functions described above – computing binding and computing global reinforcement – in different locations.

We also need to enquire further how reinforcing and salient stimuli could affect this system. The cortex receives widespread projections from dopamine neurons in the ventral tegmental area (VTA) that are activated when a stimulus with more than expected reward is received. Representations of sensory stimuli in the cerebral cortex can undergo progressive and extensive remodeling according to the behavioral importance of the stimuli (Bao et al., 2001). There are several reports that dopamine promotes neuronal oscillations. At the microcircuit level, during striatal network activity, the selective activation of either D1 or D2 receptors results in an increase of neuronal synchronization (Carrillo-Reid et al., 2011). In bullfrog retinal ganglion cells dopamine promotes synchronization (Li et al., 2012). Dopaminergic stimulation can promote, modulate or inhibit oscillatory activity depending on the structure innervated, and the frequency of the oscillation (Lee et al., 2004). For example, Cassidy et al. (2002) studied human patients following surgery for Parkinson's disease. Without medication, coherent oscillations were apparent between the subthalamic nucleus and the globus pallidus interna at <30 Hz. After exogenous dopamine stimulation the coherence frequency increased to 70–85 Hz. A neurochemical mechanism for this effect was suggested by Kuznetsova and Deth (2008). They found that dopamine, acting on D4 receptors, promotes synchronization of oscillations in cortical P cell–IN networks. This is mediated by modulation of phospholipid methylation, which alters the kinetics of potassium channels. P cell–IN networks are prominent in the claustrum. Perhaps the claustrum is involved in this type of salient activity?

For an example from another system, anxiety is reported to be associated with theta-frequency synchronization between the ventral hippocampus and the medial prefrontal cortex (mPFC) (Adhikari et al., 2010). These authors suggested that such synchronization is a general mechanism by which the hippocampus communicates with downstream structures of behavioral relevance. Other structures besides the hippocampus may do the same.

Therefore we fully agree with the conclusion by Remedios et al. (2010) and Remedios (2012) that the claustrum is involved in a variety of higher cognitive processes, including saliency-relevant ones.

ADVANTAGES OF OUR HYPOTHESIS

The advantages of our hypothesis seem to us to be:

- its simplicity of action and low computational cost. It does not necessarily need any complex mechanisms, such as axonal spike-time pattern analysis. The claustrum simply detects synchrony, processes intraclaustral synchronies, and promotes intermodal synchrony in functionally connected cortical and subcortical areas.
- it accounts for how one mechanism can exert functions that affect many "higher" brain functions.
- it gives a key role to the GABAergic IN system in claustral function.

In response to a question from a referee, "What makes the claustrum special?" we would reply that it reproduces, in a small volume, activity from all over the much larger volume of the rest of the brain. This promotes the development of multiple, fluctuating, and differential synchronous oscillations essential to the claustrum's postulated role. No other brain structure has this property.

We feel that we should mention that we have carried out an extensive exploration of other possible hypotheses in relation to the "binding" function of the claustrum, based on other fixed nerve net systems, spike-train spatiotemporal codes, and non-linear dynamics, but failed to develop any promising candidates in those fields.

PROBLEMS WITH THE HYPOTHESIS?

Duffau et al. (2007) conducted total unilateral excision of the claustrum and insula for glioma in 42 patients. They could not detect any neurological or psychological abnormalities post-operatively in any of the cases, who all returned to their normal professional lives, except for three who suffered a stroke due to vascular complications. The authors concluded that the system must have an effective compensatory system, but did not venture an opinion as to what this might be. We suggest that the answer may lie in the fact that the claustrum has extensive connections with the contralateral cortex, at least in the cat (Sherk, 1986). She states:

> "Norita (1977) found that all areas of cortex tested received input from two roughly corresponding regions in the two claustra, though the projection from the contralateral side was always considerably weaker. These findings have been confirmed by others, chiefly Macchi et al. (1981)."

EXPERIMENTS TO TEST OUR HYPOTHESIS

Central to our hypothesis is the question of how a neuron reacts if its input consists of two trains of spikes produced by two groups of afferent neurons oscillating at different frequencies. Gielen et al. (2010) conducted a theoretical and computer simulation study, and came to the conclusion that the target neuron(s) fired in a pattern that was highly phase-locked with the larger of the two afferent signals. The question then arises as to what is the result if they are of equal size. However, different types of pyramidal cell react in different ways, and there is currently no information as to which type the claustral P cells belong.

Professor Stan Gielen commented on this general problem (personal communication, 2013) as follows:

> The general structure of the claustrum as described is very similar to that of neocortical areas such as visual cortex with pyramidal cells and interneurons, where gap-junctions between interneurons facilitate oscillations within the network of interneurons. From visual cortex (areas V1, V2, V4 and many others) we know that the ratio of pyramidal cells versus interneurons is about 4:1. Moreover, there are different types of interneurons, and we gradually start to understand the possible role of various interneurons.
>
> To consider the question whether the claustrum (A) merely transmits, and distributes widely, the dominant information-laden packets it receives; or whether it can (B) amalgamate these packets and transmit the result. In my view, it might be able to do both. I will try to explain this in more detail below.
>
> First of all, I do not see how a single set of neurons with pyramidal (P) cells and interneurons (IN) might be able to process different signals encoded at different oscillatory frequencies simultaneously in parallel. Most people accept the view that

top-down effects related to attention are mediated by input to interneurons, which then start to oscillate at a particular frequency (ING-model). Then P-cells can only fire just before firing of the IN-cells, which implies that incoming oscillatory activity to the P-cells can only pass if it is in phase with the (more precisely: if its phase slightly leads) the phase of the IN-oscillations. It is quite likely that oscillatory input to the P-cells might synchronize after some time with the IN-oscillations, very similar to synchronization when a non-linear oscillator is driven by external oscillatory input. The input might synchronize with the oscillations of the driven oscillator or not depending on the difference between the frequencies of input and oscillator and depending on the strength of the coupling (see e.g. Pikovsky et al., 2001). Within this framework, the claustrum may certainly be able to transmit information (hypothesis A).

The next question is, what might happen if two different signals arrive at the P-cells, where the frequencies and phase of the two oscillatory inputs are different and also different from the oscillations of the IN neurons. Theoretically, if the frequencies of the two inputs are far apart and/or if the input to the P-cells is weak, nothing will happen: the output of the P-cells will be small and the network will not transmit any information related to the input. If the input frequencies are close to the oscillation frequency of the IN cells and if the input is sufficiently strong, then the P-cells will transmit the input (Gielen et al., 2010). I don't know exactly what might be meant by "amalgamating packets of neuronal information", but if the two inputs carry different types of information, the output might well be an amalgamate of the two inputs!

I hope this clarifies what I meant when I said that the claustrum might function satisfying both hypotheses A and B. If a group of cells receives two oscillatory inputs, one at 8 and one at 20 Hz, it will NEVER respond at a frequency somewhere in between (let's say at 14 Hz). What will happen is that the output is either at 8 or 20 Hz (some kind of winner take all, where the 8 Hz or 20 Hz depends on the strength of each of the synaptic inputs and the "natural" firing frequency of the group of output cells, or it will more or less randomly alternate between 8 and 20 Hz (like in the example of binocular rivalry).

In this context it may be interesting to know that although visual cortex mainly receives visual input, there is also a significant auditory input. If the visual and auditory input is related to the same object, responses of cells in visual cortex are much more vigorous than to the input of a single modality. Such a multi-sensory facilitation (or maybe similar events with different types of related signals) might happen in claustrum as well. This would fit with your hypothesis B, but at the same time would fit with hypothesis A if I am correct. I should add that the observation above about multi-sensory facilitation has been published in high-ranking journals, but the potential role of synchrony in this game has not been addressed yet. My guess is that neuronal oscillations are very important in this process. Moreover, it could well happen that there is a segregation of claustrum such that different parts of the claustrum transmit different inputs in parallel. In visual cortex, different parts of visual cortex may process different visual stimuli in parallel and it is highly unlikely (no one has shown so far!) that different visual inputs are processed by the same groups of neurons in parallel by neuronal oscillations at precisely the same frequency.

To respond to Professor Gielen's arguments we have suggested that, in the circuits inside the claustrum, groups of neurons oscillating at different frequencies "compete" for dominance. This involves two processes.

1. In a preliminary process, the units that will compete evolve out of activity in developing cortico-claustral-cortical circuits that integrate

two or more active areas involved in one common task (e.g. sight, sound and bodily sensations and movement in piano playing).

2. These integrated populations then compete inside the claustral syncytium for access to the final executive common path, i.e. the projection to the motor cortex, to control conscious behavior.

In this version of our hypothesis (level 1) the claustrum does not take part in processing the information carried in the packet of spike trains that form these synchronized oscillations. This information processing is all done in the cortex. The claustrum merely automatically estimates which packet is dominant, and transmits that to the motor cortex.

In another version of our hypothesis (level 2), we supposed that the claustrum does process information by 'melding' two different input frequencies into a derivative common frequency. In this scenario, one input packet carries the information, "A has occurred", and the other input packet carries the information, "B has occurred". Then the new output packet carries the bound information, "A and B have occurred." However, Dr Gielen's argument would appear to rule this possibility out.

This fits in with the "conductor of the orchestra" model. The conductor does not write or play the music, he merely chooses what is played, and who plays it. To be more exact, without the conductor the orchestra would consist of a lot of trios and quartets each playing their own pieces, trying very hard to follow what the others were doing. The conductor brings it all together. The implication for the claustrum here is that it does not just bind sensations, but it binds conscious behavior as well.

An extended answer to this problem might produce essential evidence as to the possible role of such a mechanism in the claustrum. Interesting basic data on the synchronization properties of pyramidal neurons are provided by Di Garbo et al. (2007). Relevant information might be obtained by experiments that fed axonal bursts synchronized at different frequencies into the claustrum, and recorded what form of synchronized bursts appeared in the relevant outputs from the claustrum.

In particular, the problem to be solved might be put in the following form. The claustrum consists of a fairly uniform tangled array of pyramidal cells and a wide variety of GABAergic INs, all densely interconnected by axon collaterals of the P cells, axons and dendrites of the INs, and the chemical synapse and gap junction linked syncytium formed by the INs. This complex system receives afferent axons from most brain areas. So what is the likely form of interaction between all the different synchronized oscillations, both of intrinsic and extrinsic origin, set up by this system, and how would this affect the output patterns of the P cells? Further experiments are needed to make progress in this direction.

We now turn to detailing how our hypothesis explains the new experimental data reported by Ryan Remedios (2012) in his lecture on the subject of "The claustrum and the orchestra of cognitive control" presented at the Francis Crick Memorial Conference held in Cambridge, UK. These experiments were not conducted to test our hypothesis. However, they turned out actually to do so.

At the conference, Remedios presented the results of a series of experiments he had conducted in rats using a molecular lesioning technique involving saporin (a ribosome-inactivating protein), to effectively eliminate most if not all of the claustrum. When these "aclaustral" animals were placed on the center platform of an elevated eight-arm radial maze, they did not explore the maze as did the control subjects, but, in the speaker's words, remained "frozen" at its center, with infrequent short forays into one of the arms. Likewise, rats placed in a running situation on a Rotarod continued running far longer than did the control subjects.

Our hypothesis suggests that sensory – and much cognitive – discrimination and integration (leading to an executive order) is effected by competitive synchronous gamma oscillations. This mechanism operates at a weak level within the cortex (via cortico-cortical synchronizations), and at a stronger level within the claustrum (via intraclaustral synchronizations), the connection between the two being maintained by cortico-claustro-cortical loops.

In the case of the lesioned rat "stuck" at the center of the maze, we suggest that this has nothing to do with behavioral or emotional "freezing," but that it is a case of the "donkey caught between two identical bales of hay" syndrome. The loss of the claustrum results in the rat losing its ability to choose between what are very similar arms of the maze. The weak signals, produced by the now isolated cortical system, are insufficient to trigger the executive order necessary for the rat to enter an arm of the maze. In effect, the lesioned animal can no longer make up its mind. Likewise, the running lesioned animal, by the operation of a similar mechanism, cannot decide when to stop.

In addition, with the use of blood-oxygen-level-dependent (BOLD) fMRI in these saporin-lesioned rats, Remedios reported increased widespread positive correlations across the various sensory cortical areas, as well as increased activity in the PFC. Our hypothesis also explains this finding, premised on the well-known principle that, when one mechanism in the brain is inactivated, it immediately tries to activate some other mechanism(s) to compensate for the loss. In this case, the loss of the integrating activity of the claustrum leads to an expected compensatory increase in activity in the cortico-cortical system. In this regard, we also wish to note that our hypothesis serves to explain the multiple cognitive effects of claustral activity, as detailed in this paper. It is unclear to us how an "interhemispheric bidirectional PFC-claustrum network," as detailed by Remedios in his lecture (without specifics offered as to what its function might be), would accomplish this.

Lastly, we have already commented in this chapter on the point raised by Christof Koch during the Q&A session that immediately followed the Remedios lecture, referring to the lack of salience in the stimuli used in his monkey claustrum unit physiology study (Remedios et al., 2010), which was also presented during his talk. We agree with his point that, in order to test a salience hypothesis, it is necessary to use salient stimuli, which, in the case of the monkey claustrum study (Remedios et al., 2010), would be a multisensory/multimodal task.

In a water maze test Grasby and Talk (2013) found that rats with stereotactic lesions in the anterior claustrum showed an impairment in finding a switched location of the target platform together with increased perseverance errors. The lesioned rats were not impaired in working memory or latent inhibition. The authors concluded that this result is in concordance with our hypothesis.

Another report (Koubeissi et al., 2012) detailed the result of electrical stimulation of the left claustral area, with the electrode placed 1 mm outside the claustrum (at the junction of the anterosuperior insula and the extreme capsule) in an unanesthetized human epileptic patient. The current intensities reached 18 mA at a frequency of 50 Hz. The authors reported that this resulted in behavioral arrest, blank staring, and unresponsiveness, with reduction of spontaneous respiratory movements. The patient was able to continue repetitive movements with her tongue and hands, and utter a word repeatedly during stimulation, if initiated before the stimulus train. As soon as the stimulation ceased she returned to normal, but had no memory of the incident. The similarity with the aclaustral rats described by Remedios et al. (2010) is remarkable. Electrical stimulation seems to produce a more profound effect than chemical ablation. The aclaustral rats could make a few tentative and feeble efforts to enter a wing of the maze. The patient was totally unresponsive. Our hypothesis might explain this on the following basis. In the rats the cortico-cortical pathway is still intact and could provide weak synchronization. The 50 Hz electrical stimulation of the claustrum would not only disrupt intraclaustral synchronization but might traverse the claustro-cortical projection and disrupt the cortico-cortical synchronization as well, leaving the patient with nothing. We suggest that experiments should be carried out using radial mazes and similar tests during electrical stimulation of the claustrum, together with further experiments using anti-connexins and saporin-based toxins.

The similarity of the behavior of this patient with cases of complex seizures is noteworthy. It seems possible that such clinical episodes may involve the claustrum. Our hypothesis suggests that the claustrum magnifies cortical oscillations; therefore, it might also magnify epileptic oscillations, resulting in a breakdown of claustral function (i.e. loss of the ability to respond to the environment).

CONCLUSION

This chapter presents a specific hypothesis (with variations) on the mechanism of action of the claustrum. It suggests that the claustrum consists of a large number of simple synchrony detectors that detect and react to the degree of synchrony contained in spike bursts in their selective input axons. This synchrony is facilitated by intraclaustral synchronizing interactions between the P cells and GABAergic IN cells, and is maintained in dynamic cortico-claustro-cortical reverberating circuits. We present two applications of this hypothesis. The first deals with intermodal synchronizations and the second with intramodal synchronizations. These cycles may function as the neural correlates of consciousness. In the case of the ventral claustrum this system may relate to the reinforcement (salience) value of stimuli and other higher brain functions of the claustrum. We also discuss a complementary hypothesis related to salience detection and estimation. Furthermore we delineate three levels of possible claustral activity. The first involves only identification and magnification of gamma oscillations in its afferent inflow. The second level adds integration of these oscillations, and the third involves integration of the underlying spike trains. Lastly, we develop a variant that includes ideas gleaned from the Sherman and Guillery (2011) model of thalamo-cortical circuits.

Thus our position is very close to the hypothesis proposed by Crick and Koch (2005). They described "widespread intra-claustral interactions," that may be in the form of "waves of information [that] can travel within the claustrum." This may involve, they suggest further, dendrodendritic synapses and networks of gap-junction linked neurons. They also suggested that claustral neurons "could be especially sensitive to the timing of the inputs." Our hypothesis is in concordance with the suggestion made by Crick and Koch (2005) that "rhythmic firing of claustral INs might be critical to help synchronize far-flung populations of cortical neurons." We feel that we have merely supplied some details of this process. We also support the hypothesis presented by Remedios et al. (2010) that one function of the claustrum is related to salience processing.

In a study of sparse and dense codes in the brain during exposure to language passages and music, Lloyd (2011) found that the resulting fMRI activity has properties more similar to music than to language. Crick and Koch (2005) famously likened the claustrum to the conductor of the orchestra. We agree that the evidence points to this role for the claustrum, as the conductor of an orchestra that plays a symphony woven out of a continual shimmering interplay of information-laden harmonies built out of synchronized potentials oscillating at many different frequencies.

POSTSCRIPTS

The Function of the Claustrum in a Nutshell: The Brain's War Room

Any large organization, composed of many modules that process incoming information, and which has an output in the form of intelligent behavior, will have a problem with internal information flow. Each module is busy processing its own input and providing an output, but how does one module know what another is doing, and how does the organism as a whole know what is going on, and what best to do next? Intercommunication between the modules soon leads to overload. Direct intermodular connections, however extensive, and non-linear dynamics can only go so far.

For example, in an army, one unit can make judgments about what to do next by observing what other units beside it in the line are doing, and making rational decisions. These may be right in the local context, but wrong in the context of the overall war plan. In the Napoleonic wars this is what generals in charge of brigades had to do during the time it took a galloping horse to get from Paris to Madrid. Another example is the battle of New Orleans, which was fought after the War of 1812 had ended. The advent of wireless telegraphy changed the situation. During WW2 Churchill set up a control room in the basement of 10 Downing Street. This was equipped with a large map in the center of the room, where small models of the planes, ships, tanks, etc., that were involved could be moved about in response to telegraphed reports from the front. The generals could review the whole situation at a glance, and make instant decisions about what orders to send back.

We suggest that the claustrum, with its own internal maps, is doing the same thing. It not only "binds" sensory information, as Crick and Koch proposed, but it is receives edited incoming information (evaluated for novelty and salience) from five different channels (via cortico-claustral and thalamo-claustral afferents). It then calls in further reports and analyses, when needed, from areas of cortex it activates (by means of a web of synchronized oscillations and axon bursts). Lastly, the claustral circuits integrate all this activity (via intraclaustral synchronized oscillations), to come up with a series of executive orders transmitted to the motor cortex. In other words, the claustrum may do for its brain what Churchill's War Room did for the Allied armies.

We do not imply that this model requires the infamous little green man in the War Room reading all this incoming data and making decisions based on it. The evaluation of the data and the decisions are both functions of the competitively interacting rhythms.

The Brain: Orchestra or String Quartet?

Professor Noble's discussion in this book on the conductor of the orchestra is of particular interest. Does the brain need a conductor: or is it more like a string quartet that has no conductor? As Noble says, the brain, like an orchestra, has regional specialization. But does that entail the brain having a part that tells the other parts what to do, like an orchestra does? Crick implied yes; Noble suggests not. However, when considering possible answers to this question, we suggest that we need to look more closely at what the conductor of an orchestra does to detect nuances. Let us begin by considering, at first, only the exact timing of producing notes, which is so important for good music. Here, the player is guided by the score, by what the other players are doing, and by twitches of the conductor's baton. All these are important. However, a recent study, reported by *The Economist*, compared conductors like Herbert von Karajan, who exerted rigid discipline over their flocks, with other conductors who allowed more leeway as to timing. In a blind test, expert musicians judged which produced the better music: von Karajan and his like won. So exact timing imposed by the conductor is important. But the conductor does not tell the players, during a performance, what to play. He relies on their memory and the score in front of them to do that. Of course, he is on the watch out for wrong notes, greeted with a scowl – but, during the rehearsal and performance he is tuned only to the finest distinctions of quality in the manner of how the notes are played ("Do not press so hard on your strings," and so on).

Bearing that in mind, we can look at what the claustrum may be doing in the light of the hypothesis presented in our chapter. One thing it does, we suggest, is to turn the gain up on activities in parts of the cortex that are activated by salient stimuli and are thus oscillating and working together. This is like the general wave of a baton that tells the strings to join in. Since the brain is a computer controlling a robot's behavior, and not just an orchestra producing notes, the parts activated by the claustrum have to produce computations that end in the executive orders it sends to the motor cortex to control the behavior. So the claustrum switches parts of the cortex that are doing these computations to higher gain and stronger interactions.

Secondly, it may exert, like von Karajan, rigid control of precise timings by integrating the spike-timing code that carries this information. This improves efficiency. Finally, we can ask whether it modulates the actual themes the brain is playing in any way; in other words, does it affect the actual computations the brain is carrying out, rather than merely how it carries them out? The answer may depend on the level at which the claustrum operates on the flow of information that passes through it. In our chapter we suggest three levels. At the first it may modulate only the power of the synchronized oscillations it processes. In the second level it may modulate the timing and frequencies of these oscillations. At the highest level it may integrate the temporally coded spike messages it transmits. The claustrum would not modulate the actual text passing through it at the first two levels – the conductor is not telling the players what to do. At the third level, however, it may. The details of any such operations needs far more work to explore. At present we can only sketch out possibilities such as intermodular interactions ("binding"). The scope for these "higher" computations may be increased if the claustrum is also affected by the epigenetic codes that we describe in our chapter.

Response to Comments by Sherk in this Volume

In Helen Sherk's comments (see Chapter 5) on our hypothesis she states: "One prediction of the hypothesis which can be evaluated, however, is that claustral neurons will have multimodal responses." However, we do not agree that our present hypothesis leads to this prediction. A previous hypothesis (Smythies et al., 2012a), that we have since discarded, did make this prediction. However, in our present hypothesis (Smythies et al., 2012b: and as printed in this volume) we specifically suggest that each claustral neuron receives afferent fibers from only one sensory modality. They have only unimodal responses. In our revised hypothesis these different modalities are only blended when these neurons generate intraclaustral gamma oscillations in which GABAergic syncytia play a key role. These gamma oscillations then compete for dominance and the winner activates the motor output of the claustrum, and a decision is made. In this latest version of our hypothesis we suggest that the signal detected in the claustrum by Remedios et al. (2010), in response to a non-salient stimulus, carries priming not salient information. The experiments to determine what signal results from a salient stimulus await to be done.

Secondly, Sherk says that there is no evidence for the existence of GABAergic syncytia in the claustrum. Neither is there any evidence that there are no such syncytia in the claustrum. So far, these have been looked for only in the cortex and striatum. However, we are happy to make the statement that there are FFI GABAergic syncytia and gap junctions in the claustrum, a prediction of our hypothesis to be tested by experiment.

In addition, Sherk says slowly conducting claustro-cortical fibers would be unlikely to synchronize cortical networks. In answer we refer to the paper by Burns et al. (2011), quoted in our paper in this volume, that argues that gamma oscillations cannot be used as clocks for precise temporal coding but can nevertheless synchronize different gamma-activated networks to fire simultaneously. Moreover, we quote Vicente et al. (2008) in this volume, who showed that the brain can compensate for time lags produced by long distance projections to some extent by dynamical self-organization by the distant target neurons into lag-free oscillations. We also refer the reader to the experiments carried out by Emrich et al. (2006) that demonstrate synchronized cortico-claustral gamma oscillations (see p. 313).

Lastly, Sherk says that the claustrum may be concerned with motor and cognitive processing. We fully agree that the claustrum may be closely involved in cognitive processing, and, in collaboration with the higher thalamus, with motor processing, as we present at length above.

References

Adhikari, A., Topiwala, M.A., Gordon, J.A., 2010. Synchronized activity between the ventral hippocampus and the medial PFC. Neuron 65, 257–269.

Asai, Y., Guha, A., Villa, A.E., 2008. Deterministic neural dynamics transmitted through neural networks. Neural Netw. 21, 799–809.

Babu, K.M., McCurdy, C.R., Boyer, E.W., 2008. Opioid receptors and legal highs: *Salvia divinorum* and Kratom. Clin. Toxicol. (Phila.) 46, 146–152.

Baizer, J.S., 2001. Serotonergic innervation of the primate claustrum. Brain Res. Bull. 55, 431–434.

Bao, S., Chan, V.T., Merzenich, M.M., 2001. Cortical remodelling induced by activity of ventral tegmental dopamine neurons. Nature 412, 79–83.

Baron-Cohen, S., 2008. Autism and the Asperger Syndrome. Oxford University Press, New York.

Belitski, A., Panzeri, S., Magri, C., Logothetis, N.,K., Kayser, C., 2010. Sensory information in local field potentials and spikes from visual and auditory cortices: time scales and frequency bands. J. Comput. Neurosci. 29, 533–545.

Behan, M., Haberly, L.B., 1999. Intrinsic and efferent connections of the endopiriform nucleus in rat. J. Compar. Neurol. 408, 532–548.

Beneyto, M., Prieto, J.J., 2001. Connections of the auditory cortex with the claustrum and the endopiriform nucleus in the cat. Brain Res. Bull. 15, 485–498.

Bosman, C.A., Schoffelen, J.-M., Brunet, N., Oostenveld, R., Bastos, A.M., Womelsdorf, T., et al., 2012. Attentional stimulus selection through selective synchronization between monkey visual areas. Neuron 75, 875–888.

Bruno, R., 2011. Synchrony in sensation. Curr. Opin. Neurobiol. 21, 701–708.

Budinger, E., Heil, P., Hess, A., Scheich, P., 2006. Multisensory processing via early cortical stages: connections of the primary auditory cortical field with other sensory systems. Neuroscience 143, 1063–1083.

Buffalo, E.A., Fries, P., Landman, R., Buschman, T.J., Desimione, R., 2011. Laminar differences in gamma and alpha coherence in the ventral stream. Proc. Nat. Acad. Sci. U.S.A 108, 11262–11267.

Bullock, T., 1993. *How do Brains Work?* Birkhauser Verlag, New York.

Burns, S.P., Xing, D., Shapely, R.M., 2011. Is gamma-band activity in the local field potential of V1 cortex a "clock" or filtered noise? J. Neurosci. 31 (26), 9658–9664. doi: 10.1523/JNEUROSCI.0660–11.2011.

Cappe, C., Barone, P., 2005. Heteromodal connections supporting multisensory integration at low levels of cortical processing in the monkey. Eur. J. Neurosci. 22, 2886–2902.

Cappe, C., Rouiller, E.M., Barone, P., 2012. Cortical and thalamic pathways for multisensory and sensorimotor interplay. In: Murray, M.M., Wallace, W.T. (Eds.), The Neural Bases of Multisensory Processes CRC Press, Boca Raton (FL), pp. 15–30.

Cardin, J.A., Carién, M., Meletis, K., Knoblich, U., Zhang, F., Deisseroth, K., et al., 2009. Driving fast-spiking cells induces gamma rhythm and controls sensory responses. Nature 459, 663–667.

Carrillo-Reid, L., Hernández-López, S., Tapia, D., Galarraga, E., Bargas, J., 2011. Dopaminergic modulation of the striatal microcircuit: receptor-specific configuration of cell assemblies. J. Neurosci. 31, 14972–14983.

Cassidy, M., Mazzone, P., Olivero, A., Insola, A., Tonali, P., Di Lazzaro, V., et al., 2002. Movement-related changes in synchronization in the human basal ganglia. Brain 125, 1235–1246.

Cauda, F., D'Agata, F., Sacco, K., Duca, S., Geminiani, G., Vercelli, A., 2011. Functional connectivity of the insula in the resting brain. Neuroimage 55, 8–23.

Chalk, M., Herrero, J.L., Gieselmann, M.A., Delicato, L.S., Gotthardt, S., Thiele, A., 2010. Attention reduces stimulus-driven gamma frequency oscillations and spike field coherence in V1. Neuron 66, 114–125.

Chanda, S., Xu-Friedman, M.A., 2010. Neuromodulation by GABA converts a relay into a coincidence detector. J. Neurophysiol. 104, 2063–2074.

Churchland, P.S., Sejnowski, T.J., 1999. The Computational Brain. MIT Press, Cambridge, MA.

Clarey, J.C., Irvine, D.R.F., 1986. Auditory response properties of neurons in the claustrum of the cat. Exp. Brain Res. 2, 432–437.

Clavagnier, S., Falchier, A., Kennedy, H., 2004. Long-distance feedback projections to area V1: implications for multisensory integration, spatial awareness, and visual consciousness. Cognit. Affectual Behav. Neurosci. 4, 117–126.

Colgin, L.L., Denninger, T., Fyhn, M., Hafting, P., Bonnevie, T., Jensen, O., et al., 2009. Frequency of gamma oscillations routes flow of information in the hippocampus. Nature 462, 353–357.

Collignon, O., Voss, P., Lassonde, M., Lepore, F., 2009. Cross-modal plasticity for the spatial processing of sounds in visually deprived subjects. Exp. Brain Res. 192, 343–358.

Corey, J., Scholl, B., 2012. Cortical selectivity through random connectivity. J. Neurosci. 32, 10103–10104.

Cortes, N., van Vreeswijk, C., 2012. The role of pulvinar in the transmission of information in the visual hierarchy. Front. Comput. Neurosci. 6, 29. doi: 10.3389/fncom.2012.00029.

Crick, F.C., Koch, C., 2003. A framework for consciousness. Nat. Neurosci. 6, 119–126.

Crick, F.C., Koch, C., 2005. What is the function of the claustrum? Philos. Trans. R. Soc. B 360, 1271–1279.

Dahl, C.D., Logothetis, N.K., Kayser, C., 2010. Modulation of visual responses in the superior temporal sulcus by audio-visual congruency. Front. Integ. Neurosci. 4, 10. doi: 10.3389/fnint.2010.00010.

Das, U.N., 2010. Metabolic Syndrome Pathophysiology: the role of Essential Fatty Acids. Wiley, New York.

Day-Brown, D., Wei, H., Petry, H.M., & Bickford, M.E., 2009. Characterization of claustrum-V1 connections. Poster #848.11/T15, Society for Neuroscience Annual Meeting, Chicago.

Day-Brown, J.D., Wei, H., Chomsung, R.D., Petry, H.M., Bickford, M.E., 2010. Pulvinar projections to the striatum and amygdala in the tree shrew. Front. Neuroanat. 4, 143. doi: 10.3389/fnana.2010.00143.

Di Garbo, A., Berbi, M., Chillemi, S., 2007. The synchronization properties of a network of inhibitory interneurons depend on the biophysical model. Biosystems 88, 216–227.

Divac, I., 1979. Patterns of subcortico-cortical projections as revealed by somatopetal horseradish peroxidase tracing. Neuroscience 4, 455–461.

Divac, I., Kosmal, A., Bjorklund, A., Lindvall, O., 1978. Subcortical projections to the prefrontal cortex in the rat revealed by the horseradish peroxidase technique. Neuroscience 3, 785–796.

Dockstader, C., Cheyne, D., Tannock, R., 2010. Cortical dynamics of selective attention to somatosensory events. Neuroimage 49, 1777–1785.

Doucette, W., Gire, D.H., Whitesell, J., Carmean, V., Lucero, M.T., Restrepo, D., 2011. Associative cortex features in the first olfactory brain relay station. Neuron 69, 1176–1187.

Driver, J., Noesselt, T., 2008. Multisensory interplay reveals crossmodal influences on 'sensory-specific' brain regions, neural responses and judgments. Neuron 57, 11–23.

Druga, R., 1972. Efferent projections from the claustrum (an experimental study using Nauta's method). Folia Morphol. Praha. 20 (2), 163–165.

Duffau, H., Mandonnet, E., Gatignol, P., Capelle, L., 2007. Functional compensation of the claustrum: lessons from low-grade glioma surgery. J. Neurooncol. 81, 327–329.

Edelstein, L.R., Denaro, F.J., 2004. The claustrum: a historical review of its anatomy, physiology, cytochemistry and functional significance. Cell. Mol. Biol. 50, 675–702.

Eggan, S.M., Lewis, D.A., 2007. Immunocytochemical distribution of the cannabinoid CB1 receptor in the primate neocortex: a regional and laminar analysis. Cereb.Cortex 17, 175–191.

Emrich, S.M., Ferber, S., Klug, A., Ross, B., 2006. Examining changes in synchronous oscillations associated with the segregation and persistence of forms using magnetoencephalography (MEG). Poster. Cognitive Neuroscience Society meeting, San Francisco.

Enomoto, K., Matsumoto, N., Nakai, S., Satoh, T., Sato, T.,K., Ueda, Y., et al., 2011. Dopamine neurons learn to encode the long-term value of multiple future rewards. Proc. Nat. Acad. Sci. USA 108, 15462–15467.

Falchier, A., Clavagnier, S., Barone, P., Kennedy, H., 2002. Anatomical evidence of multimodal integration in primate striate cortex. J. Neurosci. 22, 5749–5759.

Fernández-Ruiz, A., Makarov, V.A., Benito, N., Herreras, O., 2012. Schaffer-specific local field potentials reflect discrete excitatory events at gamma frequency that may fire postsynaptic hippocampal CA1 units. J. Neurosci. 32, 5165–5176. doi: 10.1523/jneurosci.4499–11.2012.

Fino, E., Yuste, R., 2011. Dense inhibitory connectivity in neocortex. Neuron 69, 1188–1203.

Fischer, J., Whitney, D., 2012. Attention gates visual coding in the human pulvinar. Nat. Commun. 3, 1051. doi: 10.1038/ncomms2054.

Flindt-Egebak, P., Olsen, R.B., 1978. Some efferent connections of the feline claustrum. Neurosci. Lett. 1 (Suppl) S159-S159.

Flint, R.D., Lindberg, E.W., Jordan, L.R., Miller, L.E., Slutsky, M.W., 2012. Accurate decoding of reaching movements from field potentials in the absence of spikes. J. Neural. Eng. 9 (4), 046006. Epub 2012 Jun 25.

Foxworthy, W.A., Clemo, H.R., Meredith, M.A., 2013. Laminar and connectional organization of a multisensory cortex. J. Comp. Neurol. 521 (8), 1867–1890. doi: 10.1002/cne.23264.

Freeman, W.J., 2001. How Brains Make up Their Minds. Columbia University Press, New York.

Friston, K.J., Shiner, T., FitzGerald, T., Galea, J.M., Adams, R., Brown, H., et al., 2012. Dopamine, affordance and active inference. PLoS Comput. Biol. 8 (1), e1002327. doi: 10.1371/journal.pcbi.1002327.

Fujioka, T., Trainro, L.J., Large, E.W., Ross, B., 2009. Beta and gamma rhythms in human auditory cortex during musical beat processing. Ann. NY Acad. Sci. 1169, 89–92.

Fukuda, M., Juháász, C., Hoechstetter, K., Sood, S., Asano, E., 2010. Somatosensory-related gamma-, beta- and alpha-augmentation precedes alpha- and beta-attenuation in humans. Clin. Neurophysi. 121, 366–375.

Fukuda, T., Kosaka, T., Singer, W., Galuske, R.A., 2006. Gap junctions among dendrites of cortical GABAergic neurons establish a dense and widespread intercolumnar network. J. Neurosci. 26, 8595–8604.

Gao, W.J., Pallas, S.L., 1999. Cross-modal reorganization of horizontal connectivity in auditory cortex without altering thalamocortical projections. J. Neurosci. 19, 7940–7950.

Gaucher, Q., Edeline, J.M., Gourévitch, B., 2012. How different are the local field potentials and spiking activities? Insights from multi-electrodes arrays. J. Physiol. (Paris) 106, 93–103.

Ghazanfar, A.A., Schroeder, C.E., 2006. Is the neocortex essentially multisensory? Trends Cogn. Sci. 10, 278–285.

Gielen, S., Krupa, M., Zeitler, M., 2010. Gamma oscillations as a mechanism for selective information transmission. Biolog. Cybern. 103, 151–165.

Gill, S.K., Ishak, M., Dobransky, T., Haroutunian, V., Davis, K.L., Rylett, R.J., 2006. 82-kDa choline acetyltransferase is in nuclei of cholinergic neurons in human CNS and altered in aging and Alzheimer disease. Neurobi. Aging 28, 1028–1040.

Gomes, A., Markov, N.T., Misery, P., Lamy, C., Giroud, P., Dehay, C., et al. 2012. Claustrocortical connectivity in the Macaque Monkey. FENS poster.

Grasby, K., Talk, A., 2013. The anterior claustrum and spatial reversal learning in rats. Brain Res. < http://dx.doi.org/10.1016/j.brainres.2013.01.014 >.

Gregoriou, G.G., Gotts, S.J., Zhou, H., Desimone, R., 2009. High-frequency, long-range coupling between prefrontal and visual cortex during attention. Science 324, 1207–1210.

Haenschel, C., Baldeweg, T., Croft, R.J., Whittington, M., Gruzelier, J., 2000. Gamma and beta frequency oscillations in response to novel auditory stimuli: a comparison of human electroencephalogram (EEG) data with in vitro models. Proc. Nat. Acad. Sci. USA 97, 7645–7650.

Hagiwara, K., Okamoto, T., Shigeto, H., Ogata, K., Somehara, Y., Matsushita, T., et al., 2010. Oscillatory gamma synchronization binds the primary and secondary somatosensory areas in humans. Neuroimage 51, 412–420. Epub 2010 Feb 10.

Hajós, M., Hoffmann, W.E., Robinson, D.D., Yu, J.H., Hajós-Korcsok, E., 2003. Norepinephrine but not serotonin reuptake inhibitors enhance theta and gamma activity of the septo-hippocampal system. Neuropsychopharmacology 28, 857–864.

Hubbard, R.S., Ramachandran, V.S., 2005. Neurocognitive mechanisms of synesthesia. Neuron 48, 509–520.

Ishizu, T., Zeki, S., 2013. The brain's specialized systems for aesthetic and perceptual judgement. Eur. J. Neurosci. 37 (9), 1413–1420. doi: 10.1111/ejn.12135.

Kaas, J.H., Lyon, D.C., 2007. Pulvinar contributions to the dorsal and ventral streams of visual processing in primates. Brain Res. Rev. 55, 285–296.

Kaiser, J., 2003. Induced gamma-band activity and human brain function. Neuroscientist 9, 475–484.

Katz, D.B., Maier, J.X., 2011. Olfactory bulb: odor signals put into context. Neuron 69, 1041–1042.

Kavounoudias, A., Roll, J.P., Anton, J.L., Nazarian, B., Roth, M., Roll, J.P., 2008. Propriotactile integration for kinesthetic perception: an fMRI study. Neuropsychologia 46, 567–575.

Kayser, C., Petkov, C.L., Logothetis, N.K., 2008. Visual modulation of neurons in auditory cortex. Cereb. Cortex 18, 1560–1574.

Kayser, C., Montemurro, N.K., Logothetis, N.K., Panzeri, S., 2009. Spike-phase coding boosts and stabilizes information carried by spatial and temporal spike patterns. Neuron 61, 4597–4608.

Kilner, J., Bott, L., Posada, A., 2005. Modulations in the degree of synchronization during ongoing oscillatory activity in the human brain. Eur. J. Neurosci. 21, 2547–2554.

Koós, Y., Tepper, J.M., 1999. Inhibitory control of neostriatal projection neurons by GABAergic interneurons. Nat. Neurosci. 2, 469–472.

Koubeissi, M.Z. , Beltagy, A., Picard, F., Edrees, M., 2012. Alteration of consciousness due to electrical stimulation of the claustrum in the human brain. Abstract 2.079, 2012. American Epilepsy Society Conference, San Diego, 11/30 to 12/4.

Kovács, I., Papathomas, T.V., Yang, M., Fehér, A., 1996. When the brain changes its mind: interocular grouping during retinal rivalry. Proc. Nat. Acad. Sci. USA 1004, 289–296.

Kraskov, A., Quiroga, R.Q., Reddy, L., Fried, I., Koch., C., 2007. Local field potentials and spikes in the human medial temporal lobe are selective to image category. J. Cogn. Sci. 19, 479–492.

Kumar, A., Rotter, S., Aertsen, A., 2010. Spiking activity propagation in neuronal networks: reconciling different perspectives on neural coding. Nat. Rev. Neurosci. 11, 615–627.

Kuznetsova, A.Y., Deth, R.C., 2008. A mode for modulation of neural synchronization by D4 receptor-mediated phospholipid methylation. J. Comput. Neurosci. 24, 14–29.

Lachaux, J.-P., Rodriguez, E., Martinerie, J., Varela, F.J., 1999. Measuring phase synchrony in brain signals. Hum. Brain Mapp. 8, 194–208.

Lacković, Z., Salković, M., Kuci, Z., Relja, M., 1990. Effect of long-lasting diabetes mellitus on rat and human brain monoamines. J. Neurochem. 54, 143–147.

Längsjo, J.W., Alkire, M.T., Kasinoro, K., Hayama, H., Maksimow, A., Kaisti, K.K., et al., 2012. Returning from oblivion: imaging the neural core of consciousness. J. Neurosci. 32, 4935–4943.

Lee, K.M., Ahn, T.B., Jeon, B.S., Kim, D.G., 2004. Change in phase synchronization of local field potentials in anesthetized rats after chronic dopamine depletion. Neurosci. Res. 49, 179–184.

LeVay, S., Sherk, H., 1981. The visual claustrum of the cat. II. The visual field map. J. Neurosci. 1, 981–992.

Li, C., Pleil, K.E., Stamatakis, A.M., Busan, S., Vong, L., Lowell, B.B., et al., 2012. Presynaptic inhibition of gamma-aminobutyric acid release in the bed nucleus of the stria terminalis by kappa opioid receptor signaling. Biological Psychiatry 71, 725–732.

Li, H., Liu, W.Z., Liang, P.J., 2012. Adaptation-dependent synchronous activity contributes to receptive field size change of bullfrog retinal ganglion cell. PLoS One 7 (3) e34336. Epub 2012 Mar 27.

Li, Z.K., Takada, M., Hattori, T., 1986. Topographic organization and collateralization of claustrocortical projections in the rat. Brain Res. Bull. 17, 529–532.

Liang, M., Mouraux, A., Ianetti, C.D., 2013. Bypassing primary sensory cortices – A direct thalamocortical pathway for transmitting salient sensory information. Cereb. Cort. 23, 1–11. doi: 10.1093/cercor/bhr363.

Lloyd, D., 2011. Mind as music. Fron. Theoret. Philosophi. Psychol. 2, 63. doi: 10.3389/fpsyg.2011.00063.

Lomber, S.G., Meredith, M.A., Kral, A., 2010. Cross-modal plasticity in specific auditory cortices underlies visual compensations in the deaf. Nat. Neurosci. 13, 1412–1418.

Macchi, G., Bentivoglio, M., Minciacchi, D., Molinari, M., 1981. The organization of the claustroneocortical projections in the cat studied by means of the HRP retrograde axonal transport. J. Comp. Neurol. 195, 681–695.

Mansour, A., Fox, C.A., Burke, S., Meng, F., Thompson, R.C., Akil, H., et al., 1994. Mu, delta and kappa opioid receptor mRNA expression in the rat CNS: an *in situ* hybridization study. J. Comp. Neurol. 350, 412–438.

Masuda, N., Aihara, K., 2002. Spatiotemporal spike encoding of a continuous external signal. Neur. Comput. 14, 1599–1628.

Meador-Woodruff, J.H., Mansour, A., Civelli, O., Watson, S.J., 1991. Distribution of D2 dopamine receptor mRNA in the primate brain. Prog. Neuropsychopharmacology Biologi. Psychi. 1991 (15), 885–893.

Melloni, L., Molina, C., Pena, M., Torres, D., Singer, W., Rodriguez, E., 2007. Synchronization of neural activity across cortical areas correlates with conscious perception. J. Neurosci. 27 (11), 2858–2865. doi: 10.1523/jneurosci.4623–06.2007.

Meng, F., Xie, G.X., Thompson, R.G., Mansour, A., Goldstein, A., Watson, S.J., et al., 1993. Cloning and pharmacological characterization of a rat kappa opioid receptor. Proc. Nat. Acad. Sci. USA 90, 9954–9958.

Minciacchi, D., Molinari, M., Bentivoglio, G., Macchi, G., 1985. The organization of the ipsi- and contralateral claustrocortical system in rat with notes on the bilateral claustrocortical projections in cat. Neuroscience 16, 557–576.

Mufson, E.J., Mesulam, M.M., 1984. Thalamic connections of the insula in the rhesus monkey and comments on the paralimbic connectivity of the medial pulvinar nucleus. J. Compar. Neurol. 227, 109–120.

Naghavi, H.R., Eriksson, J., Larsson, A., Nyberg, L., 2007. The claustrum/insula region integrates conceptually related sounds and pictures. Neurosci. Lett. 422, 77–80.

Nioche, C., Cabanis, E.A., Habas, C., 2009. Functional connectivity of the human red nucleus in the brain resting state at 3T. Am. J. Neuroradiol. 30, 396–403.

Norita, M., 1977. Demonstration of bilateral claustro-cortical connections in the cat with the method of retrograde axonal transport of horseradish peroxidase. Arch. Histol. Japon. 40, 1–10.

Nunn, J.A., Gregory, L.J., Brammer, M., Williams, S.C., Parslow, D.M., Morgan, M.J., et al., 2002. Functional magnetic resonance imaging of synesthesia: activation of V4/V8 by spoken words. Nat. Neurosci. 5, 371–375.

Olson, C.R., Graybiel, A.M., 1980. Sensory maps in the claustrum of the cat. Nature 288, 479–481.

Oram, M.W., Perrett, D.I., 1996. Integration of form and motion in the anterior superior temporal polysensory area (STPa) of the macaque monkey. J. Neurophysiol. 76, 109–129.

Padmala, S., Lim, S.-L., Passoa, L., 2010. Pulvinar and affective significance: responses track moment-to-moment stimulus visibility. Frontiers in Human Neuroscience, Sep 24;4. pii: 64. doi: 10.3389/fnhum.2010.00064.

Palva, S., Palva, J.M., 2007. New vistas for alpha-frequency band oscillations. Trends Neurosci. 30, 475–484.

Park, S., Tyszka, J.M., Allman, J.M., 2012. The claustrum and insula in Microcebus murinus: a high resolution diffusion imaging study. Front. Neuroanat. 6, 21. doi: 10.2289/fnana/2012.00021.

Pearson, R.C., Brodal, P., Gatter, K.C., Powell, T.P., 1982. The organization of the connections between the cortex and the claustrum in the monkey. Brain Res. 234, 435–441.

Pikovsky, A., Rosenblum, M., Kurths, J., 2001. Synchronization: A Universal Concept in Nonlinear Sciences. Cambridge University Press, Cambridge, UK.

Ptito, M., Fumal, A., de Noordhout, A.M., Schoenen, J., Gjedde, A., Kupers, R., 2008. TMS of the occipital cortex induces tactile sensations in the fingers of blind Braille readers. Exp. Brain Res. 184, 193–200.

Rahman, F.E., Baizer, J.S., 2007. Neurochemically defined cell types in the claustrum of the cat. Brain Res. 1159, 94–111.

Ramachandran, V.S., Anstis, S.M., 1983. Perceptual organization in moving patterns. Nature 304, 529–531.

Remedios, R., Logothetis, N.K., Kayser, C., 2010. Unimodal responses prevail within the multisensory claustrum. J. Neurosci. 30, 12902–12907.

Remedios, R., 2012. The claustrum and the orchestra of cognitive control. Francis Crick Memorial Conference, Cambridge, UK, July 7th, 2012. Available online: < http://fcm-conference.org/#talks > (accessed 18.07.13).

Rockland, K.S., Ojima, H., 2003. Multisensory convergence in calcarine visual areas in macaque monkey. Int. J. Psychophysiol. 50, 19–26.

Rosenberg, D.S., Maquiére, F., Catenoix, H., Faillenot, I., Magnin, M., 2009. Reciprocal thalamocortical connectivity of the medial pulvinar: a depth stimulation and evoked potential study in human brain. Cereb. Cortex 19 (6), 1462–1473. doi: 10.1093/cercor/bhn185.

Saalmann, Y.B., Pinsk, M.A., Wang, L., Kastner, S., 2012. The pulvinar regulates information transmission between cortical areas based on attention demands. Science 337, 753–756.

Sadowski, M., Moryś, J., Jakubowska-Sadowska, K., Narkiewicz, O., 1997. Some claustral neurons projecting to various neocortical areas show morphological differences. Folia Morpholog. (Warsaw) 56, 65–76.

Samengo, I., Montemurro, M.A., 2010. Conversion of phase information into a spike-count code by bursting neurons. PLoS One 5 (3), e9669. Mar 12.

Scalaidhe, S.P., Rodman, H.R., Albright, T.D., Gross, C.G., 1997. The effects of combined superior temporal polysensory area and frontal eye field lesions on eye movements in the macaque monkey. Behav. Brain Res. 84, 31–46.

Schilder, P., 1942. Mind: perception and Thought in Their Constructive Aspects. Columbia University Press, New York.

Sejnowski, T.J., Paulsen, O., 2006. Computational Principles. J. Neurosci. 26, 1673–1676. doi: 10.1523/JNEUROSCI.3737–05d.2006.

Sharma, J., Angelucci, A., Sur, M., 2000. Induction of visual orientation modules in auditory cortex. Nature 404, 841–847.

Sherk, H., 1986. The claustrum and the cerebral cortex. In: Jones, E.G., Peters, A. (Eds.), Cerebral Cortex, Vol. 5. Sensory-Motor Areas and Aspects of Cortical Connectivity Plenum Press, New York, pp. 467–499.

Sherk, H., LeVay, S., 1981. The visual claustrum of the cat: III. Receptive field properties. J. Neurosci. 1, 993–1002.

Sherman, S.M., Guillery, R.W., 2011. Distinct functions for direct and transthalamic cortico-cortical connections. J. Neurophysiol. 106, 1068–1077. doi: 10.1152/jn.00429.

Shinomoto, S., Kim, H., Shimokawa, T., Matsuno, N., Funahashi, S., Shima, K., et al., 2009. Relating neuronal firing patterns to functional differentiation of cerebral cortex. PLoS Comput. Biol. 5, e1000433. Epub 2009 Jul 10.

Shipp, S., 2003. The functional logic of cortico-pulvinar connections. Philos. Trans. R. Soc. B 358, 1605–1624.

Shumikhina, S., Molotchnikoff, S., 1999. Pulvinar participates in synchronizing neural assemblies in the visual cortex, in cats. Neurosci. Lett. 272 (2), 135–139.

Sim-Selley, L.J., Daunais, J.B., Porrino, L.J., Childers, S.R., 1999. Mu and kappa1 opioid-stimulated [35S]guanylyl-5'-O-(gamma-thio)-triphosphate binding in cynomolgus monkey brain. Neuroscience 94, 651–662.

Sloniewski, P., Usunoff, K.G., Pilgrim, C., 1986. Retrograde transport of fluorescent tracers reveals extensive ipsi- and contralateral claustrocortical connections in the rat. J. Comp. Neurol. 246, 467–477.

Smiley, J.F., Falchier, A., 2009. Multisensory connections of monkey auditory cerebral cortex. Hearing Res. 258, 37–46.

Smith, J.B., Alloway, K.D., 2010. Functional specificity of claustrum connections in the rat: interhemispheric communication between specific parts of motor cortex. J. Neurosci. 30, 16832–16844.

Smith, J.B., Radhakrishnan, H., Alloway, K.D., 2012. Rat claustrum coordinates but does not integrate somatosensory and motor cortical information. J. Neurosci. 32, 8583–8588.

Smythies, J., 1997. The functional neuroanatomy of awareness. Consci. Cogn. 6, 455–481.

Smythies, J., 1999. 'Reality' and 'virtual reality' mechanisms in the brain and their significance. J. Consci. Stu. 16, 69–80.

Smythies, J., Edelstein, L., Ramachandran, V., 2012a. The functional anatomy of the claustrum: the net that binds. WebmedCentral Neurosci. 3 (3) WMC003182.

Smythies, J., Edelstein, L., Ramachandran, V., 2012b. Hypotheses relating to the function of the claustrum. Front. Integr. Neurosci. 6, 53. doi: 10.3389/fnint.2012.00053.

Spector, I., Albe-Fessard, D., Hassmannova, J., 1969. Macroelectrode and microelectrode exploration of cat's claustrum. Bollettino della Società italiana di biologia sperimentale 45, 148.

Spector, I., Albe-Fessard, D., Hassmannova, J., 1974. Sensory properties of single neurons in the cat's claustrum. Brain Res. 66, 39–65.

Strumpf, H., Mangun, G.R., Boehler, C.N., Stoppel, C., Schoenfeld, M.A., Heinze, H.J., et al., 2012. The role of the pulvinar in distractor processing and visual search. Hum. Brain Mapp. doi: 10.1002/hbm.21496.

Sutoo, D., Akiyama, K., Yabe, K., Kohno, K., 1994. Quantitative analysis of immunohistochemical distributions of cholinergic and catecholaminergic systems in the human brain. Neuroscience 58, 227–234.

Tallon-Baudry, C., 2009. The roles of gamma-band oscillatory synchrony in human visual cognition. Front. Biosci. 14, 321–332.

Tian, F., Tu, S., Qiu, J., Lv, J.Y., Wei, D.T., Su, Y.H., et al., 2011. Neural correlates of mental preparation for successful insight problem solving. Behav. Brain Res. 216, 626–630.

Tononi, G., Edelman, M.T., 1998. Consciousness and complexity. Science 282 (1846), 1998. doi: 10.1126/science.282.5395.1846.

Treisman, A., Schmidt, H., 1982. Illusory conjunctions in the perception of objects. Cogn. Psychol. 14, 107–141.

Uhlhaas, P.J., Pipa, G., Lima, B., Melloni, I., Neuenschwander, S., Nikolić, D., et al., 2009. Neural synchrony in cortical networks: history, concept and current status. Front. Integr. Neurosci. 3 (17) doi: 10.3389/neuro.07.017.2009.

Van Wijk, B.C.M., Beek, P.J., Daffertshofer, A., 2012. Neural synchrony within the motor system: what have we learned so far? Front. Hum. Neurosci. doi: 10.3389/fnhum.2012.0025.

Vervaeke, K., Lorincz, A., Gleeson, P., Farinella, M., Nusser, Z., Silver, R.A., 2010. Rapid desynchronization of an electrically coupled interneuron network with sparse excitatory synaptic input. Neuron 67, 17435–17451.

Vicente, R., Gollo, L.L., Mirasso, C.R., Fischer, I., Pipa, G., 2008. Dynamical relaying can yield zero time lag neuronal synchrony despite long conduction delays. Proc. Nat. Acad. Sci. USA 105, 17157–17162. Epub 2008 Oct 28.

Vohn, R., Fimm, B., Weber, J., Schnitker, R., Thron, A., Spijkers, W., et al., 2007. Management of attentional resources in within-modal and cross-modal attention tasks: an fMRI study. Hum. Brain Mapp. 28, 1267–1275.

Volz, K.G., Schooler, L.J., Yves von Gramon, D., 2010. It just felt right: the neural correlates of the fluency heuristic. Consci. Cogn. 19, 829–837.

Von Stein, A., Sarnthein, J., 2000. Different frequencies for different scales of cortical integration: from local gamma to long range alpha/theta synchronization. Int. J. Psychophysiol. 38, 301–313.

Wilke, M., Turchi, J., Smith, K., Miishkim, M., Leopold, D.A., 2010. Pulvinar inactivation disrupts selection of movement plans. J. Neurosci. 30 (25), 8650–8659. doi: 10.1523/JNEUROSCI.0953–10.2010.

Wróbel, A., Ghazaryan, A., Bekisz, M., Bogdan, W., Kaminski, J., 2007. Two streams of attention-dependent beta activity in the striate recipient zone of cat's lateral posterior-pulvinar complex. J. Neurosci. 27, 2230–2240.

Yuan, Z., Qin, W., Wang, D., Jiang, T., Zhang, Y., Yu, C., 2012. The salience network contributes to an individual's fluid reasoning capacity. Behav. Brain Res. 229, 384–390.

Zhang, X., Hannesson, D.K., Saucier, D.M., Wallace, A.E., Howland, J., Corcoran, M.E., 2001. Susceptibility to kindling and neuronal connections of the anterior claustrum. J. Neurosci. 21 3674–368.

14

What is it to be Conscious?

Denis Noble[1], Raymond Noble[2], and James Schwaber[3]

[1]Department of Physiology, Anatomy and Genetics, University of Oxford, Oxford, UK [2]Institute for Women's Health, University College London, London, UK [3]Department of Pathology, Anatomy and Cell Biology, Thomas Jefferson University, Philadelphia, PA, USA

In *The Music of Life* (Noble, 2006), one of us argued that consciousness and the self should be viewed as processes rather than as objects.[1] That theme was repeated in more recent articles, including one which deliberately used the title *Mind over Molecule* in the context of systems biology and neuroscience (Noble, 2010). Two of the authors (DN & RN) of this article interacted extensively during the writing of those ideas. As will be clear from this chapter, all three authors share the same perspective.

In our brief contribution to this book on the functional neuroanatomy of the claustrum, we will try to clarify some of those ideas, and relate

---

[1]The distinction between "object" and "process" is an important one in this chapter. But like many ways in which we have difficulty matching language to what we conceive as reality, the distinction is not absolute. It can be context dependent. A spiral storm cloud is a process in the sense that it is formed as an attractor from many component elements. Seen from a space satellite it appears as an object, as does a spiral galaxy. At the smallest scale, fundamental particles are more like processes (with fuzzy or no boundaries) than they are like discrete particles. The particle-wave duality in physics nicely expresses the non-absolute nature of the distinction. In biology, the context depends very much on the scale from which one is viewing. From the viewpoint of cellular or organ scales, consciousness is not an object, in the sense that the necessary processes are not limited to the objects at those scales.

J. Smythies, L. Edelstein and V. Ramachandran (Eds):
The Claustrum.

353

DOI: http://dx.doi.org/10.1016/B978-0-12-404566-8.00014-3

them to some of the more specifically philosophical issues raised by the work of neuroscientists interested in the following questions:

1. How did we come to reify the self and to refer to consciousness?
2. Are the questions that are then thought to follow from this real ones?
3. Do we need to identify a region of the brain where it all "comes together"?
4. If we regard the self or consciousness as a process, does that commit us to epiphenomenalism, i.e. are they impotent, or can such processes exercise causality?

HOW DID WE COME TO REIFY THE SELF AND TO REFER TO CONSCIOUSNESS?

There is a long and extensive history in philosophy and science on questions relating to the mind, the self, and consciousness (Bennett and Hacker, 2003, 2008; Crick, 1994; Damasio, 1994; Davidson, 1970; Hacker, 2012; Hampshire, 1956; Kenny, 1969, 2009; Kim, 1993, 2000; Montefiore, 1989; Noble, 1989a, 1989b; Parfit, 1986; Ryle, 1949/2000; Sherrington, 1940; Strawson, 1959; Williams, 1976). There have been many different approaches to the questions, but it seems clear that Western thought has not always reified the self and consciousness, and that such reification may be a development of our languages as much as of our thought,[2] though the two go together of course: our language reflects as well as constrains our thought. Some traditions of Eastern thought have also avoided reification, including notably the no-self (anātman) traditions of Buddhism and Daoism. One of us has drawn on these traditions in several ways in recent work (Noble, 2006, 2009, 2011, 2012).

Minimally, we can say we are conscious.[3] We are conscious in the sense of being aware of many things and processes. It may seem a small step from that observation to the idea that some*thing* (other than the organism itself) must be the subject of such awareness. In fact it is a significant step. The transition is encouraged by most human languages since it is usual to include the subject in sentences in which we refer to awareness. "I think" could be interpreted to mean "thinking is happening," without specifying that there is any organic process other than the

---

[2] For an extensive account of the origins of consciousness, and cognate words, and the development of their meanings, see Chapter 1 in Hacker (2013).

[3] We refer here to being transitively conscious, i.e. conscious (aware) *of* something. The word is also used intransitively to mean "being awake," which is a necessary, even more minimal, condition for all other forms of awareness of things, ideas, etc., to be possible. The different forms of "conscious" are often conflated. They should not be.

animal/human involved. It is significant therefore that there are languages in which this step is not so necessary, in which it is natural and normal to omit the subject. Such languages, which include Japanese and Korean, are for this reason called pro-drop languages. "I am going" can then be simply "going" or, to be slightly more correct, "going is happening."[4] Similarly, "I think" can be expressed as "thinking is happening." The significance of this way of expressing things is then best brought out by noting that Descartes' famous philosophical statement "I think, therefore I am" (*cogito ergo sum*) could be more minimally expressed as "thinking, therefore being." "Thinking" requires that a process exists, just as "going" does, but it does not require that we should reify that process. On this way of thinking, the mistake involved in referring to consciousness lies in the last part of the word, i.e. the "ness" part. This is the reification we should avoid.

ARE THE QUESTIONS THAT ARE THEN THOUGHT TO FOLLOW FROM THIS REAL ONES?

David Hume famously looked introspectively for the self and noticed that once one had eliminated the various things of which one is aware, there is, as it were, nothing left. The self can then be regarded as a bundle of perceptions, which would not exist without those perceptions. A more modern account of the "bundle of perceptions" idea is that of Parfit (1986). The problem with this approach is that it can make the existence or otherwise of the self appear to be an empirical question. A similar problem occurs in trying to characterize the Buddhist "no-self" idea as one arising from meditative experience, as though through meditation we can, as it were, make the self disappear.[5] That would be a marvelous conjuring trick if we really could do that. Is there more than smoke and mirrors here? The way to answer this question, which is a question about whether there is a real question to be asked, is to first ask what it would

---

[4]People using such languages can, of course, deliberately insert the "I" word. In Japanese, for example, *kakimasu* (going) would become *watashi-wa kakimasu* (I am going) but that is done for emphasis or clarity and the sense is not really the same as "I am going," where there is no emphasis. Note also that the subject is not identified in the declension of the verb, as it is in, for example, the Latin *cogito*. The verbs identify only an action, not who does it. It is also important to emphasize that, in all languages, whether pro-drop or not, there must always be a subject of thinking. Our point is simply a sociological one: that some languages encourage reification, others do not. The pro-drop languages focus on the action and de-emphasize the subject.
[5]It would be better characterized as making self*ishness* disappear. One of the aims is to remove anger, animosity.

be to identify the self in this way, and hence to know what we would mean by saying that we don't find it. When we put the matter this way, it becomes obvious that we don't know what we would mean by "finding" the self. As Hacker (2012) argues, we are just as likely to find the "East Pole." There is nothing we could possibly call the "East Pole" and so it doesn't add anything to our knowledge to say that we can't find it.[6]

The "no-self" idea is then best regarded as a conceptual matter. There is no need for neuroscientists or anyone else to look for the physical seat of consciousness simply because there is no "...ness" to look for. Note that this does not mean that we cannot look for the physical basis of the *processes* involved, but it does warn us that it may be difficult to set boundaries to them. Focusing on one area of the body may be misleading. The processes may be distributed throughout our bodies and even beyond; they do not require boundaries in the way that objects require boundaries. In fact, the processes involved in intentional actions *must* usually go beyond our own bodies.

In Chapter 9 of *The Music of Life*, this approach was illustrated by imagining a debate between a neuroscientist and a philosopher concerning whether a full characterization of the physical (neuronal, muscular) processes involved in the act of pointing could count as having provided the neurophysiological basis of such an act in such a way that somehow explains away the idea of intention. Such a demonstration would be similar to the way in which the famous experiment by Libet et al. (1983) is often claimed to show that an earlier mechanical neuronal event was the *real* cause of an intention.

The problem here is that an intention is not that kind of thing, if indeed it is a thing at all. The best that the neuroscientist in the story can do is to explain the neuronal and muscular events that occur during the *movement* of pointing. That may be spectacular as neurophysiology but it is not to explain the *act* of pointing since being an act precisely requires the possibility of intention. Having an intention is a process and such processes necessarily involve social interactions. Their explanation therefore becomes one that "jumps out" of the context of neurophysiology to become an interpersonal question. To return to the story developed in *The Music of Life*, the best and only *real* explanation of the *act* of pointing was that someone wanted to know where the dog lead was. Of course, there were specific neuronal and muscular events that occurred, but looking for explanations at that level is rather like looking for the non-existent East Pole.

This view of intentionality and consciousness (if one insists on using "...ness" words in this context) bears some resemblance to the

---

[6]For a thorough-going demolition of the notion of the self, see Chapter 9.1–9.2 in Hacker (2010).

philosophy of Strawson in his book *Individuals* (1959) where he argued that one cannot ascribe consciousness to oneself without knowing how to ascribe it to others, which of course dissolves the solipsist case into a form of "East Pole" problem. This is necessarily true since what we are referring to is interpersonal, i.e. social. Once we know that the act of pointing was to show where a dog lead was to be found, we don't need to look further for the cause of that act, *as an act*. Some scientists might still insist that the *real* causes must nevertheless lie in the neuronal and muscular events. We will deal with that question later in this chapter, when we consider causal relations and epiphenomenalism.

James Watson once quipped "there are only molecules, everything else is sociology."[7] Well, yes, but it will be clear from what we have written that we would put a totally different interpretation on that statement. "Everything else" is precisely where the "secret" lies.

DO WE NEED TO IDENTIFY A REGION OF THE BRAIN WHERE IT ALL "COMES TOGETHER"?

There are two questions here that sometimes get confounded:

1. Are there are regions of the brain necessary for coordination of sensory experience and function?
2. Is there a specific region involved in pulling the function of the brain into a conscious state?

An affirmative answer to the first question does not necessitate an equally affirmative answer to the second. Indeed, an affirmative answer to the second question creates more problems than it solves. If there is a specific part of the brain that "experiences" then what is the neuronal basis of this experience? It creates an absurdity of a part of the brain that "sees" the seeing, "hears" the "hearing", "experiences" the "experience". An "eye" in the brain that "sees the seeing" would require a replication of the processes involved in being an "eye". This is not the same as requiring any part of the brain to coordinate brain function.

It may be that the brain has no specific conductor to orchestrate its activity. An orchestra may need a conductor to bring all the parts together to create the dynamics and emotion of the performance. The string section is made up of individual violins, violas, cellos; the brass of trumpets, etc. But this may not be the best way of considering brain function. Regional "specialization" is not the same as "being a violin." When neuroscientists talk of the visual or auditory cortex, are these totally dedicated to those sensory inputs? A conductor "knows the sound

---

[7] www.edge.org/conversation/the-astonishing-francis-crick

of a violin" and indicates his instruction to the violinist; but is there a part of the brain that in the same sense "knows" apart from the process itself, the playing of the violins? The brain is not an aggregate of individual sections as we might consider an orchestra – indeed, when an orchestra performs a piece is each "unaware" or uninfluenced by the function of other sections? The visual areas of the brain do what they do as part of an integrative process involving other regions. They don't produce a "violin sound" that is then added to a "cello" sound. Perception is not a jigsaw puzzle of aggregate fields; nor is it a passive process; it is dynamic and interactive.

This view is supported by the findings of neurological imaging, which show that the brain is approximately as active in the absence of performing any tasks as it is while performing tasks, albeit largely involving interactions of cortical regions other than those strongly associated with performing tasks. Since we remain just as conscious whether performing tasks or not, while being reflective or reviewing thoughts or memories, this is unsurprising. However, it undermines the idea that consciousness depends on an assemblage of activity from task-specific locales. Furthermore, when performing tasks, the default network switches and different cortical regions are recruited into different networks depending on the type of task. Yet we are conscious in all of these, again pointing to consciousness as dependent on the network interactions rather than having a causal locale.

IF WE REGARD CONSCIOUSNESS AS A PROCESS, DOES THAT COMMIT US TO EPIPHENOMENALISM?

Some parts of this section are taken, with modification, from another article (Noble, 2013), which analyzes the supervenience principle in the context of multi-scale systems biology.

A process can be viewed as an activity of a system, such as an attractor that draws the system towards one of its maintained states. There is then a question about whether such processes could coherently be viewed as epiphenomena, in some sense not so "real" as the components of the system, and possibly therefore lacking in any causal role. Consciousness has often been represented as an epiphenomenon. In this section, we will show that this idea is not correct. Note first that the problem of epiphenomenalism was created by the dualist idea that mind and body are separate objects. That view commits one to answering the question how the mind can have causal influence on the body. Rejecting that view already distances us from the problem. Modern neuroscience has largely abandoned the dualist view of mind and body, but many of its philosophical consequences still remain to haunt our ideas and interpretations.

In relation to mind, the self, and consciousness this issue can be phrased in terms of the supervenience principle, which was introduced by Davidson in 1970 (see also Kim, 1993) as a principle in mental philosophy. But it can be generalized to any theory of the relationships between events at different levels. Davidson's statement of it in relation to the mind was: "supervenience might be taken to mean that there cannot be two events alike in all physical respects but differing in some mental respects, or that an object cannot alter in some mental respects without altering in some physical respects" (Davidson, 1970). Applied to multi-level biology the principle asserts that if two states are identical at a lower level, they cannot be different at a higher level.

A very similar idea was proposed by one of us some years ago (Noble, 1967a, 1967b) as a criticism of Charles Taylor's work on teleology in *The Explanation of Behaviour* (Kenny, 1969; Taylor, 1964, 1967). At that time, the ideas were developed under the assumption that the lower levels were naturally where causal effectiveness would lie. If differences necessarily exist at lower levels when differences exist at higher levels, then it would seem that primacy in causality must be given to the lower levels.

What has developed since that time is clarification of the concept of downward causation; that is, the view that the processes of the body involve circular causality and that such interactions necessarily require causal efficacy at higher levels (see for example the set of articles in *Interface Focus* (Ellis et al., 2012)).

Downward causation from a high-level process might then be thought to break the supervenience principle. It does not, and it is important to demonstrate why. Downward causation can be defined in terms of the influence of boundary conditions determined by a higher scale (Noble, 2012). It is then clear that supervenience is always satisfied, provided that the boundaries of the system being studied are taken to include the processes determining those boundary conditions and that all events within these (even more extensive) boundaries are included. It is then trivially true that supervenience is satisfied since the lower scale has become extended to include the larger scale.

In what sense then can the causation involved be characterized as downward? The downward nature here is clearly between different scales. The issue is then a matter of definitions of scales and boundaries, and how we identify events occurring at the different scales. Moreover, since we are dealing with open rather than closed systems, we cannot know in advance how far out we should extend the boundaries in order to include everything that could determine the boundary conditions. In practice, in biology, we must deal with finite systems with boundaries open to events beyond the boundary of the system. This, after all, is what we mean by an open system. Another way to explain why downward causation does not break the supervenience principle is to note that it follows from the fact that a process at a larger scale can cause the components at a smaller scale to behave differently than they would otherwise do. This automatically guarantees that downward causation will always be reflected in differences in behavior at the lower scale.

The principle of biological relativity is then simply a statement of the fact that this does not mean that we must always regard the events at lower scales as determining those at larger scales. On the contrary, the structure of a differential equation model of an open system shows the importance of the causal influences of the boundary conditions (Noble, 2012). Without those conditions the equations would have an infinite set of solutions. It is a mistake to think that downward causation exercised through determining boundary conditions requires the supervenience principle to be broken.

It is important to note that this argument applies regardless of the level at which the causes of the boundary conditions are identified. All that is required is that those causes should depend on processes at a larger scale. The difference can be understood by noting that even events at the largest scale possible (i.e. the universe) could be claimed to depend entirely on the properties of small-scale components. In principle, though certainly not in practice, the universe could be modeled at, for example, a molecular, fundamental particle, or even lower level. Looked at from this grand perspective it is hard to see why the molecular or cellular levels should be regarded as privileged. If the justification is that we should always attribute causality to the smallest level, then we should go down to strings, or whatever we think there may be at the "lowest" level. As an explanation of living systems, this would be absurd except in the general sense in which the stuff of the universe must have whatever characteristics are necessary for life to occur.

In the context of the present chapter, the relevant boundaries must be set to include the social and environmental interactions that enable intentions to arise and be fulfilled. Intentions cannot be interpreted as functions of a brain alone.

A further important question is what counts as a causal explanation. This is where the concept of level becomes important. Thus, pacemaker rhythm, whether in the heart or in neurons, is integrated at the level of a cell. It doesn't really make sense to refer to this rhythm at a lower level since it necessarily depends on a global cell property, the electrical potential. The utility of the concept of level (as distinct from that of scale) results from its use in causal explanation.

This difference provides an important clue to the ways in which explanations at different levels can relate to each other. This was also a conclusion of the debate with Charles Taylor in *Analysis* in 1967: That the primacy of explanation at one level rather than another is a conceptual question rather than an empirical one (Noble, 1967a). Processes observed and modeled at different scales inevitably refer to different entities since they occur within different chosen boundaries. The conditions for the identification of a particular causal entity may not be satisfied at too small a scale. The system may display the characteristics of a disparate and unexplained set, and the explanation for those characteristics will then require observation at a larger scale.

Causation and causal explanation do not necessarily refer to the same processes. For example, the supervenience principle requires that there must always be differences at, say, a molecular scale if there are differences at the organism scale. Molecular causation could be said to operate at all scales. But molecular events may not provide a causal *explanation* for a higher-scale function (e.g. that of the organism in interaction with others) if the molecular events form a disparate set which we would find difficult if not impossible to organize conceptually into a schema that may provide a satisfactory explanation.

CONCLUSIONS

The main theme of this chapter is that consciousness is better viewed as process. This idea is not new of course, and many other scientists and philosophers have explored the same approach. Our main contribution to the debate is to explore some of the philosophical consequences of that view. The main one is that, once we refer to processes rather than to objects, the boundaries become fuzzy. Organisms are interaction systems. Any bodily process may, usually does, include events in the environment as well as within the organism. This is necessarily true since intentions and other functions of the conscious state must refer to social and environmental interactions as well as to processes within the body. They have causal efficacy through necessary involvement in the situational logic of those interactions.

It will be clear from this chapter that we do not take issue with the impressive neurophysiological science presented in this book. On the contrary, the evidence that the claustrum plays an important role in neurophysiological integration is increasingly strong. As we argued in the third section, this does not mean that it should be interpreted to be the seat of consciousness. In fact, we have difficulty understanding what that could possibly mean. We can identify processes that are essential for conscious states to occur, but we don't even know what a part of the brain acting as a seat of consciousness could conceivably look like. We suspect that the neurophysiological mechanisms that underpin the relevant processes are distributed and that they must include processes that are not restricted to the brain.

Acknowledgments

We acknowledge valuable email discussions with Peter Hacker.

References

Bennett, M.R., Hacker, P.M.S., 2003. Philosophical Foundations of Neuroscience. Blackwell, Oxford.

Bennett, M.R., Hacker, P.M.S., 2008. History of Cognitive Neuroscience. Wiley-Blackwell, Chichester.

Crick, F.H.C., 1994. The Astonishing Hypothesis: The Scientific Search for the Soul. Simon and Schuster, London.

Damasio, A., 1994. Descartes's Error. Putnam, New York.

Davidson, D., 1970. Mental events. In: Foster, Swanson,   Experience and Theory Duckworth, London.

Ellis, G.F.R., Noble, D., O'Connor, T., 2012. Top-down causation: An integrating theme within and across the sciences. Interface Focus 2, 1–3.

Hacker, P.M.S., 2010. Human Nature: The Categorial Framework. Wiley-Blackwell.

Hacker, P.M.S., 2012. The sad and sorry history of consciousness: being among other things a challenge to the "consciousness studies community". R. Inst. Philos. Suppl. 70, 149–168.

Hacker, P.M.S., 2013. The Intellectual Powers: A Study of Human Nature. Wiley-Blackwell.

Hampshire, S., 1956. Spinoza. Faber & Faber, London.

Kenny, A.J.P., 1969. The Five Ways. Routledge & Kegan Paul, London.

Kenny, A.J.P., 2009. Descartes: A Study of his Philosophy. St Augustine's Press, South Bend, Indiana.

Kim, J., 1993. Supervenience and Mind: Selected Philosophical Essays. Cambridge University Press, Cambridge, UK.

Kim, J., 2000. Mind in a Physical World. MIT Press, Cambridge, Mass.

Libet, B., Gleason, C.A., Wright, E.W., Pearl, D.K., 1983. Time of conscious intention to act in relation to onset of cerebral activity (readiness-potential). The unconscious initiation of a freely voluntary act. Brain 106, 623–642.

Montefiore, A.C.R.G., 1989. Intentions and Causes. In: Montefiore, A.C.R.G., Noble, D. (Eds.), Goals, No Goals and Own Goals. Unwin Hyman, London, pp. 58–80.

Noble, D., 1967a. Charles Taylor on teleological explanation. Analysis 27, 96–103.

Noble, D., 1967b. The conceptualist view of teleology. Analysis 28, 62–63.

Noble, D., 1989a. Intentional action and physiology. In: Montefiore, A.C.R.G., Noble, D. (Eds.), Goals, No Goals and Own Goals. Unwin Hyman, London, pp. 81–100.

Noble, D., 1989b. What do intentions do?. In: Montefiore, A.C.R.G., Noble, D. (Eds.), Goals, No Goals and Own Goals. Unwin Hyman, London, pp. 262–279.

Noble, D., 2006. The Music of Life. OUP, Oxford.

Noble, D., 2009. Could there be a synthesis between Western and oriental medicine? Evid. Based Complement. Alternat. Med. 6 (Supplement 1), 5–10.

Noble, D., 2010. Mind over molecule: Systems biology for neuroscience and psychiatry. In: Tretter, F., Winterer, G., Gebicke-Haerter, J., Mendoza, E.R. (Eds.), Systems Biology in Psychiatric Research Wiley-VCH, Weinheim, pp. 97–109.

Noble, D., 2011. Convergence between Buddhism and science: Systems biology and the concept of no-self (anátman). The 3rd World Conference on Buddhism and Science (WCBS). Bangkok, Thailand, pp. 1–15.

Noble, D., 2012. A theory of biological relativity: No privileged level of causation. Interface Focus 2, 55–64.

Noble, D., 2013. A biological relativity view of the relationships between genomes and phenotypes. Prog. Biophys. Mol. Biol. 111, 59–65.

Parfit, D., 1986. Reasons and Persons. Oxford University Press, Oxford.

Ryle, G., 2000. The Concept of Mind. University of Chicago Press, Chicago, (First published 1949.).

Sherrington, C.S., 1940. Man on his Nature. Cambridge University Press, Cambridge.

Strawson, P.F., 1959. Individuals. Routledge, London.

Taylor, C., 1964. The Explanation of Behaviour. Routledge & Kegan Paul, London.

Taylor, C., 1967. Teleological explanation – A reply to Denis Noble. Analysis 27, 141–143.

Williams, B., 1976. Problems of the self: Philosophical. Cambridge University Press, Cambridge, papers 1956–1972.

Selected Key Areas for Future Research on the Claustrum

Hiroyuki Okuno[1], John R. Smythies[2], and Lawrence R. Edelstein[3]

[1]Medical Innovation Center, Kyoto University Graduate School of Medicine, Tokyo, Japan [2]Center for Brain and Cognition, UCSD, 9500 Gilman Drive, La Jolla, CA, USA [3]Medimark Corporation, Del Mar, CA, USA

IMMEDIATE-EARLY GENES IN THE CLAUSTRUM

One area urgently in need of development is the role of immediate-early genes (IEGs) in the claustrum. The literature contains few references to this topic. There are several reports that c-*fos* levels in the claustrum are increased during a variety of stresses. In lactating mice subject to male threat, c-Fos (but not pCREB) expression is increased significantly in the claustrum (Gammie and Nelson, 2001). The medial prefrontal cortex differentially regulates stress-induced c-*fos* expression in the forebrain depending on the type of stressor (Figueiredo et al., 2003). Sucrose attenuated, restraint-induced c-*fos* mRNA expression is increased in the basolateral amygdala, infralimbic cortex, and claustrum (Ulrich-Lai et al., 2007). The authors suggested that their results support the hypothesis that the intake of palatable substances represents an endogenous mechanism to dampen physiological stress responses. In mice, methylphenidate induced Fos-like immunoreactivity (Fos-LI) in the mesostriatal and mesolimbocortical dopamine (DA) pathways that include the claustrum (Trinh et al., 2003). Amphetamine sensitization significantly enhanced the effects of conditioned fear on c-Fos expression in several brain regions including the claustrum (Hamamura et al., 1997), and conditioned fear-induced c-Fos expression was seen to be dramatically increased in a wide network including the claustrum (Beck and Fibiger, 1995).

J. Smythies, L. Edelstein and V. Ramachandran (Eds):
The Claustrum.

365

DOI: http://dx.doi.org/10.1016/B978-0-12-404566-8.00015-5

These reports indicate that protein synthesis is involved in early claustral responses. This constitutes a topic that has scarcely been touched upon. The IEGs are a class of genes that are rapidly and transiently induced in response to various extracellular stimuli (Christy et al., 1988; Curran and Morgan, 1985; Greenberg and Ziff, 1984). In the brain, IEG expression is mostly restricted in neurons and regulated by neuronal activities (Morgan et al., 1987; Sheng et al., 1990; Smeyne et al., 1992). Therefore, the IEG expression in the brain is believed to represent, at least in part, cortical and subcortical activities related to sensory information processing and internal representation (Guzowski et al., 1999; Lamprecht and Dudai, 1996; Parthasarathy and Graybiel, 1997).

Accumulating evidence from immunohistochemical and *in situ* hybridization studies in the claustrum has demonstrated that several IEGs are dynamically regulated in response to various stimulus conditions and behavioral paradigms. For example, administration of psychostimulants, such as cocaine and methamphetamine, rapidly induces c-*fos* and *Arc* (also called *Arg*3.1) mRNAs in the claustrum in rodents (Trinh et al., 2003; Yamagata et al., 2000). These IEGs, as well as *Nurr*1 (also called *NR*4a2), are shown to be effectively upregulated in the claustrum when animals receive physical or mental stress (e.g. forced swimming, immobilization) that induces depression-like behavior (Ons et al., 2004; Rojas et al., 2010), while some other IEGs including *egr*-1 (also called *zif* 68, *NGFI-A*, *krox* 24) and c-*jun* are rather constitutively expressed in the claustrum and their expressions are not modulated by stress (Cullinan et al., 1995). Interestingly, chronic, but not acute, social defeat stress induces prolonged expression of c-*fos* in the claustrum in mice (Matsuda et al., 1996). Fear- and anxiety-eliciting experiences also induce c-*fos* and other IEGs in the claustrum (Beck and Fibiger, 1995; Duncan et al., 1996). Further investigation about such IEG expression may help our understanding of claustrum's function, but what does IEG expression tell us about?

In the last two decades, although many of the molecular mechanisms underlying IEG regulation and functions of gene products have been uncovered, a general consensus for biological meaning of IEG expression in a given brain area has not yet been fully developed. It has been widely accepted that expression of IEGs, especially that of c-*fos*, *egr*-1 and *Arc*, reflects a recent history of neuronal activities in the brain. This is why IEG mapping is often used to visualize activated cortical areas and subcortical structures following stimuli. However, lines of evidence have indicated that IEG expression is not always correlated with solely neuronal activity. In particular, many neuromodulators and growth factors significantly contribute to IEG expression. For example, IEG expression in the hippocampus is greatly downregulated when the fornix, the major tract for cholinergic and noradrenergic inputs to the hippocampus, is transected, whereas neuronal firing properties remain unchanged (Fletcher et al.,

2006). Furthermore, infusion of brain-derived neurotropic factor (BDNF) into the dentate gyrus dramatically induces *Arc* expression (Ying et al., 2002). In rodents, both physical and mental stress induce IEGs such as c-*fos* and *Arc* in various brain areas, including the claustrum (Beck and Fibiger, 1995; Cullinan et al., 1995; Duncan et al., 1996). Because such stress induces the release of corticotropin-releasing factor (CRF), which triggers the discharge of adrenocorticotropic hormone that subsequently releases glucocorticoids, the IEG expression after stress might be directly mediated by intracellular events triggered by CRF or glucocorticoids, or both. Indeed, moderate-to-high level expression of both CRF and glucocorticoid receptors in the claustrum has been reported (Morimoto et al., 1996; Van Pett et al., 2000).

Other than dynamics and regulation of IEG expression, what do we know about IEGs, especially regarding their functions in the nervous systems? Although it is true that many IEGs still need further exploration, our knowledge about IEG functions has been progressively growing. Electrophysiological studies have demonstrated that IEGs and their products are required for synaptic plasticity of both Hebbian-type (such as long-term potentiation; LTP), and non-Hebbian type, including homeostatic neuroplasticity. Gene knockout (KO) mouse studies showed several IEGs, including c-*fos*, *egr*-1 and *Arc*, play critical roles in long-term synaptic plasticity and memory formation (for a review, see Okuno (2011)). At cellular and synaptic levels, several "effector" IEG products are involved in regulating synaptic transmission and postsynaptic signaling. For example, Arc protein is a postsynaptic protein that serves to enhance the contrast between strong and weak synapses in an activity-dependent manner (Okuno et al., 2012). Furthermore, Arc, as well as some other IEG products such as Narp and Homer1a, are implicated in a homeostatic process called synaptic scaling, in which cellular excitability gains by increasing synaptic weights when excitatory inputs to the cell are suppressed, and vice versa (Chang et al., 2010; Hu et al., 2010; Shepherd et al., 2006). Although the experimental results described above suggest the necessity of IEGs for long-lasting neuronal and synaptic plasticity, it remains unknown whether IEG expression per se is sufficient to induce the neuronal plasticity in the cell or is merely a prerequisite for neuronal plasticity and other factors or signals are required for the neuronal modification. In other words, it is still debatable whether IEG expression autonomously triggers plasticity or whether it plays a permissive role for plasticity. In the synaptic tagging and capture model, strong activity-driven, plasticity-related proteins that are putatively encoded by IEGs can be captured and utilized by weakly activated synapses in order to sustain enhancement of synaptic transmission efficacy for a long period of time (Frey and Morris, 1997). Similar to the synaptic tagging and capture, in an animal's behavioral paradigm called *behavioral tagging*, short-term memory produced by a weak training protocol can be converted to long-term memory when new IEG expression is induced by novelty exposure in a time window proximate to the training (Moncada and Viola, 2007). In this sense, IEGs may play a rather permissive role in converting short-lasting synaptic modification into a long-lasting form. The types of neuronal modifications required for the claustrum's function need to be further characterized in the future.

Finally it is worth noting that recent advances in molecular and genetic tools elegantly provide compelling evidence for the biological significance of IEG-expressing neurons in memory processes. Optogenetic or pharmacogenetic artificial activation of neuronal ensembles that expressed c-*fos* during fear-conditioning training in the hippocampus and the neocortex elicits fear responses in mice without exposure to conditioned stimuli (Garner et al., 2012; Liu et al., 2012). These results strongly suggest the critical involvement of IEG expression in circuit reorganization during memory formation processes.

ARE THERE MEMORY MECHANISMS IN THE CLAUSTRUM?

All current hypotheses pertaining to the function of the claustrum deal only with ongoing activity, for example modulating synchronized oscillations or salience signals. The observation of activity in IEGs in the claustrum points to an involvement with memory. Furthermore, the clinical observations of Koubeissi et al. (2012), described in Chapter 13, show that electrical stimulation of the claustrum in the awake human subject results in complete amnesia regarding the event. Thirdly, the likely activity of epigenetic agents such as microRNAs (miRNAs) in the claustrum suggests that protein synthesis is an important factor in claustral activity. Yet our accounts of this have all been focused on its operations at lightning speed in modulating synchronized gamma oscillations. No attention has been paid to the possibility that these operations leave some long-lasting molecular trace in the claustrum. In fact it is difficult to see what function such traces could have. In the hypothesis presented in Chapter 13, the claustrum was pictured as the scene of a complex series of two or more synchronized gamma oscillations, derived from sensory inputs, competing on a winner-takes-all basis for access to the final common motor path. Once the decision has been taken, the hypothesis suggests that the slate is wiped clean to deal with the next multisensory input from the continually changing environment. So any traces from the first episode lingering in the claustral mechanism might be expected to interfere with processing the next input. Clearly more experiments are needed during such episodes to clarify this point.

EPIGENETIC CODES IN THE CLAUSTRUM

There is now abundant evidence that different brain areas carry different and specific populations of miRNAs (Smythies and Edelstein, 2012), and that such miRNAs are transported across synapses from the pre- to the postsynaptic neuron (as well as some in the reverse direction). There is also now a long list of epigenetic factors and processes, including miRNAs, that modulate many aspects of neuronal function. (A brief summary of the evidence that supports these statements is given on pp. 370–372.) Therefore, examination of the miRNA content of postsynaptic neurons should give us some information as to the nature (modality) of the presynaptic connections of those neurons. We suggest that there exists an opportunity to apply this strategy to the claustrum. This organ is famously the target of afferent connections from practically every other region of the brain; therefore a map of each subdivision of the claustrum that identifies the differential distribution of all the candidate miRNAs should provide evidence as to the connections of that subdivision. To do this, information also needs to be gathered as to the precise miRNA content of neurons in cortical and subcortical areas belonging to different modalities. For example, suppose a particular miRNA is found that occurs in the cortex only in visual areas. Then suppose that this same miRNA was also found in a limited locality in the claustrum. This finding would provide evidence that the claustral area has a visual function.

An even more exciting research project is as follows. The miRNA content of neurons forms an epigenetic code (Smythies and Edelstein, 2012) so, if you were to take a neuron in higher polysensory cortex, this neuron will receive afferent axons from (say) three different sensory modalities – visual, somatosensory and auditory. Each of these will carry its own specific mix of miRNAs and will deliver them in packets (which we can label Vm, Sm and Am) across the synapse into the postsynaptic polysensory neuron. Therefore, the incoming mix of miRNAs that enter the postsynaptic neuron will be a mixture of Vm, Sm and Am. How much of each depends on how many axons of each modality impinge on this neuron. For example, one such neuron may have 10 visual inputs, 25 somatosensory inputs and 3 auditory inputs, which could be expressed as [10Vm + 25Sm + 3Am]. Another such neuron may receive 5 visual axons, 44 somatosensory axons and 14 auditory axons [5Vm + 44Sm + 14Am]. We can therefore generalize this and specify the make-up of the miRNA mix in trisensory cortex as [xVm + ySm + zAm]. Similar processes should govern mixtures of other epigenetic factors, such as protein transcription agents and other forms of RNA and DNA transported across synapses in polysensory cortex.

Now, exosomes transported across synapses from the presynaptic neuron into the postsynaptic neuron will be transported throughout the postsynaptic neuron (see section on exosome transport, below for evidence). In the case of a unisensory (say visual) neuron, all these genetic instructions will mix with the genetic instructions issued by the postsynaptic neuron's own genetic and epigenetic systems (i.e. its own mRNAs, MiRNAs, protein transcription factors, etc.) to modulate, in a precise and dynamic manner, details of the construction of that neuron's information processing mechanism (neuronal differentiation and specification, dendritic numbers and morphology, synapse construction, and so on; Smythies and Edelstein, 2012). This results in that neuron functioning as a visual neuron. However, if the visual input is cut off and replaced by an input of another modality (say somatosensory) then the new epigenetic instructions carried by the new input will engineer, over a short period of time, a radical change in the functional neuroanatomy of that neuron, so that it now functions as a somatosensory neuron. Likewise, in a polysensory neuron, the functional neuroanatomy of its computational machinery will be neither visual, somatosensory nor auditory, but an algebraic function of all three. Moreover there will be a great deal of variation between individual members of the same trisensory cortex owing to the varying values of x, y and z in each. It seems possible, therefore, that this system provides an epigenetic code that mediates "binding" information. The celebrated "binding problem" may be put as: how does the brain know that the color, shape and movement of a phenomenological object go together when this information is processed by three different anatomical regions of the brain? One answer may be that this information is carried by the values of x, y and z in the above equations (see Smythies and Edelstein, 2012 for details).

Bearing this in mind one can regard the claustrum as the ultimate in polysensory "cortex," since it receives input from practically every other part of the brain. Therefore a painstaking and detailed measurement of the identity and quantity of individual miRNAs in its component parts should provide invaluable information as to sensory binding.

Evidence for Transport of Exosomes Inside the Postsynaptic Neuron

There are currently three main pieces of evidence for exosome transport within postsynaptic neurons:

1. Tian et al. (2010) isolated exosomes from PC12 cells, labeled them with a lipophilic dye and an amino-reactive fluorophore. The exosomes were then incubated with PC12 cells. Live-cell microscopy revealed that the exosomes were taken up into the cells via the endocytosis pathway, enclosed in vesicles and transported to the perinuclear region.

2. Synaptic activation causes dlP-bodies (riboprotein storage bodies for RNAs) to move along dendrites to localize in distant locations (Cougot et al., 2008).
3. Using a specific technique involving TDT polymerase, Waldenström et al. (2012) took exosomes from cultured cardiomyocytes and observed that DNA was transferred from these exosomes inside target fibroblasts, where DNA was seen in the fibroblast cytosol and in nuclei themselves.

Examples of the Action of Transcription Factors in Neuronal Functional Anatomy

- In newborn neurons in the adult hippocampus, the transcription factor CREB mediates activity-driven dendritic growth and spine formation in response to a variety of signaling pathways (Magill et al., 2010).
- The transcription factors Cut and the RhoGEF Trio have been reported to control dendrite formation and morphology in Drosophila (Iyer et al., 2012; Nagel et al., 2012).
- The transcription factor Foxp4 is essential for maintaining dendritic arborization in the mouse cerebellum (Tam et al., 2011).
- The transcription factor Cux1 regulates the morphology of dendrites of pyramidal neurons in layers II-IV of the murine cerebral cortex. (Li et al., 2010).
- The transcription factor Sp4 regulates dendritic patterning during cerebellar maturation (Ramos et al., 2007).
- Neurite elongation of vertebrate neurons is regulated by homeoproteins, which are transcription factors. These have been shown to translocate through neuronal membranes (Chatelin et al., 1996) and to be transmitted to the cell nuclei.
- The transcription factor neuroD2 regulates the development and functional differentiation of hippocampal mossy fiber synapses (Wilke et al., 2012).
- The development of parvalbumin-expressing interneurons in the cerebral cortex is regulated by the transcription factors Dlx5 and Dlx6 (Wang et al., 2010).
- Addis et al. (2011) report the reprogramming of astrocytes to dopaminergic neurons using three transcription factors: ASCL1, LMX1B, and NURR1. The neurons exhibited expression profiles and electrophysiological characteristics consistent with midbrain dopaminergic neurons, notably including spontaneous pacemaking activity, stimulated release of dopamine, and calcium oscillations.

Examples of MicroRNAs that Modulate Neuronal Functional Anatomy

There is now considerable evidence to indicate that microRNAs modulate a number of neuronal functions. For example:

- spine numbers and synapse formation are controlled in an activity-regulated manner by miR-485 (Cohen et al., 2011).
- transcription of the microRNA miR355 is promoted by naturally evoked synaptic activity at the climbing fiber–Purkinje cell synapse in the mouse cerebellar flocculus (Barmack et al., 2010).
- neuronal activity regulates spine formation, in part, by increasing miR132 transcription, which in turn activates a Rac1-Pak actin remodeling pathway (Impey et al., 2010).
- microR-181a activity in primary neurons, induced by dopamine signaling, is a negative post-transcriptional regulator of GluA2 expression (Saba et al., 2012). Overexpression of this miRNA reduces GluA2 surface expression, spine formation, and miniature excitatory postsynaptic current (mEPSC) frequency in hippocampal neurons.

Optogenetic Techniques

Very recently, remarkable new investigative techniques have been developed in the field of optogenetic stimulation that promise to revolutionize claustral research.

First, following the premise that a specific memory is encoded by a sparse population of neurons, Liu et al. (2012) used a technique that directly couples the promoter of c-*fos* to the tetracycline transactivator in a process that involves training-induced neuronal activity being labeled with ChR2-EYFP, which can then be reactivated by light stimulation during testing. They identified a sparse population of neurons in the dentate gyrus of the hippocampus that were labeled during fear-conditioning associated with freezing behavior. Later reactivation of this group by light alone reproduced the freezing reaction. The authors suggested that this result reveals the engram that contributes to this conditioned behavior.

Second, Lee (2011) has developed a powerful technique that combines optogenetics and fMRI. This enables the investigator to directly trace causal communications throughout the brain across multiple synapses in vivo.

PRODUCING ACLAUSTRAL ANIMALS VIA MOLECULAR NEUROSURGERY

While much has been written regarding the vagaries of stimulating, lesioning and recording from the claustrum, which result from its irregular dimensions and close proximity to both the external and extreme capsules as well as to neighboring structures such as the insula, putamen and amygdala, Crick and Koch (2005) proposed a most logical and eminently feasible solution to this long-standing dilemma:

> Given its extended and sheet-like topography, ablating or otherwise shutting this structure down in a controlled manner – without interfering with fibres of passage or nearby regions – would require numerous, precisely targeted injections. Molecular biology could help here: if one or more genes were to be specifically and strongly expressed in neurons of the claustrum, it might be possible to silence this population by the judicious use of genetic techniques… While the shape of the claustrum makes surgical or pharmacological intervention difficult, conditional knock-in or knock-down molecular techniques are more promising. They demand the urgent identification of one or more proteins strongly expressed in the claustrum but not in adjacent regions (the insula and the putamen).

Taking their lead, Mathur et al. (2009) and Remedios (2012) sought out claustral "signatures" which would facilitate the development of an aclaustral animal. Mathur et al. (2009) approached the task via proteomics, ultimately isolating Gng2 as the most likely candidate for the development of a claustrum knockout model in rats. Pirone et al. (2012) investigated the precise localization of Gng2 and netrin G2 via double-labeling experiments in human claustrum. While Gng2 was shown to be highly expressed in the claustrum neuropil and insula (speaking for their common ontogeny), it was only lightly expressed in the external and extreme capsules, and was completely absent from the putamen. Further, it was found in neurons and glia, as well as being colocalized with glial fibrillary acidic protein (GFAP). Although a Gng2 knockout has yet to be developed, clearly, this method holds much promise.

Remedios (2012) approached the matter from an entirely different perspective – by targeting an opioid receptor known to be highly enriched in the claustrum and conjugating it to a ribosome-inactivating protein (saporin). Immunohistochemical results demonstrated that the direct-injected rats were effectively rendered aclaustral. In addition, compared with the sham-injected subjects, there were significant behavioral differences in the context of running these animals through eight-arm radial maze and rotarod tests. The aclaustral animals appeared "frozen" or indecisive when placed on the maze's center platform, while they maintained a far longer riding time on the rotarod than did the controls. This is the closest we have come to the development of an aclaustral animal, and it is rife with possibilities for the investigation of the claustrum's role in all manner of activities and processes.

References

Addis, R.C., Hsu, F.C., Wright, R.L., Dichter, M.A., Coulter, D.A., Gearhart, J.D., 2011. Efficient conversion of astrocytes to functional midbrain dopaminergic neurons using a single polycistronic vector. PLoS One 6 (12), e28719.

Barmack, N.H., Qian, Z., Yakhnitsa, V., 2010. Climbing fibers induce microRNA transcription in cerebellar Purkinje cells. Neuroscience 171, 655–665.

Beck, C.H., Fibiger, H.C., 1995. Conditioned fear-induced changes in behavior and in the expression of the immediate early gene c-fos: with and without diazepam pretreatment. J. Neurosci. 15, 709–720.

Chang, M.C., Park, J.M., Pelkey, K.A., Grabenstatter, H.L., Xu, D., Linden, D.J., et al., 2010. Narp regulates homeostatic scaling of excitatory synapses on parvalbumin-expressing interneurons. Nat. Neurosci. 13, 1090–1097.

Chatelin, L., Volovitch, M., Joliot, A.H., Perez, F., Prochiantz, A., 1996. Transcription factor hoxa-5 is taken up by cells in culture and conveyed to their nuclei. Mech. Dev. 55, 111–117.

Christy, B.A., Lau, L.F., Nathans, D., 1988. A gene activated in mouse 3T3 cells by serum growth factors encodes a protein with "zinc finger" sequences. Proc. Nat. Acad. Sci. USA 85, 7857–7861.

Cohen, J.E., Lee, P.R., Chen, S., Fields, R.D., 2011. MicroRNA regulation of homeostatic synaptic plasticity. Proc. Nat. Acad. Sci. USA 108 11650–11611.

Cougot, N., Bhattacharyya, S.N., Tapia-Arancibia, L., Bordonne, L., Filipowicz, W., Bertrand, E., et al., 2008. Dendrites of mammalian neurons contain specialized P- body-like structures that respond to neuronal activation. J. Neurosci. 28 13793–13704.

Cullinan, W.E., Herman, J.P., Battaglia, D.F., Akil, H., Watson, S.J., 1995. Pattern and time course of immediate early gene expression in rat brain following acute stress. Neuroscience 64, 477–505.

Curran, T., Morgan, J.I., 1985. Superinduction of c-fos by nerve growth factor in the presence of peripherally active benzodiazepines. Science 229, 1265–1268.

Duncan, G.E., Knapp, D.J., Breese, G.R., 1996. Neuroanatomical characterization of Fos induction in rat behavioral models of anxiety. Brain Res. 713, 79–91.

Figueiredo, H.F., Bruestle, A., Bodie, B., Dolgas, C.M., Herman, J.P., 2003. The medial prefrontal cortex differentially regulates stress-induced c-fos expression in the forebrain depending on type of stressor. Eur. J. Neurosci. 18, 2357–2364.

Fletcher, B.R., Calhoun, M.E., Rapp, P.R., Shapiro, M.L., 2006. Fornix lesions decouple the induction of hippocampal arc transcription from behavior but not plasticity. J. Neurosci. 26, 1507–1515.

Frey, U., Morris, R.G., 1997. Synaptic tagging and long-term potentiation. Nature 385, 533–536.

Gammie, S.C., Nelson, R.J., 2001. cFOS and pCREB activation and maternal aggression in mice. Brain Res. 898, 232–241.

Garner, A.R., Rowland, D.C., Hwang, S.Y., Baumgaertel, K., Roth, B.L., Kentros, C., et al., 2012. Generation of a synthetic memory trace. Science 335, 1513–1516.

Greenberg, M.E., Ziff, E.B., 1984. Stimulation of 3T3 cells induces transcription of the c-fos proto-oncogene. Nature 311, 433–438.

Guzowski, J.F., McNaughton, B.L., Barnes, C.A., Worley, P.F., 1999. Environment-specific expression of the immediate-early gene Arc in hippocampal neuronal ensembles. Nat. Neurosci. 2, 1120–1124.

Hamamura, T., Ichimaru, Y., Fibiger, H.C., 1997. Amphetamine sensitization enhances regional c-fos expression produced by conditioned fear. Neuroscience 76, 1097–1103.

Hu, J.H., Park, J.M., Park, S., Xiao, B., Dehoff, M.H., Kim, S., et al., 2010. Homeostatic scaling requires group I mGluR activation mediated by Homer1a. Neuron 68, 1128–1142.

Impey, S., Davarra, M., Lesiak, A., Fortin, D., Ando, H., Varlamova, O., et al., 2010. An activity-induced microRNA controls dendritic spine formation by regulating Rac1-PAK signaling. Cell. Mol. Neurosci. 43, 146–156. doi: 10.1016/j.mcn.2009.10.005.

Iyer, S.C., Wang, D., Iyer, E.P., Trunnell, S.A., Meduri, R., Shinwari, R., et al., 2012. The RhoGEF trio functions in sculpting class specific dendrite morphogenesis in Drosophila sensory neurons. PLoS One 7 (3), e33634.

Koubeissi, M.Z. , Beltagy, A., Picard, F., Edrees, M., 2012. Alteration of consciousness due to electrical stimulation of the claustrum in the human brain [Abstract 2.079]. American Epilepsy Society Conference, San Diego, 11/30 to 12/4.

Lamprecht, R., Dudai, Y., 1996. Transient expression of c-Fos in rat amygdala during training is required for encoding conditioned taste aversion memory. Learn. Mem. 3, 31–41.

Lee, J.H., 2011. Tracing activity across the whole brain neural network with optogenetic functional magnetic resonance imaging. Front. Neuroinformatics doi: 10.3389/fninf.2011.00021 25 October 2011.

Li, N., Zhao, C.T., Wang, Y., Yuan, X.B., 2010. The transcription factor Cux1 regulates dendritic morphology of cortical pyramidal neurons. PLoS One 5 (5), e10596.

Liu, X., Ramirez, S., Pang, P.T., Puryear, C.B., Govindarajan, A., Deisseroth, K., et al., 2012. Optogenetic stimulation of a hippocampal engram activates fear memory recall. Nature 484, 381–385. doi: 10.1038/nature11028.

Magill, S.T., Cambronne, X.A., Luikart, B.W., Liov, D.T., Leighton, B.H., Westbrook, G.L., et al., 2010. MicroRNA-132 regulates dendritic growth and arborization of newborn neurons in the adult hippocampus. Proc. Natl. Acad. Sci. USA. 107, 20382–20387.

Mathur, B.N., Caprioli, R.M., Deutch, A.Y., 2009. Proteomic analysis illuminates a novel structural definition of the claustrum and insula. Cereb. Cortex 19, 2372–2379.

Matsuda, S., Peng, H., Yoshimura, H., Wen, T.C., Fukuda, T., Sakanaka, M., 1996. Persistent c-fos expression in the brains of mice with chronic social stress. Neurosci. Res. 26, 157–170.

Moncada, D., Viola, H., 2007. Induction of long-term memory by exposure to novelty requires protein synthesis: evidence for a behavioral tagging. J. Neurosci. 27, 7476–7481.

Morgan, J.I., Cohen, D.R., Hempstead, J.L., Curran,, T., 1987. Mapping patterns of c-fos expression in the central nervous system after seizure. Science (New York, N.Y.) 237, 192–197.

Morimoto, M., Morita, N., Ozawa, H., Yokoyama, K., Kawata, M., 1996. Distribution of glucocorticoid receptor immunoreactivity and mRNA in the rat brain: an immunohistochemical and in situ hybridization study. Neurosci. Res. 26, 235–269.

Nagel, J., Delandre, C., Zhang, Y., Förstner, F., Moore, A.W., Tavosanis, G., 2012. Fascin controls neuronal class-specific dendrite arbor morphology. Development 139, 2999–3009.

Okuno, H., 2011. Regulation and function of immediate-early genes in the brain: beyond neuronal activity markers. Neurosci. Res. 69, 175–186.

Okuno, H., Akashi, K., Ishii, Y., Yagishita-Kyo, N., Suzuki, K., Nonaka, M., et al., 2012. Inverse synaptic tagging of inactive synapses via dynamic interaction of Arc/Arg3.1 with CaMKIIbeta. Cell 149, 886–898.

Ons, S., Marti, O., Armario, A., 2004. Stress-induced activation of the immediate early gene Arc (activity-regulated cytoskeleton-associated protein) is restricted to telencephalic areas in the rat brain: relationship to c-fos mRNA. J. Neurochem. 89, 1111–1118.

Parthasarathy, H.B., Graybiel, A.M., 1997. Cortically driven immediate-early gene expression reflects modular influence of sensorimotor cortex on identified striatal neurons in the squirrel monkey. J. Neurosci. 17, 2477–2491.

Pirone, A., Cozzi, B., Edelstein, L., Peruffo, A., Lenzi, C., Quilici, F., et al., 2012. Topography of Gng2- & NetrinG2-expression suggests an insular origin of the human claustrum. PLoS ONE 7, e44745. doi: 10.1371/journal.pone.0044745.

Ramos, B., Gaudillière, B., Bonni, A., Gill, G., 2007. Transcription factor Sp4 regulates dendritic patterning during cerebellar maturation. Proc. Natl. Acad. Sci. USA. 104, 9882–9887.

Remedios, R., 2012. The claustrum and the orchestra of cognitive control. Francis Crick Memorial Conference, Cambridge, UK, July 7. Retrieved from <http://fcmconference.org/#talks>.

Rojas, P., Joodmardi, E., Perlmann, T., Ogren, S.O., 2010. Rapid increase of Nurr1 mRNA expression in limbic and cortical brain structures related to coping with depression-like behavior in mice. J. Neurosci. Res. 88, 2284–2293.

Saba, R., Störchel, P.H., Aksoy-Aksel, A., Kepura, F., Lippi, G., Plant, T.D., et al., 2011. Dopamine-regulated microRNA MiR-181a controls GluA2 surface expression in hippocampal neurons. Mol. Cell. Biol. 32, 619–632. doi: 10.1128/MCB.05896-11.

Sheng, M., McFadden, G., Greenberg, M.E., 1990. Membrane depolarization and calcium induce c-fos transcription via phosphorylation of transcription factor CREB. Neuron 4, 571–582.

Shepherd, J.D., Rumbaugh, G., Wu, J., Chowdhury, S., Plath, N., Kuhl, D., et al., 2006. Arc/ Arg3.1 mediates homeostatic synaptic scaling of AMPA receptors. Neuron 52, 475–484.

Smeyne, R.J., Schilling, K., Robertson, L., Luk, D., Oberdick, J., Curran, T., et al., 1992. fos-lacZ transgenic mice: mapping sites of gene induction in the central nervous system. Neuron 8, 13–23.

Smythies, J., Edelstein, L., 2012. Transsynaptic modality codes in the brain: possible involvement of synchronized spike timing, microRNAs, exosomes and epigenetic processes. Front. Integr. Neurosci. 6, 126. doi: 10.3389/fnint.2012.00126.

Tam, W.Y., Leung, C.K., Tong, K.K., Kwan, K.M., 2011. Foxp4 is essential in maintenance of Purkinje cell dendritic arborization in the mouse cerebellum. Neuroscience 172, 562–571.

Tian, T., Wang, Y., Wang, H., Zhu, Z., Xiao, Z., 2010. Visualizing of the cellular uptake and intracellular trafficking of exosomes by live-cell microscopy. J. Cell. Biochem. 111, 488–496. doi: 10.1002/jcb.22733.

Trinh, J.V., Nehrenberg, D.L., Jacobsen, J.P., Caron, M.G., Wetsel, W.C., 2003. Differential psychostimulant-induced activation of neural circuits in dopamine transporter knock-out and wild type mice. Neuroscience 118, 297–310.

Ulrich-Lai, Y.M., Ostrander, M.M., Thomas, I.M., Packard, B.A., Furay, A.R., Dolgas, C.M., et al., 2007. Daily limited access to sweetened drink attenuates hypothalamic-pituitary-adrenocortical axis stress responses. Endocrinology 148, 1823–1834.

Van Pett, K., Viau, V., Bittencourt, J.C., Chan, R.K., Li, H.Y., Arias, C., et al., 2000. Distribution of mRNAs encoding CRF receptors in brain and pituitary of rat and mouse. J. Comp. Neurol. 428, 191–212.

Waldenström, A., Gennebäck, N., Hellman, U., Ronquist, G., 2012. Cardiomyocyte microvesicles contain DNA/RNA and convey biological messages to target cells. PLoS One 7 (4), e34653. doi: 10.1371/journal.pone.0034653 2012.

Wang, Y., Dye, C.A., Sohal, V., Long, J.E., Estrada, R.C., Roztocil, T., et al., 2010. Dlx5 and Dlx6 regulate the development of parvalbumin-expressing cortical interneurons. J. Neurosci. 30, 5334–5345.

Wilke, S.A., Hall, B.J., Antonios, J.K., Denardo, L.A., Otto, S., Yuan, B., et al., 2012. NeuroD2 regulates the development of hippocampal mossy fiber synapses. Neural Dev. 27, 7–9.

Yamagata, K., Suzuki, K., Sugiura, H., Kawashima, N., Okuyama, S., 2000. Activation of an effector immediate-early gene arc by methamphetamine. Ann. N. Y. Acad. Sci. 914, 22–32.

Ying, S.W., Futter, M., Rosenblum, K., Webber, M.J., Hunt, S.P., Bliss, T.V., et al., 2002. Brain-derived neurotrophic factor induces long-term potentiation in intact adult hippocampus: requirement for ERK activation coupled to CREB and upregulation of Arc synthesis. J. Neurosci. 22, 1532–1540.

Index

Note: Page numbers followed by "*f*", "*t*", and "*b*" refers to figures, tables, and boxes respectively.

Printed and bound by CPI Group (UK) Ltd, Croydon, CR0 4YY

08/05/2025

01865020-0001